S0-AWV-389

Nanostructured Materials

ACS SYMPOSIUM SERIES **679**

Nanostructured Materials

Clusters, Composites, and Thin Films

Vladimir M. Shalaev, EDITOR
New Mexico State University

Martin Moskovits, EDITOR
University of Toronto

Developed from a symposium sponsored by the Division of Physical Chemistry at the 213th National Meeting of the American Chemical Society,
San Francisco, CA,
April 13–17, 1997

American Chemical Society, Washington, DC

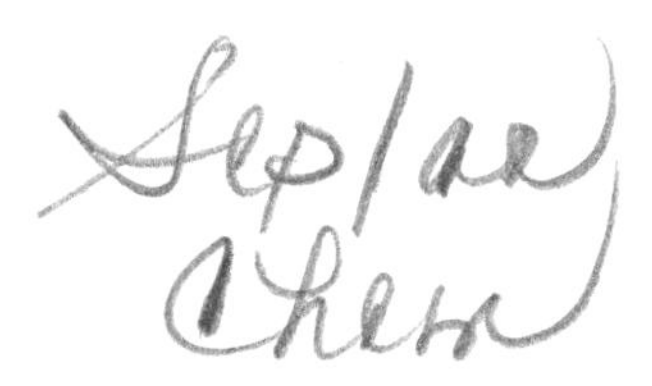

Library of Congress Cataloging-in-Publication Data

Nanostructured materials: clusters, composites, and thin films / Vladimir M. Shalaev, editor, Martin Moskovits, editor.

p. cm.—(ACS symposium series, ISSN 0097–6156; 679)

"Developed from a symposium sponsored by the Division of Physical Chemistry at the 213th National Meeting of the American Chemical Society, San Francisco, CA, April 13–17, 1997."

Includes bibliographical references and indexes.

ISBN 0–8412–3536–8

1. Nanostructure materials

I. Shalaev, Vladimir M., 1957– . II. Moskovits, Martin. III. American Chemical Society. Division of Physical Chemistry. IV. American Chemical Society. Meeting (213th: 1997: San Francisco, Calif.) V. Series.

TA418.9.N35N353 1998
620′.5—dc21 97–34825
CIP

This book is printed on acid-free, recycled paper.

PRINTED IN THE UNITED STATES OF AMERICA

Advisory Board

ACS Symposium Series

Foreword

THE ACS SYMPOSIUM SERIES was first published in 1974 to provide a mechanism for publishing symposia quickly in book form. The purpose of the series is to publish timely, comprehensive books developed from ACS sponsored symposia based on current scientific research. Occasionally, books are developed from symposia sponsored by other organizations when the topic is of keen interest to the chemistry audience.

Before agreeing to publish a book, the proposed table of contents is reviewed for appropriate and comprehensive coverage and for interest to the audience. Some papers may be excluded in order to better focus the book; others may be added to provide comprehensiveness. When appropriate, overview or introductory chapters are added. Drafts of chapters are peer-reviewed prior to final acceptance or rejection, and manuscripts are prepared in camera-ready format.

As a rule, only original research papers and original review papers are included in the volumes. Verbatim reproductions of previously published papers are not accepted.

ACS BOOKS DEPARTMENT

Contents

Preface

THE PHYSICS AND CHEMISTRY OF NANOMETER scale objects and the materials and devices derived from them is currently recognized as one of the key areas of research for the next decade. With its promise of major applications in electronics and photonics the science of mesoscopic structures has enjoyed an explosive growth in the recent past. One of the characteristics of the field is its enormous interdisciplinarity.

Physicists, chemists and engineers share equally in its establishment and often do so in isolation. This volume attempts to consolidate some of the advances currently made in nanomaterials cutting across several disciplines. It is based on the Symposium "Nanostructured Materials: Clusters, Composites, and Thin Films", held at the 1997 ACS Meeting in San Francisco.

In assembling the topics for the symposium, we tried to include both work in the area of clusters and nanoparticles and related topics which are not traditionally included in the field. These other topics provide a kind of network linking the various aspects of the field together.

We believe that the tremendous effort in the area of clusters and nanoparticles of the past decade or so, places us in a unique position to exploit the materials possibilities flowing from that research. This book provides an entree to some of those possibilities. We are grateful to the authors for participating in the Symposium and for their promptness in submitting their chapters. We would also like to thank Penny Ashcroft Moore for coordinating the symposium and the book.

VLADIMIR M. SHALAEV
New Mexico State University
Las Cruces, NM 88003

MARTIN MOSKOVITS
University of Toronto
Ontario M5S 3H6, Canada

July 22, 1997

Chapter 1

Parallel Processing of Nanocrystal Colloids: Light-Directed Assembly and Simple Devices

J. R. Heath, T. Vossmeyer, E. DeIonno, and G. Markovich

Department of Chemistry and Biochemistry, University of California—Los Angeles, 405 Hilgard Avenue, Los Angeles, CA 90095–1569

Chemical strategies for parallel preparation of nanocrystal based single electron devices and nanocrystal-patterned substrates are presented. The Langmuir-Blodgett technique is used for the preparation of ordered particle monolayers, which are incorporated in solid state capacitance devices. Single electron charging is observed in those devices. Patterned thin nanocrystal films are produced by spatially resolved photochemical activation of self-assembled organic monolayers. This method is used to form both metal and semiconductor nanocrystal patterns by binding them to exposed amine groups on the surface.

Chemical techniques for the fabricating crystalline solids in finite size have advanced rapidly, such that it is now possible to fabricate nanocrystals of virtually any class of metallic, semiconducting, and insulating materials. The unique size- and shape-dependent physical properties that are obtainable from nanocrystalline materials have made them promising candidates for a host of applications, including chemical catalysis, electronic devices, photonics-based devices, and others. Many applications, particularly those in the realm of solid-state electronics, require that particles be prepared as electrically addressable and processable patterned thin films. In this paper, we discuss two recent advances toward making such films.

In the next section, we report on parallel fabrication of single-electron capacitance devices, in which the capacitive elements are organically passivated metallic nanocrystals. The crystals are prepared as thin particle monolayer films through the use of a Langmuir trough. The Langmuir-Blodgett (LB) film is stabilized against further processing steps via solid-state chemical reactions that link the particles together. Capacitance devices fabricated in a sandwich-type configuration, exhibit single-electron (Coulomb blockade) charging effects. Capacitance/Voltage curves, taken at 77K and 300K, exhibit structure that is consistent with multiple electronic charging states. If the average particle size within the thin film is decreased, the single-electron charging energies correspondingly increase.

In the last section, we present results on the light directed assembly of metal and semiconductor nanocrystals. Self-assembled organic monolayers containing photo-active sites were prepared on a clean Si wafer. Spatially resolved surface photochemical techniques were used to selectively modify the organic monolayers, producing a pattern of chemically active amines. The amines could be subsequently functionalized to produce a pattern of some particular chemical interaction, or they could be used directly as nanocrystal binding sites. In this paper, we report on using this technique to selectively bind both semiconductor and metal nanocrystals.

Single-Electron Capacitance Devices

We have incorporated ordered, two-dimensional closest-packed phases of organically-functionalized metal nanocrystals into solid-state devices, and investigated their single electron charging properties. The device is described in the upper part of Figure 1. The nanocrystal arrays were prepared as Langmuir monolayers and subsequently transferred onto glass substrates as Langmuir Schaeffer films (*1*). The glass substrates had been patterned with a series of evaporated 0.5 mm wide aluminum stripes. The nanocrystal array was cross-linked with dithiol ligands for stabilization, similar to the technique described by Andres and coworkers (*2*). Polymethylmethacrylate (PMMA) was spin coated on top of the nanocrystal film and another striped pattern of Al electrode, perpendicular to the bottom pattern, was evaporated on the insulating polymer. One device existed at each point of electrode intersection. In this way, tens to hundreds of devices could be easily fabricated on a single 1 cm^2 glass wafer through the use of crude evaporation masks. Ultimately, the number of devices is limited only by the width and separation of the metal strips. AC impedance measurements were carried out to probe the single-electron charging characteristics of the devices under both ambient and 77 K conditions. Device capacitance as a function of the voltage applied between the Al electrodes was measured by varying a DC voltage across the electrodes while modulating it with a small amplitude sinusoidal wave and detecting the AC current with a lock-in amplifier.

Typical 77K ΔC-V curves are shown on the bottom left of Figure 1. The capacitance has a minimum around 0V bias as a result of the finite charging energy needed to load an electron onto each particle (the "Coulomb blockade"). Similar curves have been obtained for granular metal thin film devices (*3,4*). Generally, the C-V curves exhibit oscillations around 0V where the maxima correspond to a single electron (or hole, depending on polarity) charging of the particles. Ideally, the single particle curve should have a sharp "Coulomb staircase" structure, and, in fact, certain of our devices to exhibit such structure (*5*). Two factors tend to wash out such staircase structure in the current experiments. The first is a distribution in the particle size. Perhaps more important, however, is a distribution of single particle charging times during the cycle of the experiment. Either of these two phenomena will tend to smooth the observed curves after the first couple of oscillations. The difference in

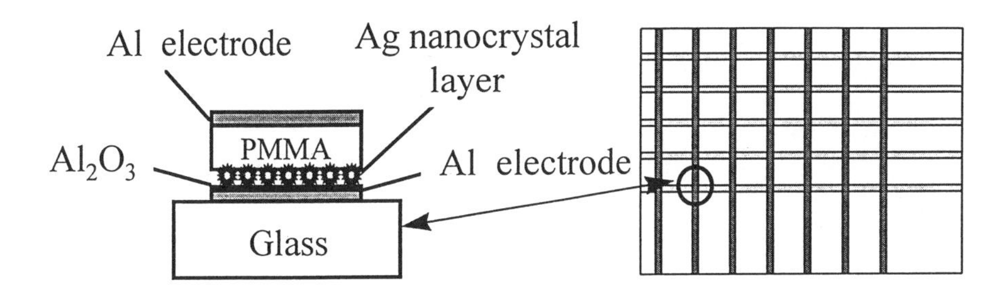

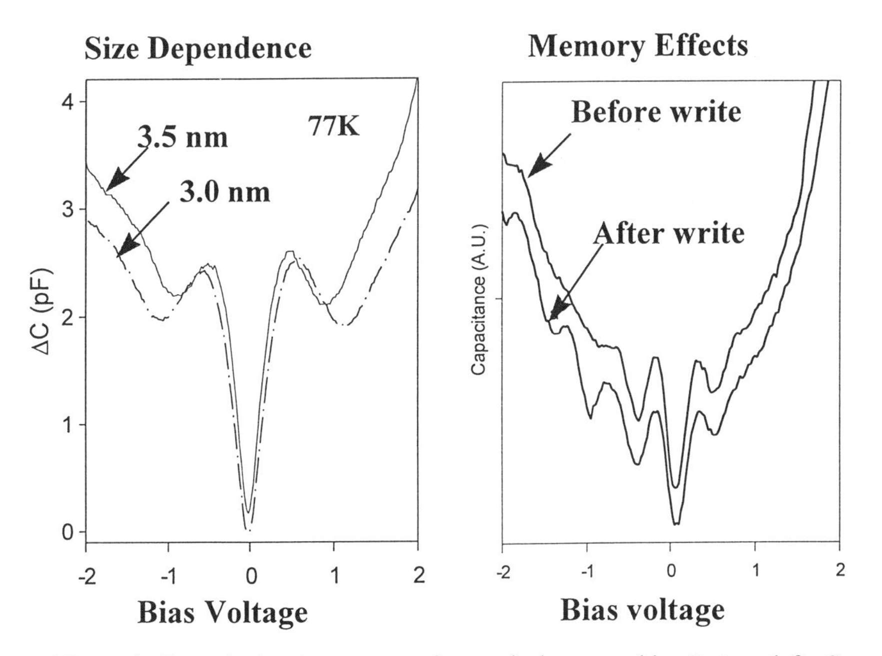

Figure 1. Top: single electron capacitance device assembly. Bottom left: Capacitance/voltage curves for two devices with different average particle sizes. Bottom right: The upper curve shows a regular C/V curve taken after biasing a device with 0V for a long period. The lower curve was taken after biasing the same device with -1.0V for 10 minutes, then switching to 0V and running a fast scan.

oscillation period between the two curves in the size dependence graph of Figure 1 originates in the different average particle size (d) in the two devices. The smaller the particle, the smaller is its capacitance ($\propto$d) and the larger is its charging energy ($\propto$1/C), which is proportional to the oscillation period.

These devices exhibit memory effects when cooled to 77K with a fixed DC bias on the electrodes. Once the voltage is removed, a 'memory' of the offset remains with the device. Evidence for this effect is presented at the bottom right of Figure 1. In this data, the C-V curve minima have split into two components. One component is unchanged, but a second component has shifted to the value of the pre-written DC voltage (= -1V) for a period of a few minutes after the DC was turned off (*3,4*).

Light Directed Assembly of Nanocrystals

Device related applications of nanocrystals often require precise spatial positioning of the particles into chemically and structurally complex environments. To do so, we have developed a strategy for the spatially resolved, light directed assembly of metal and semiconductor nanoparticles onto various substrates (*6*). Here, we describe how Pt and CdSe nanocrystalites can be assembled onto a lithographically defined micropattern of protected and deprotected amino groups attached onto a silicon substrate.

For preparing the light sensitive substrates, 3-aminopropyldimethylethoxysilane was attached onto a cleaned silicon wafer. This molecular forms a single chemical bond with the oxide surface layer on the silicon wafer, and presents and amine terminus at the air/wafer interface. These amine groups are subsequently reacted with nitroveratryloxycarbonyl glycine (NVOC-GLY) to give a NVOC terminated sample surface. NVOC is a photoremovable protecting group for primary amino groups that is used in combinatorial solid phase peptide synthesis (*7*). A pattern was photochemically defined onto the substrate surface by exposure to UV/VIS radiation ($\lambda >$ 340 nm) through a microchip mask. In order to scavenge reactive byproducts (aldehyde derivatives) of the photo-reaction which could re-react with the deprotected amino groups (*8*), a thin film of a semicarbazide hydrochloride solution in methanol was 'sandwiched' between the microchip mask and the wafer surface. This arrangement increases the yield of free amino groups as compared to the 'dry' photodeprotection – i.e. without the addition of a scavenger solution.

After photo-deprotection, the wafer was put into a solution of dodecylamine stabilized platinum particles (prepared in the same fashion as amine-terminated Au nanocrystals (*9*)) or trioctylphosphine oxide (TOPO) stabilized CdSe particles (*10,11*) in toluene. Since both types of nanoparticles can bind to free, deprotected amino groups protruding from the substrate surface, the particles assemble preferentially on areas of the wafer surface, that have been exposed to light. In order to enhance spatially resolved particle binding, we amplified the assembled pattern by coupling 1,7-diaminoheptane onto the first layer of particles attached to the surface. After this, the wafer was again exposed to the particle solution to bind more particles onto particles that have been attached to the wafer surface, before. Thus, previously bound particles themselves serve as a chemically selective 'substrate' for subsequent attachment of bifunctional ligands and further particle binding. This amplification of selective particle binding could be repeated several times. In the micrographs shown in Figure 2, the light directed assembly of Pt nanocrystals is selective (produces a cleaner pattern) than the assembly of CdSe particles.

Our initial attempts at binding Pt particles to patterned substrates initially resulted in 'noisy' patterns. In the binding process, we have found that the nanoparticle solution also plays a major role in pattern differentiation. Addition of various branched and linear alkylamines to the solution can have profound affects on particle assembly. The high contrast achieved for the Pt particle pattern was obtained by adding butylamine to the Pt/toluene solution. The CdSe pattern is a 'first generation' attempt, and we believe that enhanced differentiation for those patterns can also be achieved through the appropriate modification of the nanocrystal solution.

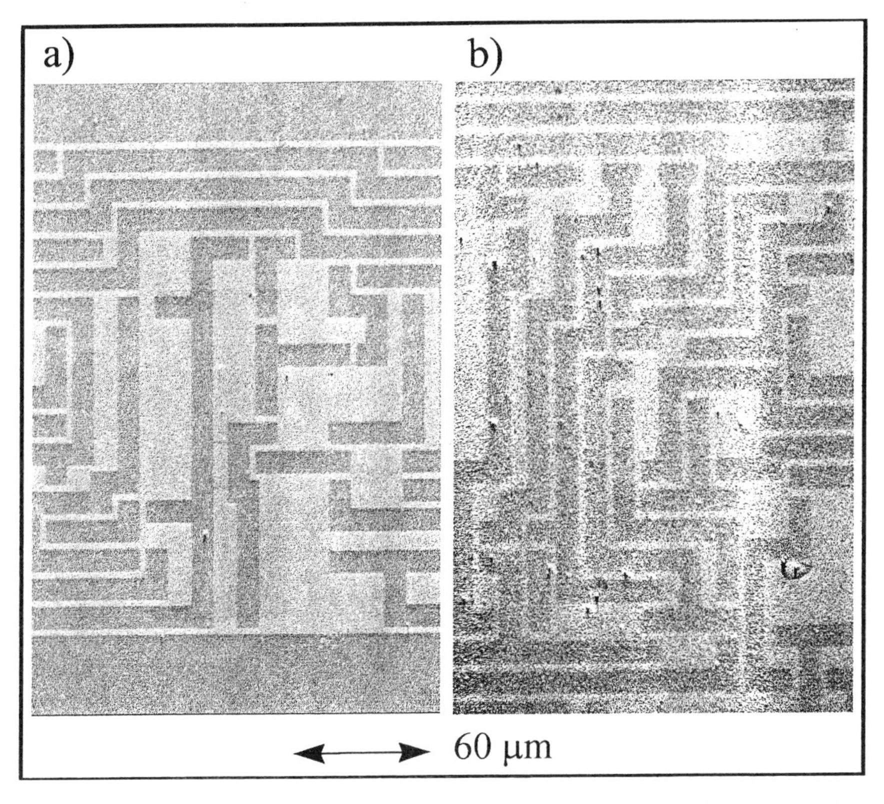

Figure 2. Light directed assemblies of Pt nanoparticles **a)** and CdSe nanoparticles **b)** on Si substrates. Each pattern was amplified three times by diaminoheptane treatment as described in the text. The lighter sections represent the areas on which the particles have assembled. The finest lines are approximately 3μm wide.

Literature Cited

1 Heath, J. R.; Knobler, C. M.; Leff, D. V. *J. Phys. Chem.* **1997**, *101*, 189.

2 Andres, R. P.; Bein, T.; Dorogi, M.; Feng, S.; Henderson, J. I.; Kubiak, C. P.; Mahoney, W.; Osifchin, R. G.; Reifenberger, R. *Science* **1996**, *272*, 1323.

3 Lambe, J.; Jaklevic, R. C. *Phys. Rev. Lett.* **1969**, *22*, 1371.

4 Cavicchi, R. E.; Silsbee, R. H. *Phys. Rev. B* **1988**, *38*, 6407.

5 Markovich, G.; Leff, D. V.; Chung, S. W.; Soyez, H. M.; Dunn, B.; Heath, J. R., *Appl. Phys. Lett.* **1997**, *70*, 3107.

6 Vossmeyer, T.; DeIonno, E.; Heath, J. R. *Angew. Chem. Int. Ed. Engl.* **1997**, *36*, 1080.

7 Fodor, S. P. A.; Read, J. L.; Pirrung, M. C.; Stryer, L.; Lu, A. T.; Solas, D., *Science* **1991**, *251*, 767.

8 Patchornik, A.; Amit, B.; Woodward, R. B. *J. Am. Chem. Soc.* **1970**, *92*, 6333.

9 Leff, D. V.; Brandt, L.; Heath, J. R. *Langmuir* **1996**, *12*, 4723

10 Murray, C. B.; Norris, D. J.; Bawendi, M. G. *J. Am. Chem. Soc.* **1993**, *115*, 8706.

11 Bowen Katari, J. E.; Colvin, V. L.; Alivisatos, A. P. *J. Phys. Chem.* **1994**, *98*, 4109.

Chapter 2

Nanometer-Scale Architecture Using Colloidal Gold

Christine D. Keating, Michael D. Musick, L. Andrew Lyon, Kenneth R. Brown, Bonnie E. Baker, David J. Pena, Daniel L. Feldheim, Thomas E. Mallouk, and Michael J. Natan[1]

Department of Chemistry, The Pennsylvania State University, University Park, PA 16802

New approaches to control the size and shape of colloidal Au nanoparticles, to tune interparticle spacing in two-dimensional Au nanoparticle arrays, and to prepare three-dimensional colloidal Au arrays exhibiting bulk-like optical and electrical properties are described. These studies derive their motivation from two sources. The first is very fundamental: progress is needed on numerous aspects of nanoparticle particle self assembly. The second is very applied: the nanostructure of noble metal films dictates a variety of exploitable bulk properties. In both cases, Au nanoparticles comprise a useful model system.

A comprehensive wish list in architecture of nanoparticulate materials might read as follows. One would like an assembly protocol that is applicable to particles of arbitrary composition, arbitrary size, and arbitrary shape. Such a protocol would be able to organize particles in a way that for a two-dimensional (2-D) material, a single (but adjustable) interparticle spacing obtains. For a 3-D material, the interlayer spacing should also be completely tunable. If desired, there should be a way to ensure registry between layers: that is, there should be a well-defined relationship between positions of particles in the nth and $(n+1)^{th}$ layers. Finally, these particle arrays should self-assemble, organizing in a prescribed fashion in response to an external stimulus (light, pH, temperature, etc.)

This is a very tall order, and even when several impressive recent accomplishments in mesoscale self-assembly are summed together (*1-7*), they pale in comparison to the hurdles yet to be traversed. To name but a few, a great deal of fundamental work is needed to learn how to control the particle size and shape of semiconductors and metals in the 20-200 nm range, to develop methods to adjust interparticle spacing in particle arrays that are not in contact or even near contact, and to understand the mechanisms by which 3-D arrays of particles assume bulk properties.

This contribution describes some progress along these lines, using colloidal Au as a model system. It is well-appreciated how bulk properties of noble metal films depend on nanometer-scale morphology (*8-9*). Accordingly, we have focused on issues in Au colloid self-assembly in some detail (*7,10-14*). Colloidal Au has a number of advantages as a model nanoparticle, amongst them size tunability from 3-150 nm (*15-16*), well-understood optical properties (*17*), and chemical stability.

[1]Corresponding author

Experimental Section

Extensive information regarding production and assembly of colloidal Au nanoparticles has been previously published (*7,10-14*). Experimental details concerning hydroxylamine-mediated particle growth and fabrication of colloidal Au rods will be described elsewhere (*18-19*).

Controlling Particle Size and Particle Shape of Colloidal Au

Although numerous routes to colloidal Au have been documented (*16,20-23*), synthesis of large colloidal Au particles (i.e. with dimensions >30 nm) poses three problems. The first is that preparations starting from $[AuCl_4]^-$ are imprecise: in other words, attempts to produce particles with a mean diameter of 65 nm just as often leads to particles with mean diameters of 55 nm or 75 nm. The second is that even when the mean diameter is acceptable, the distribution in particle size and shape is not. Larger particles lead inexorably to broad distributions in particle diameter and shape (*24*). Finally, synthetic routes to solutions of noble particles possessing high aspect ratios (i.e. the ratio of major to minor axes in prolate spheroids) are completely non-existent. Such particles have been predicted (*25*) to be extremely active for surface enhanced Raman scattering (SERS), a finding recently verified by experiment (*26*). We have investigated two new and very different routes to controlling the size and shape of colloidal Au. In the first, larger particles are grown by selective deposition of Au onto pre-formed particles. In the second, a published, template-driven electrochemical approach is used to grow extraordinarily long colloidal Au rods; conditions for growth of uniform rods as well as for rod release and resuspension have been explored.

"Seeding" By Au^{3+}/Hydroxylamine. The use of NH_2OH (and CN^-, for that matter) in electroless metal plating is driven by its tendency to reduce Au^{3+} only at metal surfaces and not in solution (*27*). We reasoned that if such properties carried over to colloidal Au nanoparticles, it would be possible to take solutions of pre-formed, monodisperse colloidal Au nanoparticles and carefully increase their size. A similar (but less successful) strategy was articulated in the European Patent literature for production of >100 nm colloidal Au from 2.5 nm "seeds" (*28*)– hence the name "seeding".

The basic experiment is carried out by mixing solutions containing 12-nm diameter colloidal Au, $HAuCl_4$, and NH_2OH (*18*). Different reagent permutations all lead to the same result: a growth in the effective diameter of the particles in solution. The process can be followed in real time by uv-vis: the surface plasmon band at ≈520 nm grows noticeably in magnitude within seconds of mixing. Importantly, control experiments indicate that in the absence of colloidal Au, mixing of NH_2OH and $HAuCl_4$ do not lead to formation of any Au nanoparticles over time scales from minutes to hours (*18*). A critical benefit to this approach is that by limiting the amount of $HAuCl_4$ added, a solution containing a known number of particles of one size can be grown to another, larger size. Thus, starting with 17 nM solutions (in particles) of 12-nm diameter colloidal Au, six stepwise growths were carried out, and in each case, the particle monodispersity was improved relative to particles of the same mean diameter grown directly by citrate reduction (*15, 24*) of $HAuCl_4$ (*18*).

Figure 1a shows the optical spectrum in H_2O and 1b the TEM image of a representative set of colloidal Au particles prepared by repetitive seeding. The uv-vis spectrum is relatively uninformative, exhibiting a λ_{max} typical for 30-40 nm colloidal Au. In contrast, the TEM image is noteworthy in two respects. First, the vast majority of particles are more uniformly-sized and -shaped than typical colloidal Au sols in this size regime. Second, a minority of particles are very irregularly shaped, with colloidal rods the most prominent motif. These rods make up a few percent of the total particle population, and as they are not present in earlier steps of the multi-step synthesis, they must result from a crystal-face selective NH_2OH reduction process. The exact nature of this reaction remains to be elucidated, but is clearly of importance, especially in light of the extreme activity of rod-like Ag particles in SERS (*26*). A final point to be gleaned from the data in Figure 1 is the weakness of uv-vis as a characterization method for colloidal Au: the optical spectrum provides no easily-extracted information on the size or shape of *minority* colloidal Au particles.

Template-Directed Growth of Rods. A quite different approach to synthesis of colloidal rods makes use of published work on electrochemical deposition of metals into cylindrical cavities of a host template (*29-32*). For example, vapor deposition of a metal onto one side of a polycarbonate membrane allows it to be used as electrode for spatially-selective metal ion reduction in the cavities. Colloidal metal rods of diameters from 10-400 nm and of lengths from 0.2-13 μm have been grown by this approach. What has not been investigated systematically is the current efficiency for metal deposition, the percentage of pores filled, the correlation between charge passed and rod length, techniques for rod extraction, and structural integrity of extracted rods– in other words, everything that is necessary to evaluate the utility of template-directed nanoparticle growth as a means to produce *free* particles.

We have investigated these parameters in some detail for the polycarbonate membrane system; Figure 2 is a TEM image of 30-nm diameter free-standing colloidal Au rods following release from a polycarbonate membrane. The colloidal Au rods exhibit a very uniform diameter and a very non-uniform length. The factors controlling this phenomenon will be described elsewhere (*19*), but it is gratifying to see that the rods can in fact be removed from the membrane host and manipulated in solution.

Controlling Interparticle Spacing in 2-D

If one allows nanoparticles coated by an organic film (either via molecular adsorption or covalent surface modification) to approach as closely as possible, the edge-to-edge interparticle spacing ends up being twice the length of the organic modifier. Domains of crystalline nanoparticle arrays have been prepared in this fashion (*1-5*). A much more challenging problem is to control interparticle spacing when the particles are not allowed to touch. For example, can an arbitrarily large 2-D hexagonal lattice of 20-nm diameter particles with a 30-nm edge-to-edge spacing be constructed? Clearly, such an architecture cannot be prepared by exploitation of molecular van der Waals' interactions, since rigid, thin, 15-nm long molecules are not available. We are pursuing two approaches to controlling interparticle spacing in 2-D. The first is a brute force approach, in which a large number of surfaces with variable particle size and variable particle coverage are rapidly generated (*13,33*). The second exploits the kinetics and thermodynamics of Au colloidal particle binding to organosilane-functionalized surfaces (*7,34*).

Combinatorial Approach. By sequential, stepped immersions of a organosilane-modified substrate into solutions of colloidal Au and a Ag^+ plating solution, respectively, it is possible to generate a "combinatorial" surface, exhibiting large numbers of different nanostructures on the same surface (*13*). In our initial foray, samples with continuous gradients were generated using linear translation rates. More recently, we have switched to time$^{1/2}$ ($t^{1/2}$)-based translation rates, exploiting the $t^{1/2}$-dependent kinetics of particle binding (*7*), and have investigated the excitation-

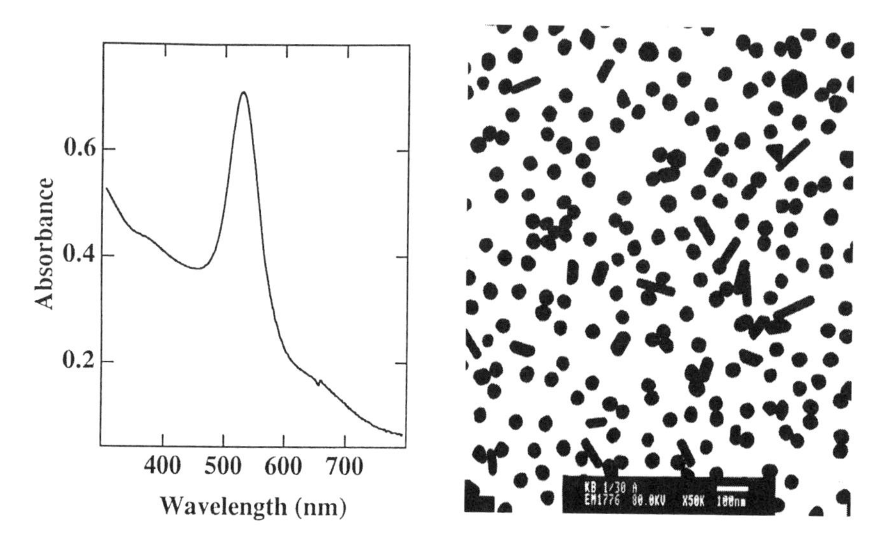

Figure 1. Optical spectrum (left) and TEM image (right) for colloidal Au particles prepared by repetitive seeding of 12-nm diameter colloidal Au with $NH_2OH/[AuCl_4]^-$.

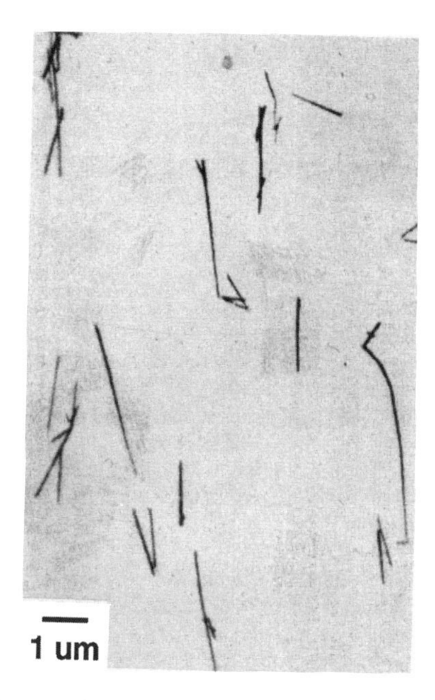

Figure 2. TEM image of 30-nm diameter Au rods following release from a polycarbonate membrane. Scale bar = 1 µm.

wavelength dependence of the SERS enhancement. Figure 3 shows a representative SERS intensity map for a surface comprising a 14-nm diameter Au gradient (30-minute maximum exposure) and a Ag overcoat gradient (40 minutes total immersion in Ag^+). The excitation wavelength was 647.1 nm, and the resulting SERS intensities for adsorbed *trans*-4,4'-bispyridylethylene were background-corrected and normalized against 30 mW of 514.5 nm excitation. The scale bar below the image correlates shading with SERS intensity. The data show that for this particular sample, maximum intensity was obtained at the lowest Au coverages and intermediate Ag coverages. Interestingly, the maximum SERS signal is found in other locations on the same sample when different excitation wavelengths are used (data not shown). For this sample, to achieve the optimal particle spacing for 647.1-nm excitation will require at least one more iteration, so that the maximum signal is found not at the perimeter: only when alternative nanostructures in all directions lead to lower signals can a nanostructure be called optimal (for a given particle size). It remains to correlate the nanostructure [as determined by atomic force microscopy (AFM)] to optical spectra and more importantly, to SERS theory: work along these lines is in progress (*33*).

Thermodynamic Approach. An alternate approach to controlling nanostructure in 2-D is to adjust the interparticle repulsion via surface charge, and to prepare particle arrays under thermodynamic control. We previously reported that binding of 12-nm diameter colloidal Au to 3-aminopropyltrimethoxysilane (APTMS)-coated glass exhibited at early times a sticking probability (p) of nearly one, and at longer times a sticking probability near zero that was governed by interparticle repulsion (*7*). This could be described as a "kinetic" approach to nanoarchitecture, in that interparticle spacing was governed in part by particle diffusion to the substrate. We reasoned that if p could be dramatically lowered without altering the equilibrium constant for adsorption, Au particles would bind in a more regular fashion, with the spacing governed solely by thermodynamic considerations.

One route to lowering p is by controlling the state of protonation of the amine group of APTMS, trivially accomplished by changing the pH. Figure 4 shows plots of 12-nm diameter Au particle coverage vs. time at pH 5.0, 3.5, and 9.0. Particle coverages were obtained by uv-vis, since Au colloidal films (both monolayers and multilayers) obey Beer's law over very large ranges of coverage. Also shown are non-linear least squares fits to the particle coverage (solid lines) at early times, from which p– the only adjustable fit parameter– was extracted. The data show that good fits to $t^{1/2}$ diffusion are obtained at all three values of pH. More importantly, p varies from nearly one at low-to-intermediate pH to a factor of ≈30 lower at pH 9. These data are consistent with the known negative charge on colloidal Au (*16*); as the local concentration of $-NH_3^+$ is decreased, p is reduced. At very long times, the particle coverage at pH 9 approaches that at pH 5 (the pH of as-prepared colloidal Au), indicating that the thermodynamics of adsorption (i.e. the equilibrium constant K) have not been substantially altered. These preliminary data are very encouraging, suggesting that with more exact control of pH, surfaces exhibiting ultralow p and high might be attainable. The nanostructure of such surfaces, especially those made with very monodisperse particles, will be quite interesting.

Controlling Particle Spacing in 3-D

When, if ever, do collections of Au nanoparticles adopt bulk electrical, optical, and/or dielectric properties, and how can these properties be tuned by controlling particle spacing in 3-D? It must first be pointed out that all three of these properties are of current interest for Au nanoparticulate films, either for fundamental studies or specific

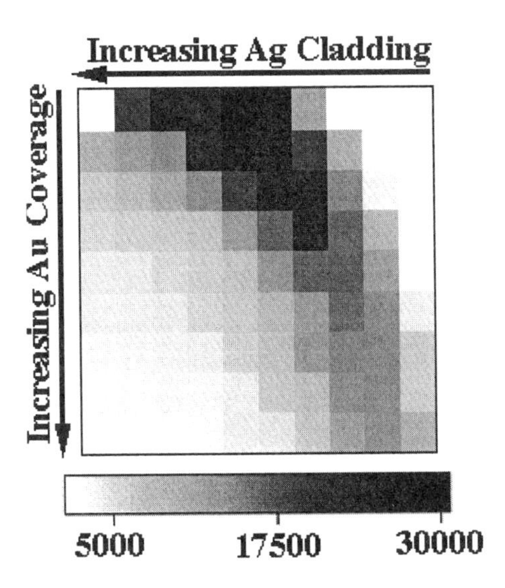

Figure 3. Background-subtracted, normalized (to 30 mW, 514.5-nm excitation) SERS intensity map of a 4-cm^2 sample (30 minute 14-nm diamter colloidal Au, 40-minute Ag^+ exposures; 1610 cm^{-1} stretch of *trans*-4,4'-bispyridylethylene; average of two, 2-second integrations on 2 mm x 2 mm area). Intensities = counts/s; λ_{ex} = 647.1, 5 cm^{-1} bandpass).

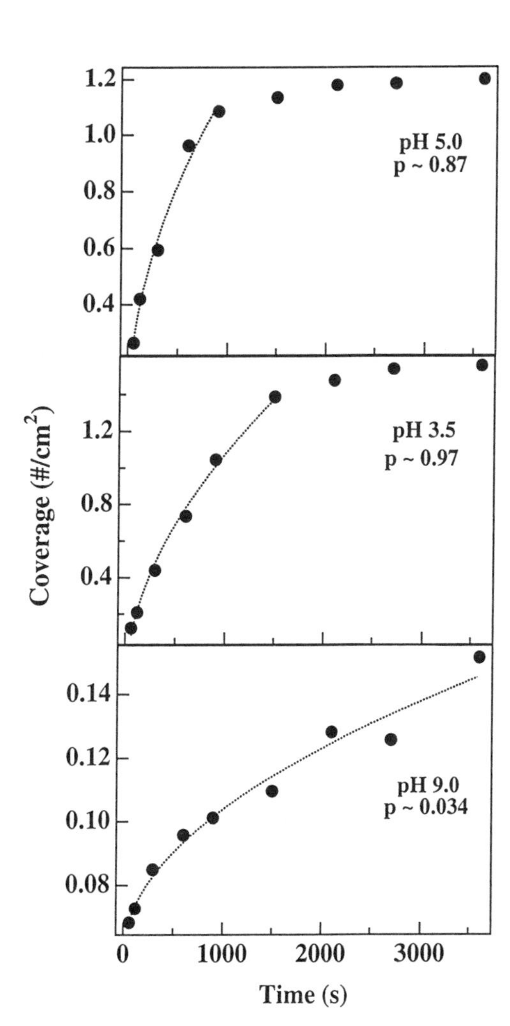

Figure 4. Particle coverage vs. time for 13-nm diameter colloidal Au on 3-aminopropyl-trimethoxysilane-coated glass slides at pH 5, 3.5, and 9, and non-linear least square fits (solid lines) as described in (*7*) .

applications (*35-37*). Brust et al. and Bethell et al. have reported stepwise assembly of Au colloid multilayers, formed by repetitive deposition of alkylthiol-coated Au nanoparticles onto a substrate (*38,39*). However, no information on nanostructure was provided, and correlation between optical and electrical properties was lacking. Such information is critical to nanometer-scale architecture in 3-D, which depends on knowing the geometric relationship between adjacent layers of particles.

Au Colloid Multilayers. Accordingly, we undertook a careful examination of Au colloid multilayer formation, starting from Au colloid monolayers. This stepwise approach consisted of repetitive dips in solutions of bifunctional organic crosslinker (followed by a wash step) and into solutions of colloidal Au (and another wash step) (*14,40*). Using 11-nm colloidal Au particles and two-carbon linkers (2-mercaptoethanol, 2-mercaptoethylamine), highly conductive surfaces can be prepared by 10-12 iterations of stepwise assembly onto 11-nm diameter Au colloid monolayers. Using uv-vis-near infrared spectrophotometry, AFM, two-point and/or four-point resistance measurements, and atomic absorption spectrometry, quantitative correlations between nanostructure and bulk properties have been made (*14,40*). Figure 5 shows a representative tapping-mode AFM image and a uv-vis-near IR transmission spectrum for a Au colloid multilayer after 5 immersion cycles into 2-mercaptethanol/11-nm diameter colloidal Au. The AFM image reveals pyramid-shaped mounds of particles, with occasional gaps between the bases. In the z-dimension (normal to the surface), the pyramid-shaped peaks are 50-70 nm above the lowest point imaged, consistent with up to six stacked particles. [It should be noted that the particle widths are inaccurate due to AFM tip convolution (*41*)]. This sample is only weakly transparent to visible and near-ir light, as evidenced by the low, wavelength-independent transmittance from 400 nm to 2.5 μm. In general, we have found that the % transmittance decreases with increasing particle coverage. At the same time, there is a dramatic decline in resistivity; while this sample had a measured resistance of 350 kΩ, increasing the particle coverage by a factor of two (to ≈70 x 10^{11}/cm^2) drops the resistance by 5 orders of magnitude (*14*). The resulting films are extraordinarily good conductors, even though the volume fraction of Au is only ≈0.5 (*40*).

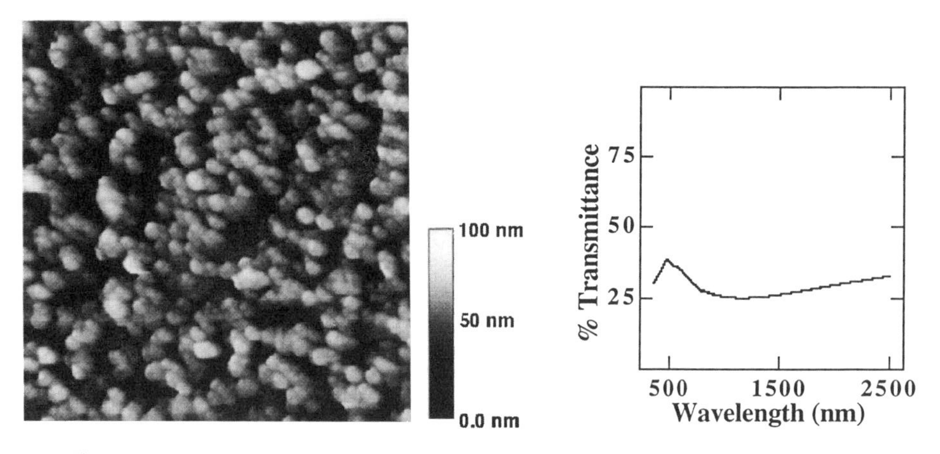

Figure 5. (Left) Tapping-mode AFM image (1 μm x 1 μm) of 2-mercaptoethanol-linked, 11-nm diameter Au colloid multilayer (monolayer + five treatments). (Right) Uv-vis near-infrared transmission spectrum of the same sample. Particle coverage = 38.7 x 10^{11} particles/cm^2.

Prospectus

The field of nanoparticulate materials assembly is still very much in its infancy, with the basic physics, chemistry, and engineering still being worked out. Though our accomplishments in this area have been modest, we are encouraged by the ease with which colloidal Au building blocks can be manipulated. And while we are fully aware that construction principles for colloidal Au may not necessarily be applicable to other nanoparticulate materials, we believe that their complete articulation is a worthwhile endeavor.

Acknowledgments

Support from NSF (CHE-9256692, CHE-9626326, CHE-9627338), NIH (GM55312-01), EPA (R825363-0100), USDA (96-35102-3840), and the Henkel Corporation/ACS Division of Colloid and Surface Chemistry (through a graduate fellowship to CDK) is gratefully acknowledged. DLF and TEM acknowledge support from DARPA. Acknowledgment is also made to the Electron Microscopy Facility for the Life Sciences in the Biotechnology Institute at The Pennsylvania State University.

Literature Cited

(1) Andres, R.P.; Bielefeld,J. D.; Henderson, J. I.; Janes, D. B.; Kolagunta, V. R.; Kubiak, C. P.; Mahoney, W. J.; Osifchin, R.G. *Science* **1996**, *273*, 1690-1693.

(2) Murray, C. B.; Kagan, C. R.; Bawendi, M. G. *Science* **1995,** *270*, 1335-1338.

(3) Harfenist, S. A.; Wang, Z. L.; Alvarez, M. M.; Vezmar, I.; Whetten, R. L. *J. Phys. Chem.* **1996**, *100*, 13904-13910.

(4) Leff, D. V.; Brandt, L.; Heath, J. R. *Langmuir* **1996**, *12*, 4723-4730.

(5) Alivisatos, A. P.; Johnsson, K. P.; Peng, X.; Wilson, T. E.; Loweth, C. J.; Bruchez, M. P., Jr.; Schultz, P. G. *Nature* **1996**, *382*, 607-611.

(6) Chang, S.-Y.; Liu, L.; Asher, S. A. *J. Am. Chem. Soc.* **1994**, *116*, 6739-6744.

(7) Grabar, K. C.; Smith, P. C.; Musick, M. D.; Davis, J. A.; Walter, D. G.; Jackson, M. A.; Guthrie, A. P.; Natan, M. J. *J. Am. Chem. Soc.*, **1996**, *118*, 1148-1153.

(8) Moskovits, M. *Rev. Modern Phys.* **1985**, *57*, 783-826.

(9) Brandt, E. S.; Cotton, T. M. In *Investigations of Surfaces and Interfaces-Part B*; 2nd ed.; B. W. Rossiter and R. C. Baetzold, Eds.; John Wiley & Sons: New York, 1993; Vol. IXB, Chapter 8, pp. 633-718.

(10) Freeman, R. G.; Grabar, K. C.; Allison, K. J.; Bright, R. M.; Davis, J. A.; Jackson, M. A.; Smith, P. C., Walter, D. G.; Natan, M. J. *Science* **1995**, *267*, 1629-1632.

(11) Grabar, K. C.; Freeman, R. G.; Hommer, M. B.; Natan, M. J. *Anal. Chem.* **1995**, *34*, 735-743.

(12) Grabar, K. G.; Allison, K. A.; Baker, B. E.; Bright, R. M.; Brown, K. B.; Keating, C. D.; Freeman, R. G.; Fox, A. F.; Musick, M. D.; Natan, M. J. *Langmuir* **1996**, *12*, 2353- 2361.

(13) Baker, B. E.; Kline, N. J.; Treado, P. J.; Natan, M. J. *J. Am. Chem. Soc.* **1996**, *118*, 8721-8722.

(14) Musick, M. D.; Keating, C. D.; Keefe, M.; Natan, M. J. *Chem. Mater.* **1997**, accepted for publication.

(15) Frens, G. *Nature Phys. Sci.* **1973**, *241*, 20-22.
(16) *Colloidal Gold: Principles, Methods, and Applications*; Hayat, M. A., Ed.; Academic Press: San Diego, 1989; Vol. 1-3.
(17) Bohren, C. F.; Huffman, D. R. *Absorption and Scattering of Light by Small Particles*; John Wiley & Sons: New York, NY, 1983.
(18) Brown, K. R.; Walter, D. W.; Natan, M. J., manuscript in preparation.
(19) Lyon, L. A.; Feldheim, D. L.; Mallouk, T. E.; Natan, M. J., manuscript in preparation.
(20) Chan, Y. N. C.; Schrock, R. R. *Chem. Mater.* **1992**, *4*, 24-27.
(21) Esumi, K.; Sato, N.; Torigoe, K.; Meguro, K. *J. Colloid Interface Sci.* **1992**, *149*, 295-298.
(23) Fukumi, K.; Chayahara, A.; Kadono, K.; Sakaguchi, T.; Horino, Y.; Miya, M.; Fujii, K.; Hayakawa, J.; Satou, M. *J. Appl. Phys.* **1994**, *75*, 3075-3080.
(22) Krasnansky, R.; Yamamura, S.; Thomas, J. K.; Dellaguardia, R. *Langmuir* **1991**, *7*, 2881-2886.
(24) Goodman, S. L.; Hodges, G. M.; Trejdosiewicz, L. K.; Livingston, D. C. *J. Microsc.* **1981**, *123*, 201-213.
(25) Zeman, E. J.; Schatz, G. C. *J. Phys. Chem.* **1987**, *91*, 634-643.
(26) Emory, S. R.; Nie, S. *Science* **1997**, *275*, 1102-1106.
(27) Stremsdoerfer, G.; Perrot, H.; Martin, J. R.; Clechet, P. *J. Electrochem. Soc.* **1988**, *135*, 2881-2886.
(28) Schutt, E.G. Eur. Patent Application #90317671.4, filed September 25, 1990.
(29) Martin, C. R. *Science* **1994**, *266*, 1961-1966.
(30) Hornyak, G. L.; Patrissi, C. J.; Martin, C. R. *J. Phys. Chem. B* **1997**, *101*, 1548-1555.
(31) Preston, C.K.; Moskovits, M. *J. Phys. Chem.* **1993**, *97*, 8495-8503.
(32) Routkevitch, D.; Bigioni, T.; Moskovits, M.; Xu, J. M. *J. Phys. Chem.* **1996**, *100*, 14037-14047.
(33) Baker, B. E.; Natan, M. J., unpublished results.
(34) Keating, C. D.; Musick, M. D., Natan, M. J., unpublished results.
(35) Fukumi, K.; Chayahara, A.; Kadono, K.; Sakaguchi, T.; Horino, Y.; Miya, M.; Fujii, K.; Hayakawa, J.; Satou, M. *J. Appl. Phys.* **1994**, *75*, 3075-3080.
(36) Yagil, Y.; Deutscher, G. *Appl. Phys. Lett.* **1988**, *52*, 373-374.
(37) Vezzoli, G. C.; Chen, M. F.; Caslavsky, *J. Nanostruct. Mat.* **1994**, *4*, 985-1009.
(38) Bethell, D.; Brust, M.; Schiffrin, D. J.; Keily, C. *J. Electroanal. Chem.* **1996**, *409*, 137-143.
(39) Brust, M.; Walker, M.; Bethell, D.; Schriffin, D. J.; Whyman, R. *J. Chem. Soc., Chem Comm.* **1994**, 801-802.
(40) Musick, M. D.; Keating, C. D.; Keefe, M. H.; Natan, M. J., manuscript in preparation.
(41) Grabar, K. C.; Brown, K. R.; Keating, C. D.; Stranick, S. J.; Tang, S.-L.; Natan, M. J. *Anal. Chem.* **1997**, *69*, 471-477.

Chapter 3

Heterosupramolecular Chemistry

The Self-Assembly and Self-Organization of Nanoscale Devices and Materials in Solution

Lucy Cusack, S. Nagaraja Rao, and Donald Fitzmaurice[1]

Department of Chemistry, University College—Dublin Dublin 4, Ireland

TiO_2 nanocrystallites have been prepared by hydrolysis of titanium tetraisopropoxide in the presence of a stabiliser modified by incorporation of a diaminopyridine moiety. The latter selectively bind, by complementary hydrogen bonding, viologens modified by incorporation of a uracil moiety. Bandgap excitation of a semiconductor nanocrystallite results in electron transfer to a bound viologen. TiO_2 nanocrystallites have also been prepared by hydrolysis of titanium tetraisopropoxide in the presence of a stabiliser modified by incorporation a uracil moiety. The latter are selectively bound by TiO_2 nanocrystallites prepared in the presence of a stabiliser modified by a diaminopyridine. Subsequent self-organisation yields an ordered array, or superlattice, of these semiconductor nanocrystallites. Some implications of these and related findings for the bottom-up assembly of nanometer scale devices and materials in solution are considered.

Conventionally, the term supermolecule is applied to non-covalently linked molecular components provided their intrinsic properties remain largely unperturbed and there exists a well defined supramolecular function or property (*1*).

With the fruitful application of supramolecular concepts throughout chemistry however, has come the need for a more inclusive definition. Accordingly, the term supermolecule is increasingly applied to covalently linked molecular components provided, as above, their intrinsic properties remain largely unperturbed and there exists a well defined supramolecular function or property (*2*).

Recent work directed toward development of a systematic chemistry of molecular and condensed phase components requires a more inclusive definition still (*3,4*). Accordingly, the term heterosupermolecule is applied to non-covalently or covalently linked molecular and condensed phase components whose intrinsic properties are largely unperturbed and which possess a well defined heterosupramolecular function or property.

In this Chapter we will consider two examples of heterosupramolecular chemistry. Firstly, the self-assembly of a condensed phase electron donor and a molecular electron acceptor in solution to form a heterosupramolecular dyad whose associated heterosupramolecular function is light-induced vectorial electron transfer (*4*).

[1]Corresponding author

Secondly, the self-assembly and self-organisation of condensed phase components in solution to form a superlattice (*5*). Examples of related studies from other laboratories will also be considered (*6*). Finally, the potential of such studies for the bottom-up assembly of nanometer scale devices in solution will be considered.

Heterosupramolecular Dyad

Described is self-assembly of a modified TiO_2 nanocrystallite and a modified viologen in solution to form the heterosupramolecular dyad denoted TiO_2-(I+III) in Scheme I. Also described, is characterisation of the associated heterosupramolecular function, namely light-induced vectorial electron transfer.

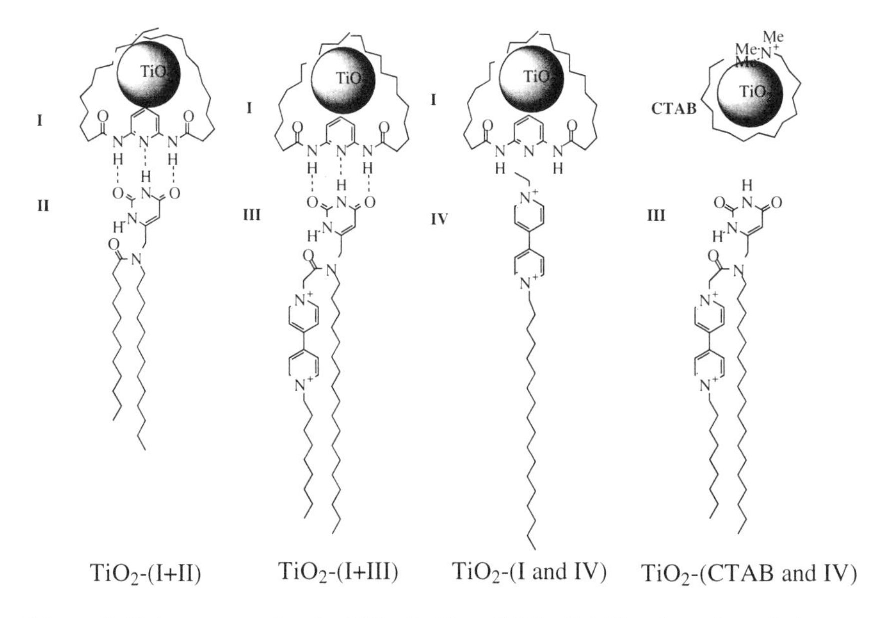

Scheme I: Heterosupermolecules TiO_2-(I+II) and TiO_2-(I+III) and condensed phase and molecular component mixtures TiO_2-(I and IV) and TiO_2-(CTAB and III).

Preparation of Molecular and Condensed Phase Components. Preparation of the molecular and condensed phase components shown in Scheme I has been described in detail elsewhere (*4*). Briefly, the molecular components (I, II, III and IV) were prepared following methods reported by Lehn and co-workers (*7*), while the condensed phase components (TiO_2-I and TiO_2-CTAB) were prepared following methods reported by Fendler and co-workers (*8*). It is noted that the resulting TiO_2 (anatase) nanocrystallites were 22±2 Å in diameter and that the molar particle concentration was 3×10^{-5} mol dm^{-3}.

Self-assembly of a Heterosupramolecular Dyad. It has been demonstrated by Lehn and co-workers that diaminopyridine and uracil, appropriately derivatised by addition of alkane chains, self-assemble in chloroform by formation of an array of three complementary hydrogen bonds (*4,7,9*). The strategy adopted therefore was to modify the stabiliser used to prepare a colloidal dispersion of TiO_2 by incorporation of a diaminopyridine moiety so that the resulting nanocrystallites were capable recognising and selectively binding substrates incorporating a uracil moiety. The molecular electron acceptor, namely viologen, was modified by incorporation of a uracil moiety.

Shown in Figure 1 are the ^{1}H NMR spectra recorded for TiO_2-I, III and a 1:1 mixture of TiO_2-I and III, denoted TiO_2-(I+III), in chloroform-d/acetone-d_6. The amidic proton resonances of TiO_2-I are observed at ∂ 8.67. The amidic and imidic proton resonances of III are observed at ∂ 9.35 and ∂ 9.70 respectively. The basis for these assignments have been discussed in detail for TiO_2-(I+II) (*4*). For TiO_2-(I+III), the above resonances are observed at ∂ 8.76, ∂ 9.36 and ∂ 9.88 respectively. The measured down field shifts in the resonances assigned to the amidic protons of I and the imidic proton of III are consistent with self-assembly, by complementary hydrogen bonding, of the heterosupramolecular complex TiO_2-(I+III).

^{1}H NMR spectra were also recorded for a 1:1 mixture of TiO_2-I and IV, denoted TiO_2-(I and IV), and a 1:1 mixture of TiO_2-CTAB and III, denoted TiO_2-(CTAB and III), in chloroform-d/acetone-d_6. For TiO_2-(I and IV), the amidic proton resonances of TiO_2-I are observed at ∂ 8.67 both prior to and following addition of IV. This observation is consistent with the assertion that TiO_2-I and IV do not self-assemble to form a heterosupermolecule by complementary hydrogen bonding. Similarly, for TiO_2-(CTAB and III), the imidic proton resonance of III is observed at ∂ 9.79 both prior to and following its addition to TiO_2-CTAB. This observation is, again, consistent with the assertion that TiO_2-CTAB and III do not self-assemble to form a heterosupermolecule by complementary hydrogen bonding.

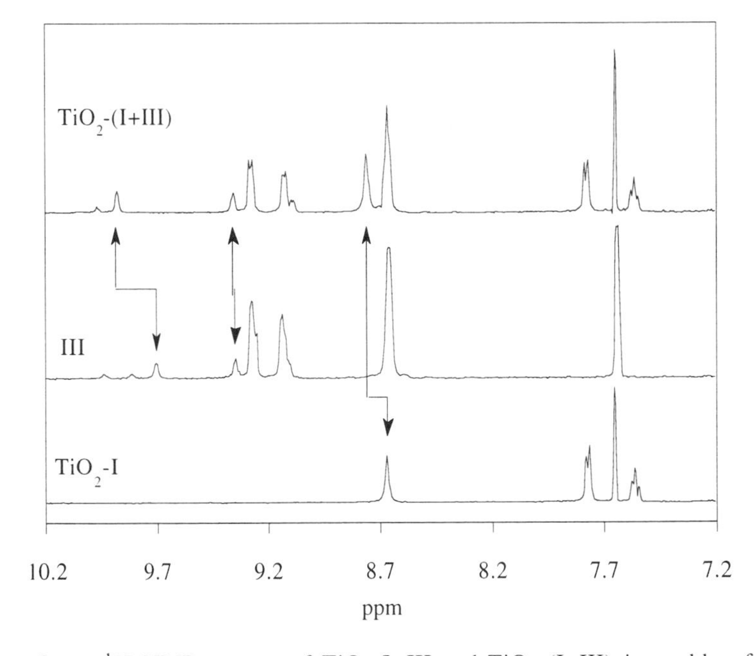

Figure One : ^{1}H NMR spectra of TiO_2-I, III and TiO_2-(I+III) in a chloroform-d/acetone-d_6 mixture (1:1 by volume) at 20°C. The nanocrystallite concentration was $3x10^{-5}$ mol dm^{-3} while the concentration of I and III was $4x10^{-3}$ mol dm^{-3}. Under these conditions about half the molecules of I present are adsorbed at a nanocrystallite surface.

In short, both a TiO_2 nanocrystallite prepared in the presence of a stabiliser incorporating a diaminopyridine moiety and a viologen incorporating a uracil moiety are necessary to self-assemble the heterosupermolecule TiO_2-(I+III) in solution.

Light-Induced Electron Transfer in Heterosupramolecular Dyad. Based on the above findings it was expected that bandgap excitation of the TiO_2 nanocrystallite in TiO_2-(III) would result in electron transfer to a viologen, but that bandgap excitation of the TiO_2 nanocrystallite in either TiO_2-(I and IV) or TiO_2-(CTAB and III) would not. Transient and steady-state absorption experiments were performed to test this expectation.

Transient Optical Absorption Spectroscopy of TiO_2-(I+III). The μs absorption transient measured following bandgap excitation of TiO_2-I in deaerated chloroform/acetone is shown in Figure 2. The corresponding ms transient (not shown) was also measured. Concerning the above we note the following : Firstly, the μs transient is not measurably dependent on the extent to which TiO_2-I is deaerated. Secondly, while the ms transient for deaerated TiO_2-I decays to about 50% of its initial amplitude between pulses, the same transient for aerated TiO_2-I decays fully. Thirdly, while the visible spectrum measured following bandgap irradiation of deaerated TiO_2-I agrees with that reported for photogenerated electrons trapped in a TiO_2 nanocrystallite (*10,11*), no spectrum is measured for aerated TiO_2-I under the same conditions. On this basis, the μs transient in Figure 2 is assigned to long-lived electrons trapped in the TiO_2 nanocrystallite of TiO_2-I (4).

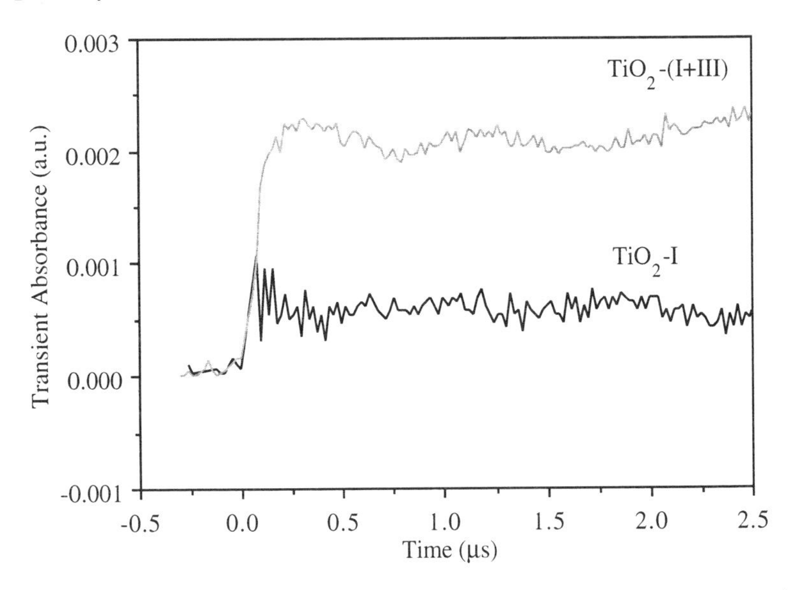

Figure Two : Transient absorption at 600 nm for TiO_2-I and TiO_2-(I+III) in a deaerated chloroform/acetone mixture (1:1 by vol.) at 25°C following bandgap excitation at 355 nm (2 mJ per pulse). The nanocrystallite concentration was 3×10^{-5} mol dm^{-3} while the concentration of I and III was 4×10^{-4} mol dm^{-3}. Under these conditions about four fifths of the molecules of I present are adsorbed at a nanocrystallite surface.

The μs absorption transient measured following bandgap excitation of TiO_2-(I+III) in deaerated chloroform/acetone is also shown in Figure 2. The corresponding ms transient (not shown) was also measured. Concerning the above we note the following : Firstly, the μs transient is not measurably dependent on the extent to which TiO_2-(I+III) is deaerated. Secondly, while the ms transient for deaerated TiO_2-(I+III) rises initially within the laser pulse and slowly during 10 ms, the ms transient for aerated TiO_2-(I+III) rises only within the laser pulse. Thirdly, while the ms transient for deaerated TiO_2-(I+III) decays to about 80% of its initial amplitude, the same transient for aerated TiO_2-(I+III) decays fully. Fourthly, while the visible spectrum measured following bandgap irradiation of deaerated TiO_2-(I+III) agrees well with that reported for the radical cation of viologens (*12*), and as expected is offset from the baseline due to the presence of long-lived electrons trapped in the TiO_2 nanocrystallites of TiO_2-I (*10,11*), no spectrum is measured for aerated TiO_2-(I+III) under the same conditions. On this basis, the μs transient in Figure 2 is assigned to radical cation of hydrogen bonded III and to long-lived electrons trapped in the TiO_2 nanocrystallite of TiO_2-(I+III) (*12*). Also on this basis, the slow component of the μs transient for deaerated TiO_2-(I+III) is assigned to the radical cation of III formed by diffusion to TiO_2-I (*13*).

Transient Optical Absorption Spectroscopy of TiO_2-(I and IV) and TiO_2-(CTAB and III). The μs absorption transient measured following bandgap excitation of TiO_2-(I and IV) in deaerated chloroform/acetone is shown in Figure 3. The corresponding ms transient (not shown) was also measured. Concerning the above we note the following : Firstly, the μs transient is not measurably dependent on the extent to which TiO_2-(I and IV) is deaerated. Secondly, while the ms transient for deaerated TiO_2-(I and IV) rises within the laser pulse and slowly during 10 ms, the ms transient for aerated TiO_2-(I and IV) rises only within the laser pulse. Thirdly, while the ms transient for deaerated TiO_2-(I and IV) decays to about half its maximum amplitude, the same transient for aerated TiO_2-(I and IV) decays fully and agrees well with that for aerated TiO_2-I. Fourthly, while the visible spectrum measured following bandgap irradiation of degassed TiO_2-(I and IV) agrees well with that reported for the radical cation of viologens (*12*), and is offset from the baseline due to the presence of long-lived electrons trapped in the TiO_2 nanocrystallites of TiO_2-I (*10,11*), no spectrum is measured for aerated TiO_2-(I and IV) under the same conditions. On this basis, the μs transient is assigned to electrons trapped in a TiO_2 nanocrystallite and to the radical cation (slow component) of IV formed by diffusion to the surface of TiO_2-I (*13*).

The μs absorption transient measured following bandgap excitation of TiO_2-(CTAB and III) in deaerated chloroform/acetone is also shown in Figure 3. The corresponding ms transient (not shown) was also measured. Qualitatively, these spectra agree with those measured for TiO_2-(I and IV). On this basis, the μs transient in Figure 3 is assigned is assigned to electrons trapped in a TiO_2 nanocrystallite and to the radical cation of III formed by diffusion to the surface of TiO_2-CTAB (*14*).

Efficiency. As the optical absorption at 355 nm of the nanocrystallite in TiO_2-I nm is 0.1 a.u, the pulse energy is 2 mJ, the cross sectional area for irradiation is 0.4 cm^2 and assumed reflection losses are 20%, it is estimated that one electron-hole pairs is generated in every thirteenth TiO_2 nanocrystallite (*4*). From the initial amplitude of the μs transient in Figure 2 for deaerated TiO_2-(I+III) and the known extinction coefficient for the reduced form of viologen (*12*), it is estimated that the charge separation efficiency is about 6% (*10,11*). From the initial amplitudes of the μs transients for degassed TiO_2-(I and IV) and TiO_2-(CTAB and III), equal to that for degassed TiO_2-I, it is clear that no radical cations of IV and III, respectively, are formed within the laser pulse although, as for TiO_2-(I+III), they are subsequently formed on the ms time-scale by diffusion.

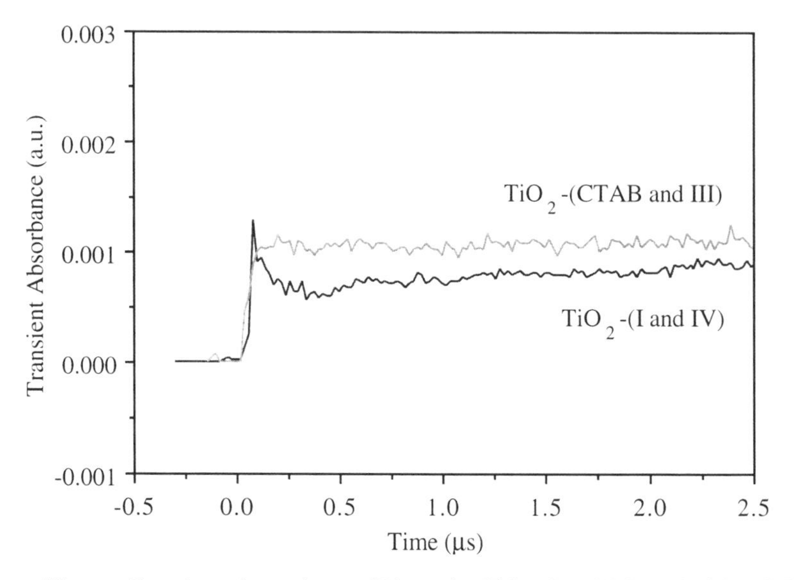

Figure Three : Transient absorption at 600 nm for TiO_2-(I and IV) and TiO_2-(CTAB and III) in a deaerated chloroform/acetone mixture (1:1: by vol.) at 25°C following bandgap excitation at 355 nm (2 mJ per pulse). The nanocrystallite concentration was $3x10^{-5}$ mol dm^{-3} while the concentration of I and III was $4x10^{-4}$ mol dm^{-3}. Under these conditions about four fifths of the molecules of I present are adsorbed at a nanocrystallite surface.

Summary. TiO_2-I containing a diaminopyridine moiety recognises and selectively binds the modified viologen III containing a uracil moiety. Light-induced vectorial electron flow is observed for the resulting donor-acceptor complex shown in Scheme II. It is noted that, there is ample precedent for electron transfer over long distances in supermolecules and their organised assemblies (*15*). In the absence of a uracil moiety the modified nanocrystallite TiO_2-I does not recognise or selectively bind the viologen IV. Similarly, in the absence of a diaminopyridine moiety the modified nanocrystallite TiO_2-CTAB does not recognise or selectively bind the modified viologen III. In neither case is direct light-induced electron transfer to the viologen moiety observed.

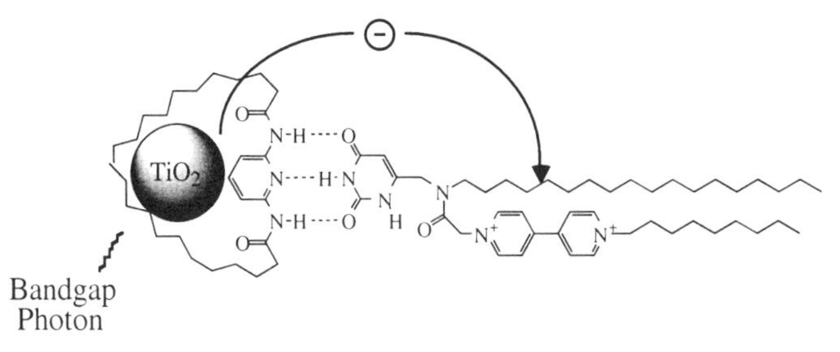

Scheme II: Light-induced vectorial electron flow is observed following bandgap excitation of the constituent nanocrystallite of the heterosupermolecule TiO_2-(I+III).

Heterosupramolecular Assembly

In light of the findings outlined above, it was expected that two appropriately modified nanocrystallites would recognise and selectively bind each other as shown in Scheme III. Specifically, that a TiO_2 nanocrystallite prepared in the presence of a stabiliser incorporating a diaminopyridine moiety, TiO_2-I, would recognise and selectively bind a TiO_2 nanocrystallite prepared in the presence of the stabiliser incorporating a uracil moiety, TiO_2-II. It was also hoped that the resulting heterosupermolecules, denoted TiO_2-(I+II)-TiO_2, would self-organise to form a superlattice.

Preparation of Molecular and Condensed Phase Components. Preparation of the molecular and condensed phase components shown in Scheme III has been outlined above and described in detail elsewhere (*4*).

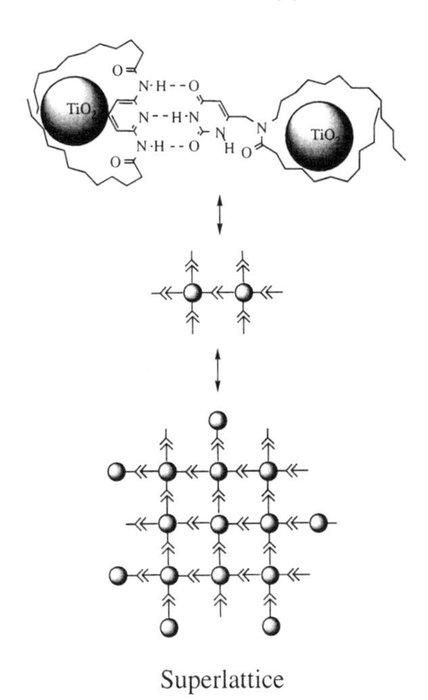

Scheme III: Self-assembly of the heterosupermolecule TiO_2-(I+II)-TiO_2 and its subsequent self-organisation to form a superlattice.

Self-assembly and Self-Organisation of a Heterosupramolecular Assembly. Shown in Figure 4 are ^1H NMR spectra of TiO_2-I, TiO_2-II and a 1:1 mixture of TiO_2-I and TiO_2-III, denoted TiO_2-(I+II)-TiO_2, in chloroform-d. Upon mixing of TiO_2-I and TiO_2-II to form TiO_2-(I+II)-TiO_2, the resonance at ∂ 7.53 assigned to the amidic protons in TiO_2-I and the resonance at ∂ 8.02 assigned to the imidic proton in TiO_2-II are shifted down-field by less than 0.3 ppm to ∂ 7.64 and ∂ 8.29 respectively. Following 10 days ageing however, the resonances assigned to the amidic protons in TiO_2-I and the imidic proton in TiO_2-II are shifted down-field to ∂ 8.23 and ∂ 9.52 respectively. It is clear therefore, that TiO_2-I and TiO_2-II self-assemble by complementary hydrogen bonding to form the heterosupermolecule TiO_2-(I+II)-TiO_2 shown in Scheme III(*5*), but that the above hydrogen bonds are formed on a time-scale of days and are weaker than those formed immediately upon mixing I and

II (*4*). It should be noted that this latter observation strongly supports the assertion that I and II are mostly at the surface of a TiO_2 nanocrystallite (*4*).

The average diameter of the aggregates present in TiO_2-I, TiO_2-II and TiO_2-(I+II)-TiO_2 as determined by dynamic light scattering is plotted in Figure 5. As expected, some aggregation is observed in both TiO_2-I and TiO_2-II during 10 days. Specifically, the average aggregate diameter in TiO_2-I increases from an initial value of less than 10 nm to 50 nm, while in TiO_2-II it increases from an initial value of less than 10 nm to 27 nm. In the case of TiO_2-(I+II)-TiO_2 however, the extent of aggregation during 10 days is significantly greater. Specifically, the average aggregate diameter increases from an initial value of less than 10 nm, to a value of 260 nm after 3 days and a final value of greater than 1000 nm after 10 days. Low-resolution electron micrographs of samples prepared from the dispersions aged for 10 days confirm that, in the case of TiO_2-I and TiO_2-II there is limited aggregation. However, low resolution micrographs confirm that, in the case of TiO_2-(I+II)-TiO_2 there is extensive aggregation leading to the formation of mesoaggregates possessing diameters in the range 500 nm - 1500 nm.

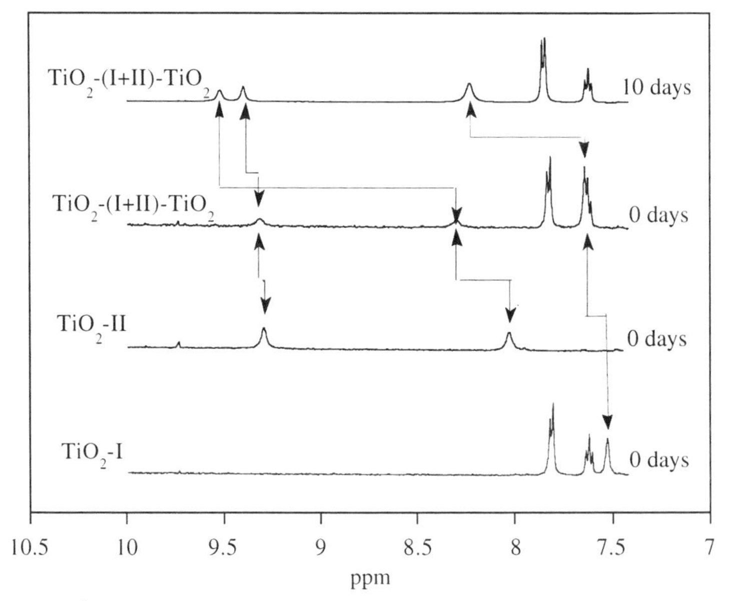

Figure Four : ^1H NMR spectra in chloroform-d at 20°C of TiO_2-I, TiO_2-II after 0 days and of TiO_2-(I+II)-TiO_2 after 0 and 10 days. The nanocrystallite concentration was $3x10^{-5}$ mol dm^{-3} while the concentration of I and III was $4x10^{-4}$ mol dm^{-3}. Under these conditions about four fifths of the molecules of I and II present are adsorbed at a nanocrystallite surface.

The light scattering and electron microscopy studies described above support the view that TiO_2-I and TiO_2-II self-assemble, by complementary hydrogen bonding as shown in Scheme III, to form the heterosupermolecule TiO_2-(I+II)-TiO_2. However, they also clearly demonstrate that these heterosupermolecules subsequently self-organise to form mesoaggregates. Stating that the heterosupermolecule TiO_2-(I+II)-TiO_2 self-organises to form a mesoaggregate implies an ordering of the constituent nanocrystallites.

That this is the case, or that a superlattice is indeed formed, is clearly seen from the medium and high resolution electron micrographs of a mesoaggregate shown in Figure 6. From the corresponding high-resolution electron micrograph in Figure 6, it is also clear that the superlattice consists of nanocrystallites that are 21±2 Å in diameter and organised in planes that are separated by 6±2 Å. This, in turn, is consistent with an observed onset for absorption of 350±10 nm and the presence of the modified stabilisers I and II at the surface of the constituent nanocrystallites of the superlattice (*16*).

Summary. It has been demonstrated that the methodology whose development has accompanied the self-assembly of heterosupermolecules in solution may be extended to permit the self-assembly and self-organisation of heterosupramolecular assemblies, also in solution. Specifically, it has been demonstrated that a TiO_2 nanocrystallite prepared in the presence of a stabiliser incorporating a diaminopyridine moiety recognises and selectively binds a TiO_2 nanocrystallite prepared in the presence of a stabiliser incorporating a uracil moiety to form the heterosupermolecule TiO_2-(I+II)-TiO_2, and that the resulting heterosupermolecules self-organise to form a superlattice or heterosupramolecular assembly.

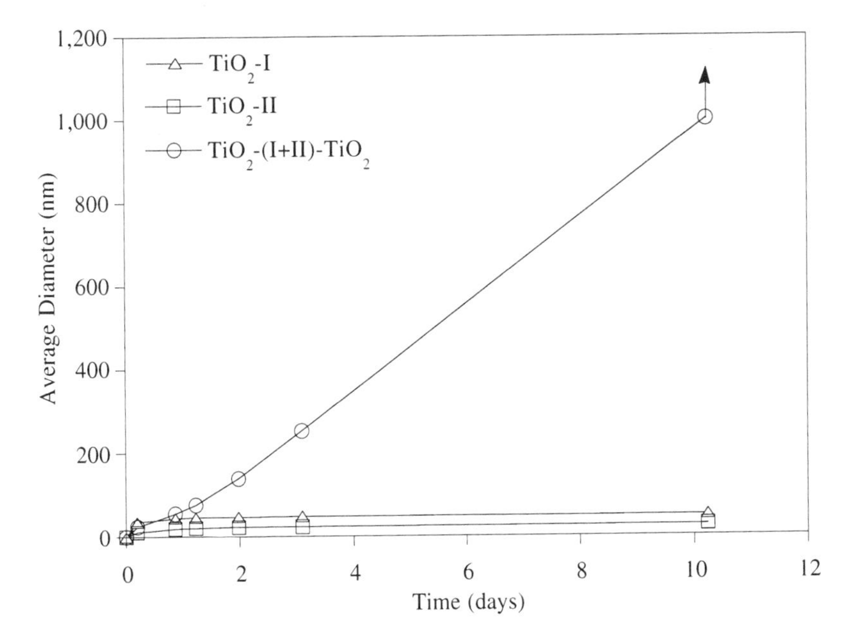

Figure 5: Average aggregate diameter in TiO_2-I, TiO_2-II and TiO_2-(I+II)-TiO_2 (1:1 mixture by vol.) during 10 days ageing in the dark at 25 ^oC. The nanocrystallite concentration was $3x10^{-5}$ mol dm^{-3} while the concentration of I and III was $4x10^{-4}$ mol dm^{-3}. Under these conditions about four fifths of the molecules of I and II present are adsorbed at a nanocrystallite surface.

Heterosupramolecular Chemistry

The self-assembly and self-organisation of complex nanostructures in solution is an important objective of materials chemistry and physics. The importance of this objective is a result of a desire to be able to programme the bottom-up assembly of nanometer scale devices in solution. It is in this context, that we have sought to develop a systematic chemistry, termed heterosupramolecular chemistry, of non-covalently and covalently linked condensed phase and molecular components whose intrinsic properties largely persist but which possess well defined heterosupramolecular functions or properties (*4*,*5*).

The difficulties encountered to date, and those likely to be encountered in the future, in developing a heterosupramolecular chemistry are justified in detail below. Briefly, however, the development of such a chemistry is expected to greatly facilitate the programmed bottom-up assembly of nanometer-scale devices in solution. A detailed justification of this latter assertion is given below.

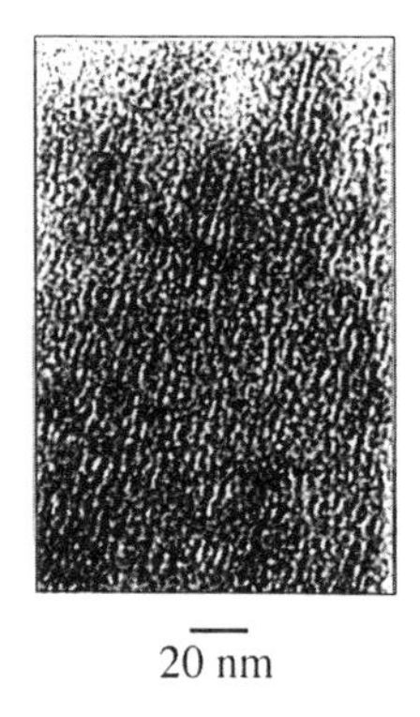

20 nm

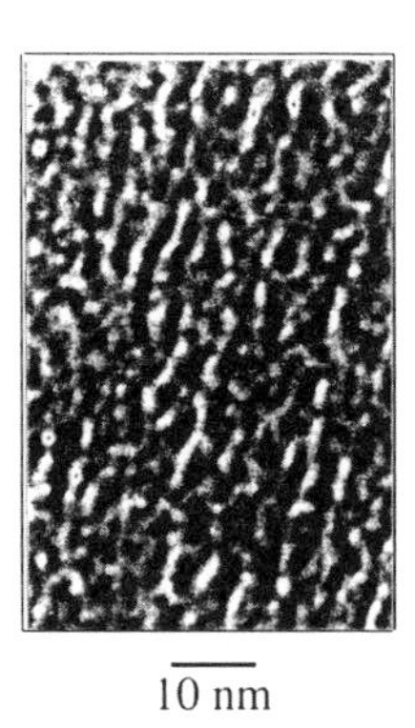

10 nm

Figure Six : Medium and high resolution electron micrographs of TiO_2-(I+II)-TiO_2 mesoaggregate in Figure 5 after 10 days. The nanocrystallite concentration was $3x10^{-5}$ mol dm^{-3} while the concentration of I and III was $4x10^{-4}$ mol dm^{-3}. Under these conditions about four fifths of the molecules of I and II present are adsorbed at a nanocrystallite surface.

The use of both condensed phase and molecular components will permit heterosupermolecules and heterosupramolecular assemblies possessing novel functions and properties to be self-assembled and self-organised in solution. In this context, recent advances in the preparation of nanocrystallites of a wide range of materials possessing well defined sizes, surface properties and crystal structures have been important (*17*). Also of importance have been the development of strategies for linking molecules, typically capping groups or sensitiser molecules, to the surface of these nanocrystallites (*17*,*18*). Further, conventional supramolecular chemistry continues to provide an ever increasing number of receptor-substrate pairs that can be used to self-assemble the condensed phase and molecular components of a heterosupermolecule or heterosupramolecular assembly (*19*).

The incorporation of condensed phase components in a heterosupermolecule or heterosupramolecular assembly facilitates practical and efficient modulation of function or property. Specifically, as the function or property of the heterosupermolecule or heterosupramolecular assembly is dependent on the bulk properties of the constituent condensed phase components, modulation of a bulk property will therefore modulate the function or property of the heterosupermolecule or heterosupramolecular assembly. That this is so, has been demonstrated for a number of cases to date and is expected to be generally true (*4*).

The use of both condensed phase and molecular components will permit heterosupermolecules and heterosupramolecular assemblies possessing novel architectures to be self-assembled and self-organised in solution. That this will be necessary is increasingly apparent. For example, Grätzel has described a stable and efficient regenerative photoelectrochemical cell for the conversion of solar energy to electrical energy (*18*). Innovative aspects of the Grätzel cell include the use of ruthenium complexes as sensitisers whose absorption spectra overlap well the solar emission spectrum and the use of 10 μm thick nanoporous-nanocrystalline semiconductor films with a surface roughness of greater than 1000 as photoanodes. Central to the efficient operation of the Grätzel cell is the fact that the sensitiser molecules, as a consequence of their being adsorbed directly at the photoanode, are effectively stacked and the probability of an incident visible photon being absorbed is close to unity. It is important to note that while the above light-harvesting strategy is clearly based on that of green plants, its practical implementation utilises a heterosupramolecular assembly and that implementation of the same strategy using only condensed phase or molecular components has not proved possible.

Another desirable feature is that a heterosupermolecule or heterosupramolecular assembly be self-assembled and self-organised in solution using a parallel algorithm. That is, that subcomponents of the heterosupermolecule or heterosupramolecular assembly be self-assembled at the same time in the same solution and that they be subsequently assembled to yield the required device or material. While it is clear that the use of both molecular and condensed phase components does not reduce the difficulties presented by the need to meet this requirement, neither does it add to them. In this context however, we note recent work of Mirkin and Alivisatos in which the self-assembly of Au nanocrystallites have been directed by use of non-self-complementary oligomers of DNA (*6*). These studies can be usefully discussed as heterosupramolecular chemistry.

Finally, for heterosupramolecular chemistry to fulfil its promise it will be necessary to be able to scale-up the process of self-assembly and self-organisation in solution. In this respect the use of DNA receptor-substrate pairs offers the prospect of utilising related commercial technologies (*6*).

In the future it is possible that solid state and synthetic chemistry will evolve into a seamless continuum of activities which may be usefully discussed in the framework of a systematic heterosupramolecular chemistry.

Literature Cited

[1] (a) Lehn, J.-M. *Angew. Chem. Int. Ed. Engl.* **1988**, *27*, 89. (b) Cram, D. J. *Angew. Chem. Int. Ed. Engl.* **1988**, *27*, 1009.

[2] (a) Lehn, J.-M. *Supramolecular Chemistry*; VCH: New York, 1995; Chapter 8. (b) Balzani, V.; Scandola, F. *Supramolecular Photochemistry*; Ellis Horwood: New York, 1991; Chapter 3.

[3] (a) Marguerettaz, X.; O'Neill, R.; Fitzmaurice, D. *J. Am. Chem. Soc.* **1994**, *116*, 2628. (b) Marguerettaz, X.; Fitzmaurice, D. *J. Am. Chem. Soc.* **1994**, *116*, 5017. (c) Marguerettaz, X.; Redmond, G.; S. Nagaraja Rao; Fitzmaurice, D. *Chem. Eur. J.* **1996**, *2*, 420.

[4] (a) Cusack, L.; Rao, S. N.; Wenger, J.; Fitzmaurice, D. *Chem. Mater.* **1997**, *9*, 624. (b) Cusack, L.; Rao, S. N.; Fitzmaurice, D. *Chem. Eur. J.* **1997**, *3*, 202. (c) Cusack, L.; Marguerettaz, X.; Rao, S. N.; Wenger, J.; Fitzmaurice, D. *Chem. Mater.* **1997**, *9*, 0000.

[5] (a) Cusack, L.; Rizza, R.; Gorelov, A.; Fitzmaurice, D. *Angew. Chem. Int. Ed. Eng.* **1997**, *36*, 848.

[6] (a) Murray, C.; Kagan, C.; Bawendi, M. *Science* **1995**, *270*, 1335. (b) Mirkin, B.; Letsinger, R.; Mucic, R.; Storhoff, J. *Nature* **1996**, *382*, 607. (c) Alivisatos, P.; Peng, X.; Wilson, T.; Johnson, K.; Loweth, C.; Bruchez, M.; Schultz, P. *Nature*

1996, *382*, 609. (d) Peng, X.; Wilson, T.; Alivisatos, P.; Schultz, P. *Angew. Chem. Int. Ed. Engl.* **1997**, *36*, 145.
[7] Brienne, M.-J.; Gabard, J.; Lehn, J.-M. *J. Chem. Soc. Chem. Comm.* **1989**, 1868.
[8] Kotov, N.; Meldrum, F.; Fendler, J. *J. Phys. Chem.* **1994**, *98*, 2735.
[9] (a) Feibush, B.; Fiueroa, A.; Charles, R.; Onan, K.; Feibush, P.; Karger, B. *J. Am. Chem. Soc.* **1986**, *108*, 3310. (b) Feibush, B.; Saha, M.; Onan, K.; Karger, B.; Giese, R. *J. Am. Chem. Soc.* **1987**, *109*, 7531. (c) Hamilton, A.; Van Engen, D. *J. Am. Chem. Soc.* **1987**, *109*, 5035. (d) Bisson, A.; Carver, F.; Hunter, C.; Waltho, J. *J. Am. Chem. Soc.* **1994**, *116*, 10292.
[10] (a) Henglein, A. *Ber. Bunsenges. Phys. Chem.* **1982**, *86*, 246. (b) Duonghong, D.; Ramsden, J.; Grätzel, M. *J. Am. Chem. Soc.* **1982**, *104*, 2977. (c) Kolle, U.; Moser, J.; Grätzel, M. *Inorg. Chem.* **1985**, *24*, 2253. (d) Rothenberger, G.; Moser, J.; Grätzel, M.; Serpone, M.; Sharma, D. *J. Am. Chem. Soc.* **1985**, *107*, 8054.
[11] There is apparently general agreement that the spectrum of trapped holes does not extend to wavelengths longer than about 450 nm : (a) Bahnemann, D.; Henglein, A.; Lilie, J.; Spanhel. L. *J. Phys. Chem.* **1984**, *88*, 709. (b) Micic, O.; Zhang, Y.; Cromack, R.; Trifunac, A.; Thurnauer, M. *J. Phys. Chem.* **1993**, *97*, 7277.
[12] (a) Kok, B.; Rurainski, H.; Owens, O. *Biochem. Biophys. Acta* **1965**, *109*, 347. (b) Trudinger, P. *Anal. Biochem.* **1970**, *36*, 222. (c) Wantanabe, T.; Honda, K. *J. Phys. Chem.* **1982**, *86*, 2617.
[13] Frei, H.; Fitzmaurice, D.; Grätzel, M. *Langmuir* **1990**, *6*, 198.
[14] Cusack, L.; Marguerettaz, X.; Rao, S. N.; Fitzmaurice, D. manuscript in preparation. The detailed reasons for the slower rate of decay of the transient assigned to long-lived trapped electrons in TiO_2-(CTAB+III) as compared with TiO_2-(I+IV) will be discussed here and supported by detailed NMR studies.
[15] (a) Chidsey, C. *Science* **1991**, *251*, 919. (b) Finklea, H.; Hanshew, D. *J. Am. Chem. Soc.* **1992**, *114*, 3173. (c) Cleland, G.; Horrocks, B.; Houlton, A. *J. Chem. Soc. Faraday Trans.* **1995**, *91*, 4001.
[16] Serpone, N; Lawless, D.; Khairutdinov, R. *J. Phys. Chem.* **1995**, *99*, 16646.
[17] Alivisatos, P. *J. Phys. Chem.* **1996**, *100*, 13226.
[18] (a) O'Regan, B.; Grätzel, M. *Nature* **1991**, *353*, 737. (b) Nazeeruddin, M. K.; Kay, A.; Rodicio, I.; Humphry-Baker, R.; Müller, E.; Liska, P.; Vlachopoulos, N.; Grätzel, M. *J. Am. Chem. Soc.* **1993**, *115*, 6382. (c) McEvoy, A.; Grätzel, M. *Sol. Energy. Mater. Sol. Cells* **1994**, *32*, 221. (d) Hagfeldt, A.; Graetzel, M. *Chem. Rev.* **1995**, *95*, 49.
[19] Philp, D.; Stoddart, J. F. *Angew. Chem. Int. Ed. Engl.* **1996**, *35*, 1154.

Chapter 4

Self-Organization of Spherical Nanoparticles in Two- and Three-Dimensinal Superlattices

Interconnected Network of Nanomaterial

M. P. Pileni

Laboratoire SRSI, Unité de Recherche Associée, Central National de la Recherche Scientifique 1662, Université Pierre et Marie Curie, 4 Place Jussieu, 75005 Paris, France
Commisariat à l'Energie Atomique, Dutch State Mines (Netherlands)— DRECAM Service de Chimie Moléculaire CE Saclay, 91 191 Gif sur Yvette Cedex, France

Syntheses of nanosized particles have been performed by using reverse micelles as microreactors. The size of the nanoparticles are controlled by the size of the micelles. The particles are coated and dispersed in hexane. By deposition of the particles on a carbon grid, 2D and 3D superlattices organized in hexagonal and face centered cubic structure respectively. Two size selection methods are presented. Elongated aggregates forming an interconnected network of copper metallic particles are build up by using oil in water micelles.

The fabrication of assemblies of perfect nanometer-scale crystallites (quantum crystal) identically replicated in unlimited quantities in such a state that they can be manipulated and understood as pure macromolecular substances is an ultimate challenge in modern materials research with outstanding fundamental and potential technological consequences (1). These potentialities are mainly due to the unusual dependence of the electronic properties on the particle size. Optical properties are one area where nanoparticles have markedly different properties from the bulk phase (2, 3).

The preparation and characterization of these nanomaterials have thus motivated a vast amount of work (1). One of the methods used to control the size and/or the polydispersity of the particles is the synthesis in reverse micelles (4, 5). The achievements of an accurate control of the particle size, their stability and a precisely controllable reactivity of the small particle are required to allow attachment of the particles to the surface of a substrate or to other particles

without leading to coalescence and hence losing their size-induced electronic properties. It must be noted that manipulating a nearly monodispersed nanometer size crystallites with an arbitrary diameter presents a number of difficulties.
There are a number of reasons for forming films of inorganic particles attached or embedded just under the surface. Moreover, the ability to assemble particles into well-defined two-and-three dimensional spatial configurations should produce interesting properties as new collective physical behavior (6). The development of general procedure for the fabrication of "quantum" crystal is a major challenge of future research. One of the approach to obtain 2D and 3D structures is to assemble nanoparticles and in ordered arrays. This requires hard sphere repulsion, controlled size distribution, inherent Van der Walls attraction between particles and dispersion forces. The polydispersity in particle size prevents construction of such well-defined two-or-three dimensional structures.
Recently, in our laboratory spontaneous arrangements either in a monolayer organized in a hexagonal network or three dimentional FCC arrangements of particle has been observed (7-9). Similar arrangements with CdSe semiconductors CdSe (10) and metallic particles as Gold or Silver (11-13) have been reported.
In this paper we present two methods to construct 2D and 3D nanoparticles. The monolayer arrangement is a hexagonal network whereas the 3D organization is arranged as face centered cubic crystal. Nanomaterial arranged in interconnected network is presented.

Self organization in 2D and 3D structures.

The preparation of colloidal silver and silver sulfide nanoparticles involves making the functionalized surfactant silver bis di(2-ethylhexyl) sulfosuccinate called Ag(AOT). Particles are obtained by mixing two sodium di(2ethkhexyl) sulfosuccinate, Na(AOT) reverse micellar solutions at a given water-content, W = $[H_2O]/[AOT]$. The first one is made with Ag(AOT), Na(AOT) and the second one either hydrazine or sulfide with 100% of Na(AOT). The details of these syntheses are given on reference 14 and 15. In both cases, a control in the particle size is obtained. The increase in the particle size depends on the producted material. With silver particles,$(Ag)_n$, the size varies from 2nm to 6nm (14) whereas with $(Ag_2S)_n$, it varies from 2nm to 10nm (15). To stabilize the particles and to prevent their growth, 1 µl/ml of pure dodecanethiol is added to reverse micellar system containing the particles. This induces a selective reaction at the interface, with covalent attachment, between thio derivatives and silver atoms (16). The micellar solution is evaporated at 60°C and a solid mixture made of dodecanethiol-coated nanoparticles and surfactant is obtained. To remove the AOT and excess dodecanethiol surfactant a large amount of ethanol is added. The particles are dried and dispersed in heptane. Surprisingly, the behavior differs for $(Ag_2S)_n$ and $(Ag)_n$ nanoparticles. With $(Ag_2S)_n$, only a small fraction of the particles is dispersed in heptane and a size selection takes place. Hence, the

extraction induces a large size selection with a strong decrease in the size distribution which drops from 30% to 14%. With $(Ag)_n$ a slight size selection occurs and the size distribution drops from 43% to 37%.
In both cases, when the particles are dispersed either in reverse micellar solution or in heptane, the solutions are optically clear. This permits to follow the colloidal particles by spectroscopy.

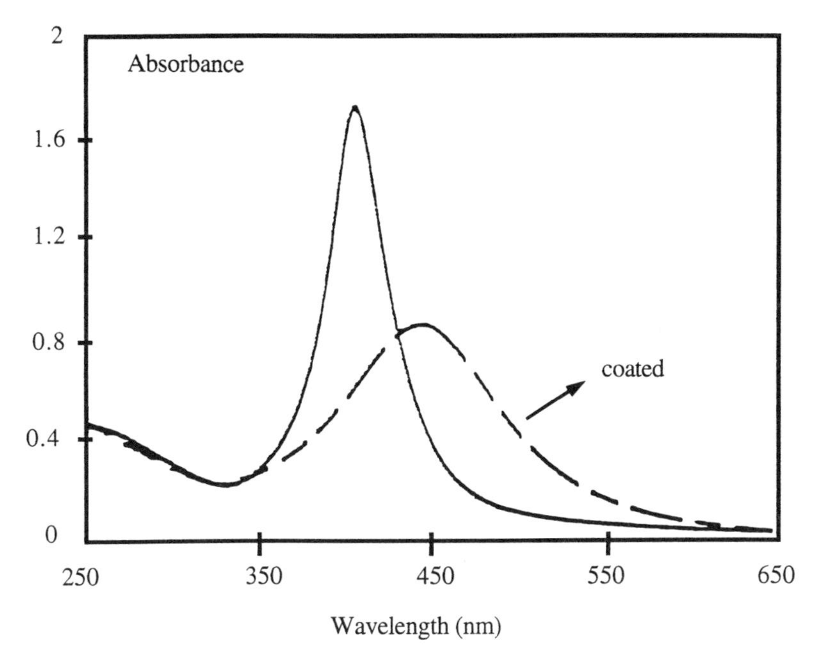

Figure 1. Absorption spectra of $(Ag)_n$ coated and not coated nanosized particles.

The plasmon peak in the $(Ag)_n$ absorption spectrum decreases with decreasing the particle size (14). This confirms the prediction obtained by simulation (17). The attachment of dodecanthiol on the silver particles can be monitored by UV-Vis Spectroscopy: Figure 1 shows a markedly drops in the extinction coefficient and a red shift in the plasmon band when dodecanthiol addition is added to micellar solution whereas no change in the UV absorption is observed below 300 nm. These behaviors are due to a change in the free electron density with changes in the surface plasmon band of silver particles (17, 18). Similar behavior has been observed by NaSH addition to silver colloids in aqueous solution (19)]. Conversely, the $(Ag_2S)_n$ nanometer particles absorption spectrum is not structured and does not change with the particle size. A blue shift compared to the optical band edge of bulk silver sulfide (1240 nm) is observed.
Because of the fact that a size selection takes place during the extraction of $(Ag_2S)_n$ particles from the micelles, self assemblies in 2D and 3D dimensions are

obtained. Conversely, for $(Ag)_n$ particles a size selected precipitation has to be performed. In the following paragraph the two self organizations are presented.

Silver sulfide, $(Ag_2S)_n$, self assemblies. By using a dilute solution of $(Ag_2S)_n$ nanocrystallites (particle volume fraction, ϕ, equal to 0.01%), monolayers of particles are formed. The particles are organized in a hexagonal network in very large domains (Fig. 2A) with 2nm as an average inter-particle distance. The monolayer domain is very large and forms long ribbons (9). The length and width of the ribbons depend on the particle size. Whatever this size, the ribbons are always very long. Direct observations on the TEM allow to estimate the length of the ribbons to be about 100mm. Figure 2B shows a monolayer observed by TMAFM, (Atomic Force Microscopy in Tapping Mode). This image shows formation of monolayer in a very large domain, with holes regularly distributed on the surface. From a cross section of the monolayer (Fig. 2C), the average depth of the holes is found to be 6nm which corresponds to the average diameter of the particles. The cross sections of the monolayer confirm the formation of rather compact arrangement of nanoparticles.

At high particle concentration (about 10^{-3}M), the solution remains optically clear. Supports used for TEM. and TMAFM experiments are immersed in the solution for two hours and then dried at room temperature. TEM images reveal the formation of relatively large aggregate area (Fig.3A). A similar pattern over a large area is obtained by TMAFM experiments (Fig 3B) as observed by TEM (Fig. 3A). The size and the average distribution of the islands are similar. A cross section of these islands shows the formation of a large aggregate having a truncated pyramid form.

Magnification an aggregate shows that it consists of nanosized particles which appear not to be placed at random but be collectively organized. It is clear from TEM observations that $(Ag_2S)_n$ layers are usually overlapping indicating a three dimensional structure. The particles are highly oriented in a four fold symmetry, attributed to the orientation of the particles in the {001} plane of face centered cubic (f.c.c.) structure. This behavior is observed for various particle sizes (Fig. 4). Nanocrystal made of particles differing by their size are obtained.

Self assemblies made with silver metallic nanoparticles. With $(Ag)_n$ particles, most of the thiododecane coated particles are redispersed in heptane. The size polydispersity is rather large (Fig.5 A and B) and is reduced through the size selected precipitation, SSP, technique. This method is based on the mixture of two miscible solvents differing by their ability to dissolve the surfactant alkyl chains. The silver coated particles are highly soluble in hexane and poorly in pyridine. Thus, a progressive addition of pyridine to hexane solution containing the silver coated particles are performed. At a given pyridine volume (which corresponds to roughly 50 %), the solution becomes cloudy and a precipitate appears. This corresponds to agglomeration of the largest particles as a result of

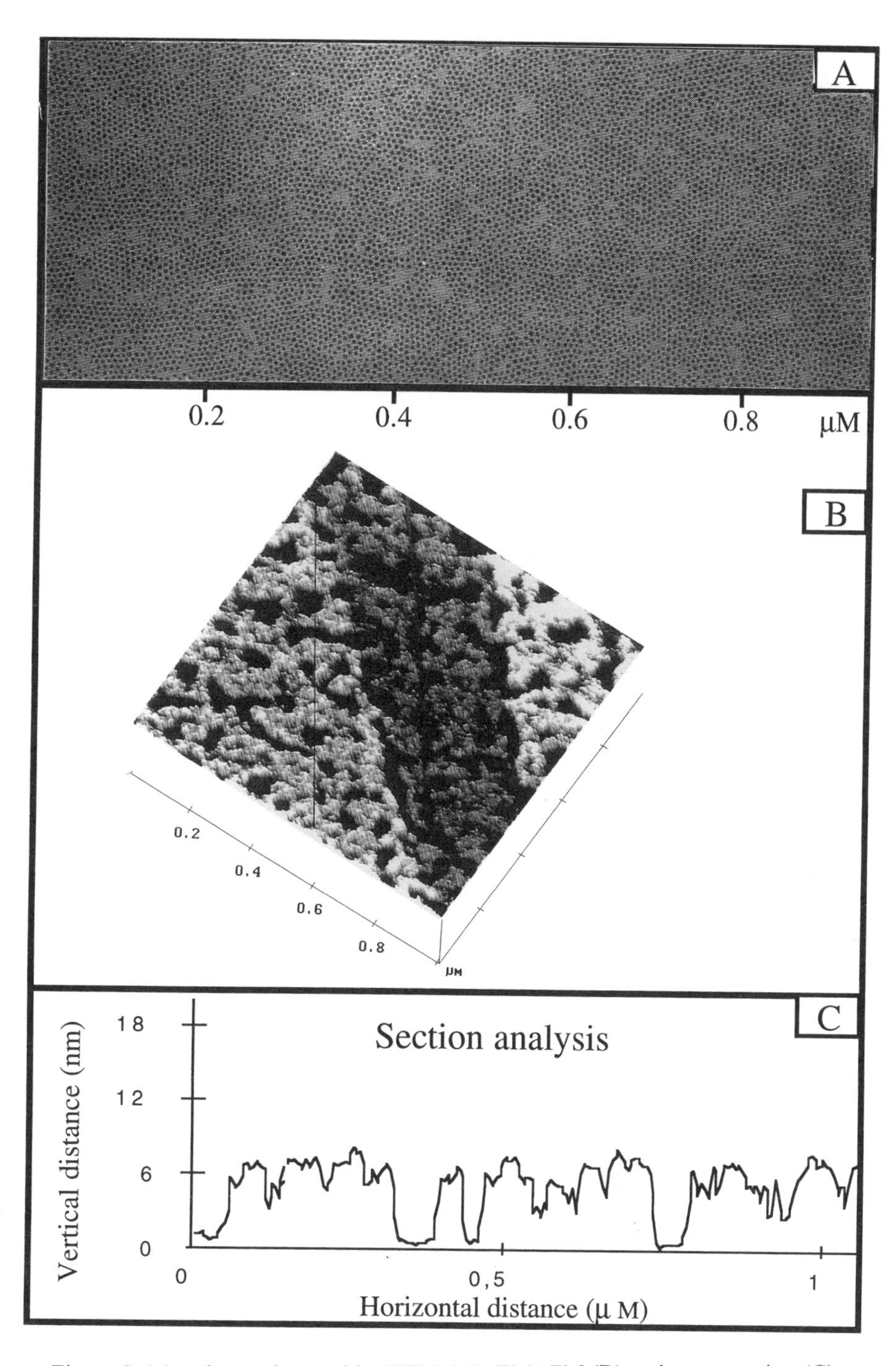

Figure 2. Monolayer observed by TEM (A), TMAFM (B) and cross section (C)

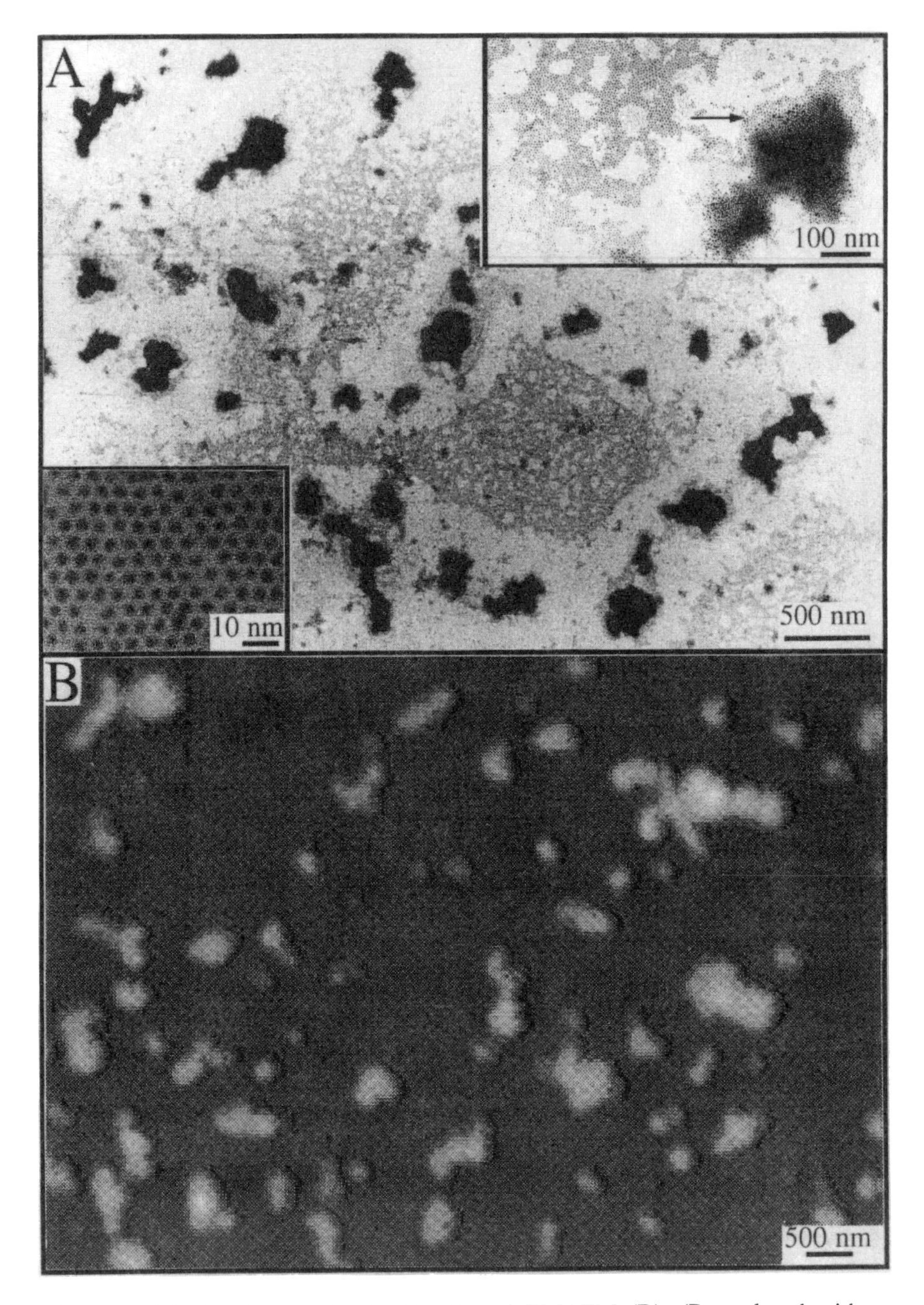

Figure 3. Islands observed by TEM (A) and TMAFM (B). (Reproduced with permission from reference 8. Copyright 1996 Wiley-VCH.)

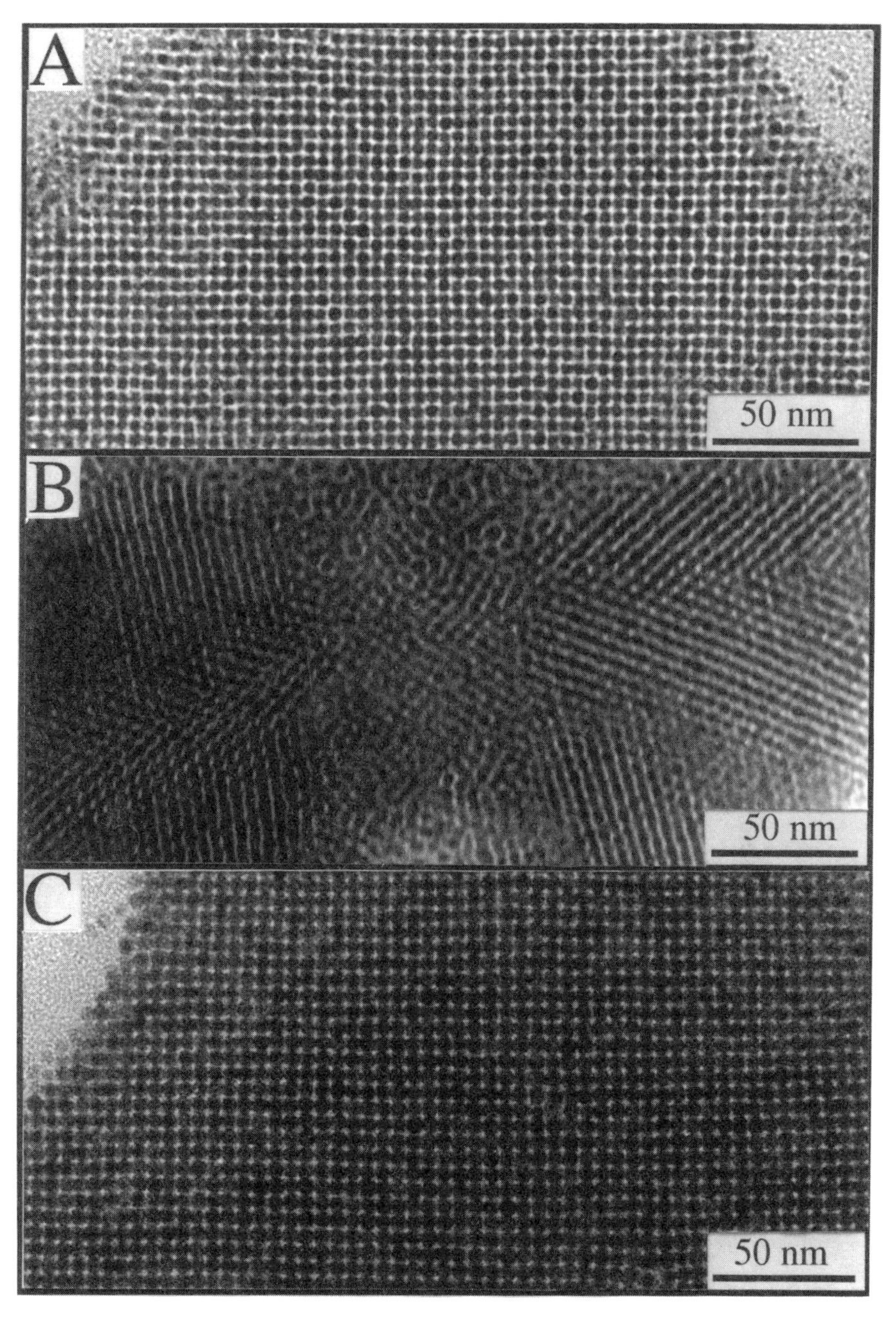

Figure 4. Magnification of crystal made of particles having 3 nm (A), 4 nm (B) and 6 nm (C) diameters, respectively.

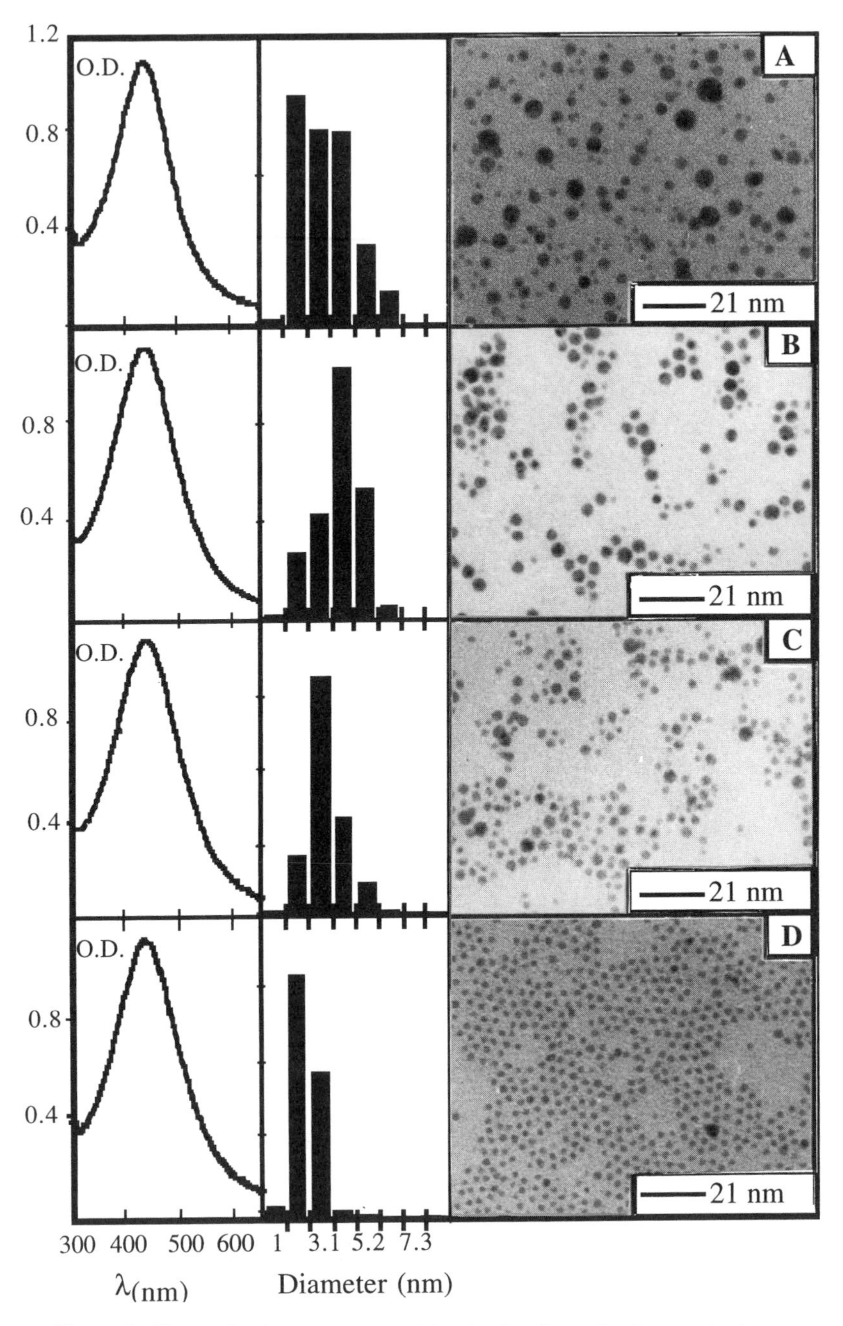

Figure 5. Change in the average particle size by size selection method.

their greater Van der Waals interactions. The solution is centrifugated and an agglomerated fraction rich in large particle is collected, leaving the smallest particles in the supernatant. The agglomeration of the largest particles is reversible and the precipitate, redispersed in hexane and forms a homogeneous clear solution. It is observed an increase in the average diameter and a decrease in the size distribution compared to what it has been observed before the size selection. This procedure is also performed with the supernatant, which contains the smallest particles. It is repeated several times and a strong decrease in the average particle size and its distribution is observed. When the size distribution is small enough, a self organization is observed. The silver nanosized particles form on carbon grid a hexagonal network with 2nm as an average distance between particles (Figure 5 C). To built a 3D self organization, we keep the particles obtained after one SSP. Particles having an average diameter equal to 4.1 nm are dispersed in hexane. By leaving the carbon grid in the solution, the TEM pattern is completely covered by a monolayer made of particles. They are organized in a hexagonal close packed network. In some region of the TEM pattern, contrast differences are observed. This is attributed to crystal growth in 3D. Increasing the immersion time until complete evaporation of the solvent yields to formation of large aggregates. Figure 6 A shows a rather high aggregate orientation around a large hole or ring. The average distance between the oriented aggregates varies from 20 to 60 nm. The aggregate magnification confirms that it consists of $(Ag)_n$ nanoparticles. The aggregate average size is the range of 0.03 to 0.55 mm2. High magnification of one of these aggregates shows that the particles are arranged in two different symmetries. Figure 6B shows the formation of a polycrystal. Its magnification (Fig.6 C), shows either a hexagonal or a cubic arrangement of nanoparticles. The transition from one structure to another is abrupt and there is a strong analogy with "atomic" polycrystals with a small grain called nanocrystals. Each domain or grain has a different orientation. This clearly shows that the stacking of nanoparticles is periodic and not random. The "pseudo-hexagonal" structure corresponds to the stacking of a {110} plane of the f.c.c structure. On the same pattern it is observed (Fig.6C) a four fold symmetry which is again characteristic of the stacking of {001} planes of the cubic structure. This cannot be found in the hexagonal structure. As matter of fact, no direction in a perfect hexagonal compact structure for which the projected positions of the particles can take this configuration. This is confirmed by TEM experiments performed at various tilt angles where it is always possible to find an orientation for which the stacking appears to be periodic. Hence, by tilting a sample having a pseudo hexagonal structure, a four fold symmetry is obtained. From these informations, it is concluded that the large aggregates of silver particles are formed by stacking of monolayers in a face centered cubic arrangement.

Interconnected network made of nanoparticles.

Surfactants having a positive curvature, above a given concentration usually called the critical micellar concentration, c.m.c, self assemble to form oil in water

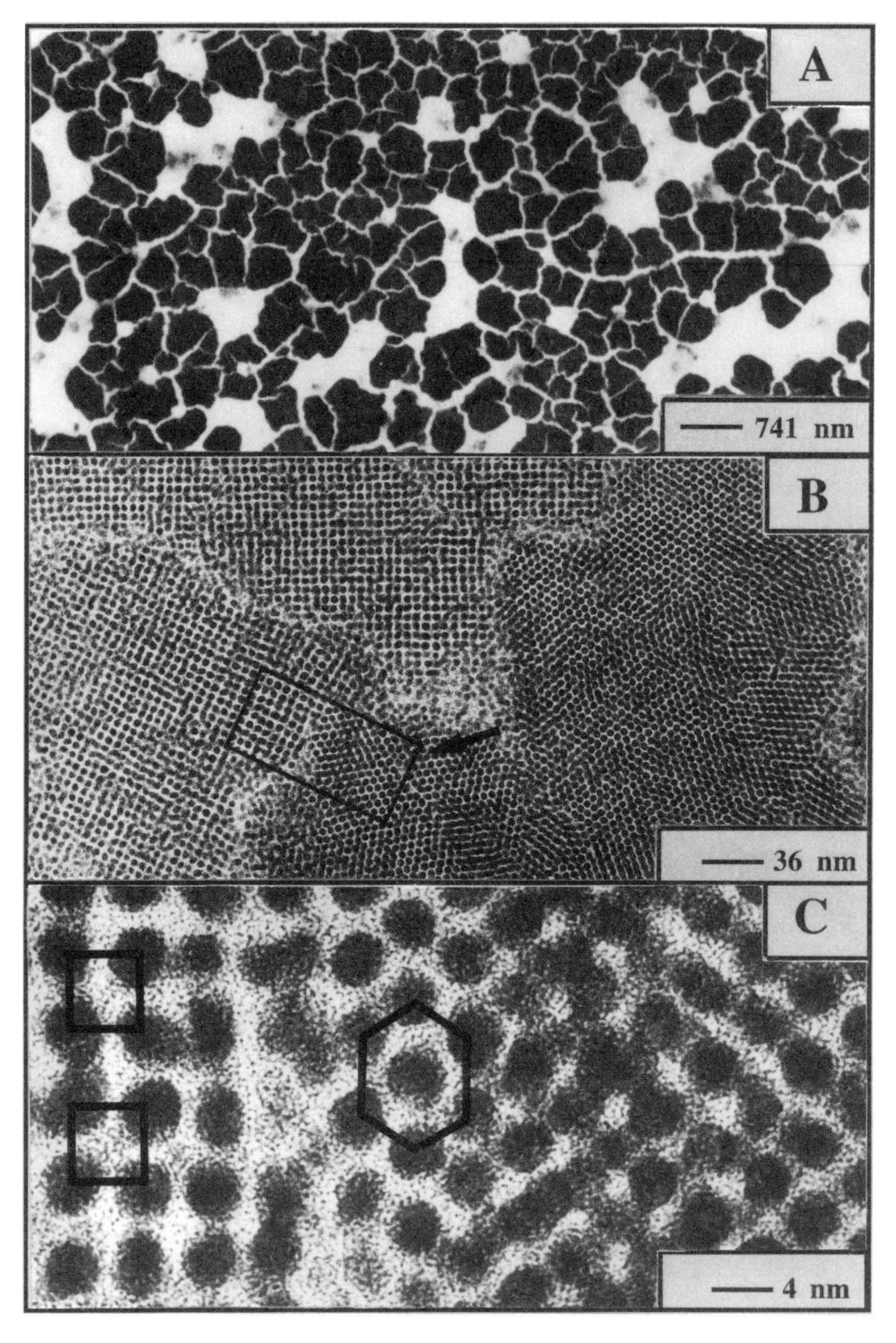

Figure 6: Progressive enhancement of $(Ag)_n$ aggregates

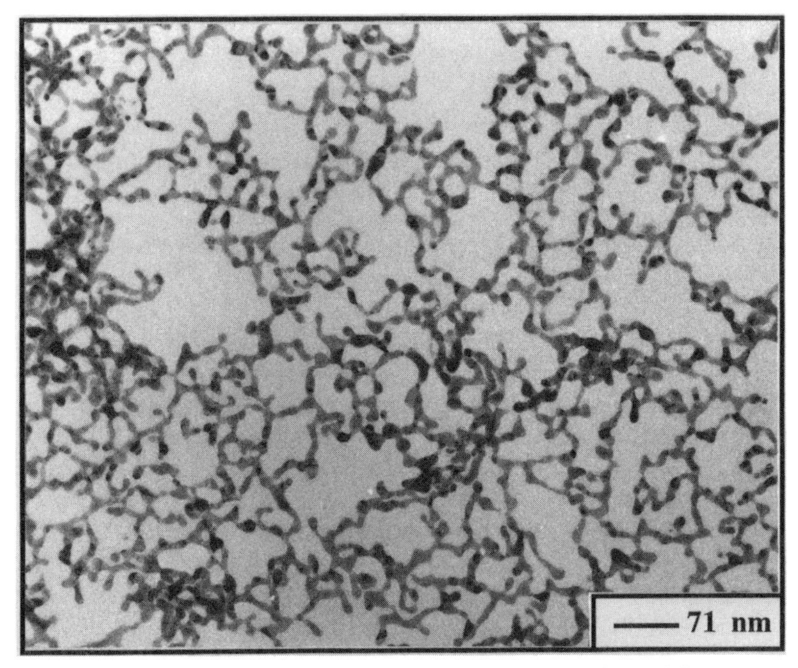

Figure 7. Interconnected network of copper metallic particles.

aggregates called normal micelles. Copper dodecyl sulfate, $Cu(DS)_2$, characterized by a c.m.c equal to $1.3.10^{-3}$M has been used to make nanoparticles. Synthesis of nanoparticles is performed by addition of a reducing agent as sodium borohydrate to the $1.3.10^{-3}$M $Cu(DS)_2$ micellar solution. Copper metallic particles forming an interconnected network over a very large network (Fig.7) is observed (20). Enhancement of the network shows a change in the material shape. The chemical reduction is performed in solution and permits to record the absorption spectrum of the $(Cu)_n$ interconnected network. It is observed a red shift in the spectrum compared to what is observed with spherical particles (21). From our knowledge this is the first example which confirms the predictions from which a red shift of the absorption spectrum is expected where the length of elongated material increases..

Acknowledgments: I would like to thank my coworkers:
F. Billoudet, S.Didierjean, J.L. Lindor, Dr. I Lisiecki, Dr.L.Motte, Dr. C.Petit, A. Taleb.
Special thanks are due to A. Filankembo, who has been in charge of the figures.
The Atomic Force Microscopy in Tapping Model has been performed in collaboration with Dr. E.Lacaze from Groupe de Physique des solides, U. R. A. C. N. R. S. 17, Université P. et M. Curie (Paris VI) and D. Diderot (Paris VII) in Paris.

References.

1- Special issue: Nanostructured Materials, Chem.Mater.; **1996**, 8,number 5
2- Brus, L.E. J. Chem. Phys., **1983**, 79, 5566.
3- Wokaun, A.; Gordon, J. P; Liao, P.F. Phys. Rev. Lett. **1982**, 48, 957.
4 - Pileni, M.P. ; J. Phys. Chem. **1993**, 97, 6961
5 - Pileni, M.P. ; Langmuir. 1997,in press
6- Heitmam, D.; Kotthaus, J.P.; Phys Today , **1993**, 56
7- Motte, L.; Billoudet, F. and Pileni, M.P.J. Phys. Chem. **1995**, 99, 16425
8- Motte, L.; Billoudet, F., E. Lacaze, and Pileni, Advanced of Materials, **1996**, 8, 1018.
9- Motte, L.; Billoudet, F., E. Lacaze, J. Douin and Pileni, M.P.; J. Phys. Chem. **1997**, 101,
10- Murray, C.B.; Kagan, C.R.; Bawendi, M.G.; Science, **1995**, 270, 1335
11- Whetten, R.L.; Khoury, J.T.; Alvarez, M.M., Murthy, S.;Vezmar, I.; Wang, Z.L.; Cleveland, C.C.; Luedtke, W.D. and Landman, U. Advanced Materials, **1996**, 8, 429
12- Brust, M.; Bethell, D.; Schiffrin, D.J. and Kiely, C.J. Advanced Materials, **1995**, 7, 9071
13- Harfenist, S.A.; Wang, Z.L.; Alvarez, M.M.; Vezmar, I. and Whetten, R.L.; J.Phys.Chem.; **1996**, 100, 13904
14- Petit,C.; Lixon,P. and Pileni, M.P. J. Phys. Chem.; **1993**, 97, 12974
15- Motte,L. Billoudet,F. and Pileni, M.P. J. Materials Science.**1996**, 31, 38
16- C. D. Bain, J. Evall, G. M. Whitesides, J. Am. Chem. Soc.; **1989**, 111, 7155
17- Charlé, K.P.; Frank, F. and Schulze, W.; Ber. Bunsenges. Phys. Chem.; **1984,** 88, 354
18- Kreibig, U.; J. Phys. F: Metal. Phys. **1974**, 4, 999.
19 -Mulvaney, P.; Linnert, T. and Henglein, A.; J. Phys. Chem.; **1991**,95, 7843.
20- Lisiecki I., Billoudet F.and Pileni M.P.; J. Phys. Chem.; **1996**, 100, 4160
21- Lisiecki I. and Pileni M.P.; J. Amer. Chem. Soc.; **1993**, 115, 3887

Chapter 5

Microstructured Particles for Electrorheological Applications

Ping Sheng[1], Wing Yim Tam, Weijia Wen, Hongru Ma, and M. M. T. Loy

Department of Physics, The Hong Kong University of Science and Technology, Clear Water Bay, Kowloon, Hong Kong

By carrying out first principles calculations, particles fabrication and experiments, we demonstrate the enhancement of the electrorheological effect through the use of microstructured composite particles, consisting of a spherical dielectric core coated with a thin layer of metal and an additional outer layer of insulating material with high dielectric constant. It is argued that through the choice of the core material and the outer layer material, the outer layer thickness, and the particle size, there can be a path of continuous improvement on the electrorheological fluid properties that makes possible the eventual fulfillment of the diverse application requirements.

Electrorheological (ER) fluids are a class of materials whose rheological properties are controllable through the application of an electric field. The most common type of ER fluids is a colloid of solid dielectric particles dispersed in a fluid. Here the change in the rheological properties is known to be accompanied by a structural transformation whereby the initially dispersed solid particles are aggregated into dense columns spanning the sample and aligned along the direction of the applied electric field. These dense columns are the source of ER fluids' increased viscosity when shearing occurs perpendicular to the electric field. Physically, the field-induced rheological transformation may be understood as follows. Due to the difference in the dielectric constants of the solid particles and the liquid, each particle is polarized in the presence of an applied electric field. In the lowest-order approximation, this results in dipole-dipole interaction between the particles. To lower the interaction energy would mean the re-arrangement of the solid particles into a periodic structure, i.e., a solid. Based on the dipole-dipole interaction model, Tao (*1*) has calculated the lowest-energy periodic structure to be that of body-centered tetragonal (BCT). This result has later been verified by more accurate calculations (*2-5*). The columns seen during the application of the electric field represent the superstructure built up from the locally BCT- ordered particles.

[1]Corresponding author

Electrorheological fluids are known to hold great potential for many applications such as in shock-absorbing vibration dampers, clutches, and robotics (*6,7*). In practice, however, large scale utilization of ER fluids has yet to be realized for lack of ER systems with the required yield stress, stability of the ER effect over extended period, ER dynamic range, and acceptable wear resistance, among other conditions. In this paper, we propose a novel ER system consisting of doubly coated particles in a dielectric fluid. Such particles offer a much broader range of material and controllable parameter options to both enhance the ER effect and fulfill many of the application constraints. In particular, the choice of the core material, which controls the density of the particle, can directly impact the sedimentation problem. The metallic coating, which can be very thin, would enhance the ER effect as shown below. The right choice of the outer dielectric layer material would serve to electrically insulate the particle, provide mechanical wear protection, as well as further enhance the ER effect through its high dielectric constant. Furthermore, it is shown that the outer coating thickness and the particle size are controllable design parameters which may be optimized in accordance with application requirements. We have successfully fabricated some preliminary versions of such particles. Below we present heuristic argument, first-principles calculations, and experimental results demonstrating the expected improvement in yield stress (*8*). More importantly, by coupling physical understanding of the ER mechanism with the experimental fabrication and characterization of the particles, we wish to show that there can be a path of continuous improvement through which optimal ER fluid performance may be achieved.

An intuitive understanding of the ER effect can be obtained by capacitive electrostatic energy argument. Consider two conducting spheres of radius R in close proximity to each other, with a separation δ between their surfaces at the point of closest approach (*9*). Provided $R >> \delta$, the mutual capacitance has the form $C \propto R\ell n(R/\delta)$. Under an applied electric field $\vec{E}$ along the axis joining the centers of the two spheres, the leading order term of the free energy density is proportional to C. If now we apply a shear perpendicular to the electric field, the line segment joining the centers of the two spheres will be both tilted with an angle θ with respect to the electric field direction, as well as lengthened. If the new separation is denoted d, then $(2R+d)\cos\theta = 2R+\delta$ so that $d \approx \delta + R\theta^2$ for θ small. Within the small angle approximation, the shear stress is given by

$$-\frac{1}{R}\frac{\partial C}{\partial \theta} \propto \frac{R}{\delta}\sqrt{\frac{d-\delta}{R}} \quad ,$$

which is noted to have a peak value. The static yield stress is defined by the peak value of the shear stress, given by the condition of $d = 2\delta$, so that the static yield stress $\sim \sqrt{R/\delta}(\varepsilon_\ell E^2/8\pi)$, where the quantity in bracket is the unit of energy density in the present problem, with ε_ℓ denoting the liquid dielectric constant. This heuristic argument tells us that in order to obtain a large ER effect in terms of the yield stress, we should strive to optimize the insulating layer coating thickness so that it would prevent shorting between the conducting spheres on the one hand and provide wear resistance and give a large yield stress on the other. Since the effect depends on the ratio between R and δ, it is also predicted that the particle size will have an effect. Also, a large dielectric constant for the coating material (or the liquid) is a plus, as it increases the capacitance of the system.

To put our understanding of the ER fluids system on a more rigorous basis, it is desirable to carry out first principles calculations (*2*). The basis of such a theoretical framework is the observation that the operating (applied electric field) frequencies of the ER fluids are generally below 10^5 Hz, whereas the dielectric particle size is below 100 microns due to the fact that larger particles can not be suspended effectively in fluids. That means comparison between the electromagnetic wavelength (on the order a km) and the scale of inhomogeneities in an ER fluid would firmly place the system in the long-wavelength limit, where the effective medium approximation is accurate so that the electrical response of the system may be characterized by an effective complex dielectric constant $\bar{\varepsilon}$. The electrostatic Gibbs free energy of the system is then given by

$$F = -\mathrm{Re}(\bar{\varepsilon}_{zz} E^2 / 8\pi) \quad , \tag{1}$$

where $\bar{\varepsilon}_{zz}$ is the zz component of the effective dielectric tensor, z being the applied field direction, $E = -\partial\phi / \partial z$ is the applied electric field, and ϕ being the electric potential. From Eq. (1), it is easy to see that as far as the electrostatic free energy is concerned, the equilibrium state is one where $\mathrm{Re}(\bar{\varepsilon}_{zz})$ is maximized with respect to the positions of the solid particles. We denote that state the electrostatic ground state of the ER fluid. The ER mechanical properties, such as the shear modulus and the static yield stress in the high field regime, may be obtained by perturbations away from the electrostatic ground state.

For the calculation of $\bar{\varepsilon}$, we use the Bergman-Milton representation (*10-13*). For simplicity, we first consider a two-component system consisting of spherical solid particles of radius R and complex dielectric constant $\varepsilon_1 = \kappa_1 + i4\pi\sigma_1 / \omega$ dispersed in a fluid characterized by $\varepsilon_2 = \kappa_2 + i4\pi\sigma_2 / \omega$, where κ denotes the real part of the dielectric constant, σ the conductivity, and ω the angular frequency. The case of coated particles, to be described later, represents a generalization of the two-component system formalism. The electrostatic problem to be solved is given by

$$\nabla \cdot \varepsilon(\vec{r}) \nabla \phi = 0 \quad . \tag{2}$$

By expressing $\varepsilon(\vec{r}) = \varepsilon_1 \eta(\vec{r}) + \varepsilon_2 [1 - \eta(\vec{r})] = (\varepsilon_1 - \varepsilon_2)\eta(\vec{r}) + \varepsilon_2$, where $\eta(\vec{r})$ is the characteristic function for the solid component, defined as having the value 1 at those spatial points occupied by the solid particles, and zero otherwise, Eq. (2) may be rewritten as

$$\nabla \cdot [1 - \frac{1}{s} \eta(\vec{r})] \nabla \phi = 0 \quad , \tag{3}$$

where $s = \varepsilon_2 / (\varepsilon_2 - \varepsilon_1)$ is the only relevant material parameter in the problem. The formal solution to Eq. (3), given the condition of $\Delta\phi / \ell = E = 1$ in the z direction, can be expressed in the operator notation as

$$\phi = -(1 - \frac{1}{s}\Gamma)^{-1} z = -s \frac{z}{s - \Gamma} \quad , \tag{4}$$

where

$$\Gamma = \int d\vec{r}'\eta(\vec{r}')\nabla' G_o(\vec{r} - \vec{r}') \cdot \nabla' \quad (5)$$

is an integral-differential operator, with $G_o(\vec{r} - \vec{r}') = 1/4\pi|\vec{r} - \vec{r}'|$ denoting the Green's function for the Laplace equation. By defining the inner product operation as

$$\langle \phi | \psi \rangle = \int d\vec{r}'\eta(\vec{r}')\nabla'\phi^* \cdot \nabla'\psi, \quad (6)$$

it becomes possible to write the effective dielectric constant as

$$\frac{\bar{\varepsilon}_{zz}}{\varepsilon_2} = -\frac{1}{V}\int d\vec{r}\left(1 - \frac{1}{s}\eta\right)\frac{\partial\phi}{\partial z}$$

$$= 1 + \frac{1}{s}\langle z|\phi\rangle\frac{1}{V}, \quad (7)$$

where V denotes the sample volume. From Eqs. (4) and (7), it follows that the effective dielectric constant is given by the Bergman-Milton representation (*10-13*):

$$\frac{\bar{\varepsilon}_{zz}}{\varepsilon_2} = 1 - \frac{1}{V}\sum_n \frac{|\langle z|\phi_n\rangle|^2}{s - s_n} = 1 - \sum_n \frac{f_n^z}{s - s_n}$$

$$= 1 - \frac{1}{V}\sum_{n,m} \langle z|\psi_n\rangle\langle\psi_n|\frac{1}{s - \Gamma}|\psi_m\rangle\langle\psi_m|z\rangle, \quad (8)$$

where s_n and ϕ_n are the nth eigenvalue and eigenfunction of the operator Γ, and $\{\psi_n\}$ is any arbitrary complete basis set, e.g., that composed of eigenfunctions of individual spheres, which are the actual basis used in our calculations. The matrix inversion operation in Eq. (8) is noted to be required only for a particular diagonal matrix element and not for the whole matrix. As a result, the answer may be obtained efficiently by using the recursive Green's function method (*14*).

The remarkable feature about the representation, Eq. (8), is that the geometric information is separated from the material information, in contrast to approaches that involve the direct numerical solution of the Laplace equation. This separation means that as far as the dielectric properties are concerned, whereas the material information appear only in s, the microstructural information are given by the spectral function, i.e., by its value $|\langle z|\phi_n\rangle|^2/V = f_n^z$ and the location of the poles s_n, both of which are known to be real. Furthermore, it has been proved that s_n must lie in the interval [0,1] (*10-13*). Once the spectral function is obtained, it becomes simple to calculate the effective properties that can be derived from energy considerations, including those due to the conductivities and their associated frequency dependencies, as these factors appear only in s. If either ε_1 or ε_2 is complex and frequency dependent, then the resulting s, and consequently $\bar{\varepsilon}_{zz}$, will be complex and frequency dependent. Whereas the imaginary part of $\bar{\varepsilon}_{zz}$ characterizes the overall electrical dissipation of the system, the frequency dependence of the real part of $\bar{\varepsilon}_{zz}$ is what gives rise to the frequency dependencies of the yield stress and the shear modulus (*2*). It should be emphasized that the present formulation is rigorous and *includes all the multipole interactions and (self-consistent) local field effects.* Furthermore, it treats the real and imaginary parts of the dielectric constants as two sides of the same coin, which is the complex dielectric constant.

Provided the electrostatic ground state has a unique local spatial structure, the target configurations for the maximum $\bar{\varepsilon}_{zz}$ can be limited to periodic structures. We have performed numerical calculations of $\bar{\varepsilon}_{zz}$ to five-decimal accuracy for six periodic structures--the body-centered tetragonal (BCT), the face-centered cubic, the hexagonal, the body-centered cubic, the simple cubic, and the diamond. It was found that at any given concentration of the particles, body-centered tetragonal has the largest $\bar{\varepsilon}_{zz}$ and face-centered-cubic is a close second, with the rest decreasing in the order given above. The fact that BCT is the favored structure is in agreement with prior calculations based on dipole interactions (*1, 3-5*). It should be noted that a direct implication of our theory, that $\bar{\varepsilon}_{zz}$ is maximized under a strong electric field, is the nonlinear electrical response of the overall system.

An important point to be emphasized is that the lattice structure derived from the ER effect is stabilized only by the hard core repulsion between the solid spheres. That is, even at equilibrium there is an electrostatic contraction pressure on the particles. As a result, the ER lattice structure has no bulk modulus in the usual sense because the first derivative of the electrostatic potential (as seen by the particles) does not vanish along those distortion paths that would pull the spheres apart. However, for those distortion paths that maintain at least some of the contacts between any given spheres and its neighbors (such as the shear distortion defined below) the definition of a modulus becomes possible.

In order to calculate the shear modulus and the yield stress, it is necessary to perturb the system away from its lowest electrostatic ground state. For the BCT structure, shearing in a direction perpendicular to the z-axis means not only a tilt of the c-axis away from the electric-field direction by an angle θ, but also a distortion (*15*) in the lattice constants c and a given by $c/R = 2/\cos\theta$, $a/R = [8-(c^2/2R^2)]^{1/2}$, shown schematically in the inset to Figure 1. As a result, under shear the volume fraction of solid spheres in the BCT structure is also θ dependent, given by $p_o(\theta) = 4\pi\cos^3\theta/3(8\cos^2\theta-2)$. For θ small, $\bar{\varepsilon}_{zz}(\theta)$ may be expanded about its optimal value as

$$\mathrm{Re}[\frac{\bar{\varepsilon}_{zz}(\theta)}{\varepsilon_2}] = \mathrm{Re}[\frac{\bar{\varepsilon}_{zz}(0)}{\varepsilon_2}] - \frac{1}{2}\mu\theta^2 + -- , \quad (9)$$

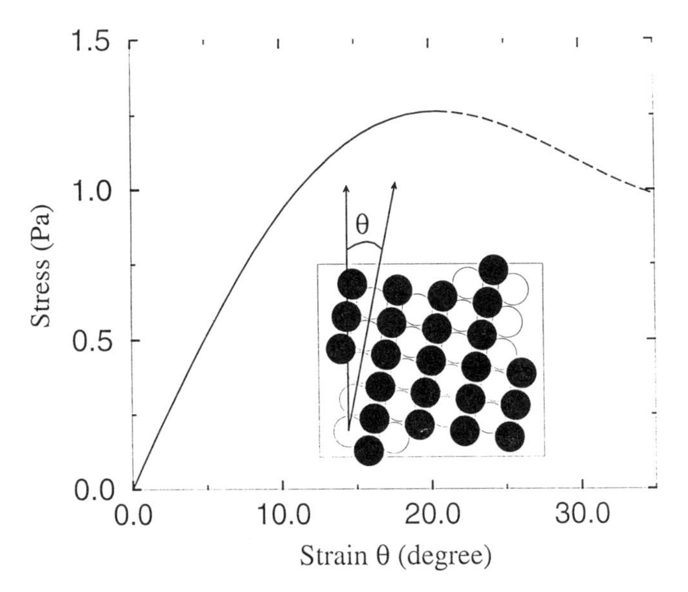

Figure 1. Calculated stress, in units of Pascal, plotted as a function of strain, i.e., the tilt angle θ as defined in the inset. The dashed line indicates the unstable regime. The maximum stress is defined as the static yield stress. The parameters used in the calculation are $\varepsilon_1 = 8.4 + 0.43i$, $\varepsilon_2 = 2.71$, E=1.32KV/mm, and $p = 0.22$. The inset shows the geometry of the particles under shear. Reprinted with permission from Reference 2, Phys. Rev. Lett. 77, 1996, 2499, Ma et al.

where μ is the shear modulus in units of $|\varepsilon_2|E^2/8\pi$. It may be expressed as

$$\mu = \mathrm{Re}\{\frac{2p\varepsilon_2}{|\varepsilon_2|p_o(0)}[F_x(s) - F_z(s) + \frac{\partial F_z(s)}{\partial(\theta^2)}]|_{\theta=0} + \frac{p}{3p_o(0)}[\frac{\bar{\varepsilon}_{BCT}(0) - \varepsilon_2}{|\varepsilon_2|}]\}, \quad (10)$$

with p denoting the volume fraction of solid spheres, $F_z(s) = \langle z|(s-\Gamma)^{-1}|z\rangle / V$, $F_x(s) = \langle x|(s-\Gamma)^{-1}|x\rangle / V$, and $\bar{\varepsilon}_{BCT}(0)$ being the θ=0 effective dielectric constant of the BCT structure. Due to the vertical nature of the structure, i.e., either in columnar or phase-separation state (column diameter approaching ∞), $\bar{\varepsilon}_{BCT}(0)$ is accurately related to the overall effective dielectric constant by the relation

$$\bar{\varepsilon}_{zz}(0) = \frac{p}{p_o(0)}\bar{\varepsilon}_{BCT}(0) + [1 - \frac{p}{p_o(0)}]\varepsilon_2. \quad (11)$$

For the calculation of the static yield stress, it is necessary to go beyond the small θ expansion. Numerical evaluation of the stress-strain relation, i.e., $\mathrm{Re}\{|\varepsilon_2|^{-1}(\partial\bar{\varepsilon}_{zz}(\theta)/\partial\theta)\}$ versus θ, is found to display a maximum as shown in Fig. 1. By definition, value of the maximum stress is the static yield stress (in units of $|\varepsilon_2|E^2/8\pi$), as further strain would make the stress decrease, i.e., the structure becomes unstable.

In the above the calculations of the shear modulus and the static yield stress are considered for the BCT ground state. In actual experiments, it is often the case that columns are formed, as mentioned earlier. However, the difference in the free energies of the column state and the true ground state (due to the electrostatic surface energy of the columns) is small (at most 0.1% of the total electrostatic energy) (*2*). As a result, it is expected that the calculated shear modulus or yield stress should be relatively accurate for the experimentally encountered column state as well.

It is possible to derive physical upper bounds on the effective dielectric constant, the shear modulus and the static yield stress from the Bergman-Milton representation. The key observation here is that whenever the value of s coincides with one of the eigenvalues of the operator Γ(with a non-zero f_n), divergence in the electrical and mechanical responses can occur. However, if we limit ourselves to the values of κ_1 and κ_2 that are positive, then the real part of s will be either less than zero or greater than 1, i.e., outside the range of s_n. The only two places where there can be close approach between s and s_n are 0 and 1, the boundary points for the value of s_n. It is known that in a periodic structure, $s_n \to 0$ as $\delta \to 0$, where 2δ is the closest approach between the surfaces of two spheres. We therefore have two elements in the determination of the effective electrical and mechanical responses of the system: the ratio $\varepsilon_1/\varepsilon_2$ and the dimensionless parameter R/δ. To get the upper bounds, we set $|\varepsilon_1 / \varepsilon_2| \to \infty$ so that $s = 0$. We also assume the periodic structure to be that of BCT. That means for the purpose of getting the upper bounds, the responses of the system can be reduced to functions of R / δ only. Since there is a physical minimum in the value of δ, i.e., atomic separation, we thus have upper bounds in terms of R / δ. By carrying out the effective dielectric constant, the shear modulus, and the static yield stress calculations under the conditions specified above, we obtain the following three upper bounds:

(1) The effective dielectric constant is bounded by $[p / p(0)]\varepsilon_2[3.95\ln(R / \delta) - 5.35]$. Here $p(0) = 2\pi / 9$ is the volume fraction of spheres in the BCT structure.

(2) The shear modulus is bounded by $[p / p(0)]1.9(R / \delta)|\varepsilon_2|E^2 / 8\pi$.

(3) The static yield stress is bounded by $[p / p(0)]1.38\sqrt{R / \delta}\,|\varepsilon_2|E^2 / 8\pi$.

Taking $p = p(0)$, $R = 20\mu$, $\delta = 1\,\mathring{A}$, $E = 1$ kV/mm, and $\varepsilon_2 = 2.5$, we get 8 KPa and 4 MPa, respectively, for the upper bounds on the static yield stress and the shear modulus. It should be noted that the functional dependence of $\sqrt{R / \delta}\,(|\varepsilon_2|E^2 / 8\pi)$ in (3) is exactly what we have derived heuristically earlier using the capacitance argument (with $\varepsilon_2 = \varepsilon_\ell$). Thus both physical arguments and upper bounds derived from first principles identify the same parameters for improving the ER yield stress.

For the coated particles system, the above formalism is generalized as follows. First, the dielectric constant is now given by $\varepsilon(\vec{r}) / \varepsilon_\ell = 1 - \Sigma_i^3 \eta_i(\vec{r}) / s_i$, where $s_i = \varepsilon_\ell / (\varepsilon_i - \varepsilon_\ell)$, ε_ℓ denotes the liquid dielectric constant (replaces ε_2 in the two-component case), ε_i is the dielectric constant of the ith solid materials (i=1,2,3), and $\eta_i(\vec{r})$ is the characteristic function of solid material component i, taking the value 1 in the region of component i, and zero otherwise. With the same imposed voltage condition as before, the effective dielectric constant is given by

$$\frac{\bar{\varepsilon}}{\varepsilon_\ell} = 1 - \frac{1}{V} \sum_{i=1}^{3} \frac{1}{s_i} \int d\vec{r} \, \frac{\partial \phi(\vec{r})}{\partial z} \eta_i(\vec{r}) \, , \tag{12}$$

where $|\phi\rangle = [1 - \Sigma_i \Gamma_i / s_i]^{-1} |z\rangle$ is the formal solution to Eq. (1), with

$$\Gamma_i = \int d\vec{r}' \eta_i(\vec{r}') \nabla' G(\vec{r}, \vec{r}') \cdot \nabla' \; . \tag{13}$$

The electrostatic ground state for the coated particles remains BCT, just as in the two-component case. The calculation of either the shear modulus or the static yield stress follows the same procedure as before.

We now describe the fabrication process for the dielectric/metal (DM) doubly-coated particles (DMP) (*8*). The core particles are solid glass spheres commercially available in various sizes. We used spheres with diameters 1.5μ ($\pm 0.1\ \mu$) and 50μ ($\pm 7.5\mu$). Conventional electroless plating process (*16*) was used to deposit the inner conducting nickel layer. The outer insulating layer material used was titanium oxide, chosen for its hardness, resistance to wear, and high dielectric constant. This outer layer was deposited using a sol-gel process (*17*). Thickness control of the layers were achieved by solution concentrations and process durations. It is crucial to avoid particle coagulation during the process, and this was achieved by adding excess sugar solution to form a buffer, which could subsequently be removed by heating. The sugar solution, in addition, formed a foam structure during heating which further separated the particles. After heating (500C for 10 hours) to remove all organic components, the DMP's were collected as ashes and dispersed in silicon oil to form the DMP ER fluid. The heating also serves to properly anneal the TiO_2 coating, with the very desirable properties of high dielectric constant, excellent adhesion and hardness. We have taken cross sectional electron microscope pictures at various stages of the coating process for monitoring purpose. A 1.5 μ particle with only the nickel coating is shown in Figure 2(a). It is seen that the metallic coating was on the order of 100 Å in thickness with excellent uniformity. Figure 2(b) shows a doubly-coated particle at the completion of the process (after heating.) In contrast to Fig. 2(a), the TiO_2 coating has a much greater variation in thickness, ranging from 100 Å to 250 Å in this case. Figure 2(c) shows, at higher magnifications, a 50 μ doubly-coated particle showing the annealed crystalline structure of the TiO_2 outer coating. Comparison between Figures 2(a) and 2(b) or 2(c) also indicated the disappearance of the metal/glass boundary clearly visible in Figure 2(a). This is most likely due to diffusion of the nickel atoms during the heating process after the TiO_2 coating process. However, as we will see in yield stress measurements below, this conducting layer, while not visible in the picture, is sufficiently continuous to fulfill its mission of expelling the electric field (from the interior of the particles) and led to behavior markedly different from particles without such an inner conducting coating. As a further check, we have performed compositional analysis of the particles. The results are in general agreement with the above picture. A parallel plate viscometer was used to measure the yield stress of the DMP/silicon oil ER system. To remove any traces of water from the system, all samples went through a 24 hours heating process at 120 °C prior to each measurement. The yield stress was measured with a 50Hz AC field applied between the parallel plates, and using samples with particle/fluid fraction fixed to 0.2. Figures 3(a) and 3(b) show the measured static yield stress for a DMP/silicon oil ER system using the 1.5 μ and 50 μ diameter particles, respectively, as a function of electric field strength. Static yield stress over 2000Pa was obtained at 2kV/mm for the DMP system. As comparisons, we also showed the static yield stress measured for the bare glass particles as well as glass particles coated only with TiO_2 without inner metallic coatings, (see also Figure 3(c) with expanded scale for Figure 3(a)). The ER enhancement of the DMP system was over two orders of magnitude. It is seen that the DMP system followed generally the expected electric field squared dependence. These DMP's were found to be extremely robust and can sustain both the high electric field and high stress which are prerequisites for any application. The size dependence is seen to be in general agreement with the intuitive picture presented above. In Figures 3(a) and 3(b) are also plotted theoretical predictions calculated from TiO_2 and Silicone oil dielectric constant values of 85 and 2.5, respectively, with the thickness of the TiO_2 coating thickness indicated besides each of the theory curves (solid lines). The metal dielectric constant is taken to be infinite, which also makes its thickness immaterial. Good agreements are seen for both particle sizes. Using the same theoretical model, we have also calculated the yield stress for bare glass particles and TiO_2 coated particles (without the inner metallic coating). The results, together with the experimental results for such particles, are shown in Figure 3(c) for the 1.5 μ

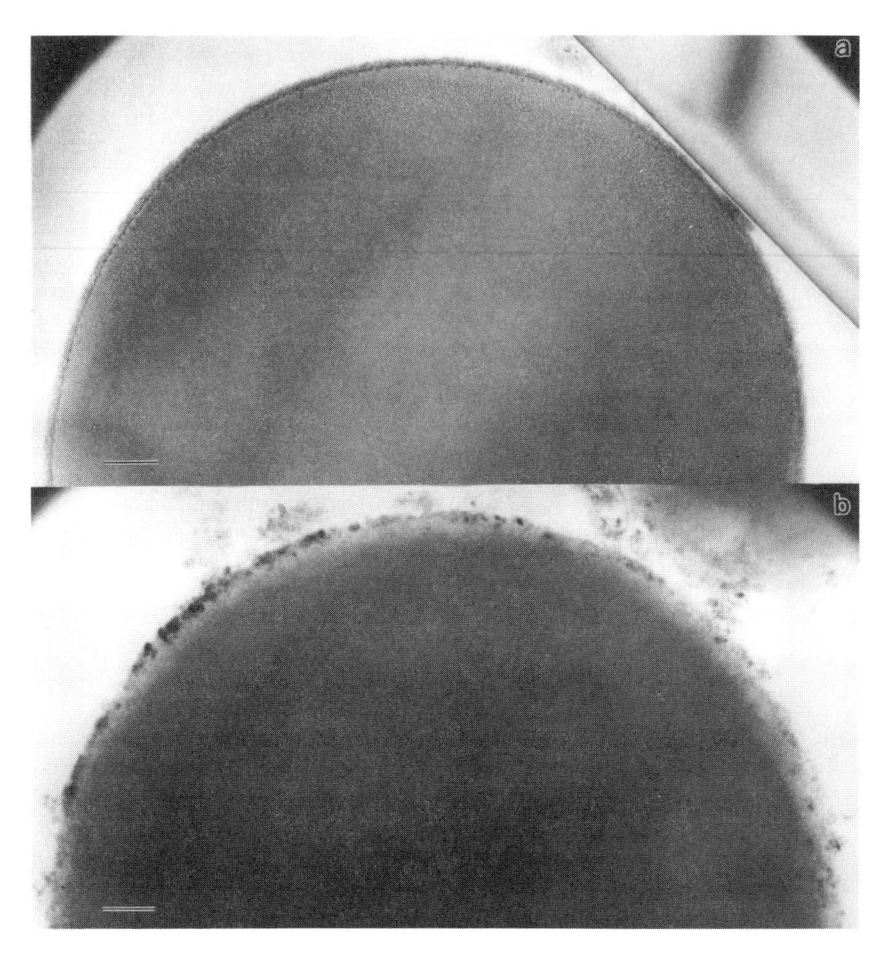

Figure 2. Cross sectional electron micrographs for (a) 1.5μ nickel coated, (b) 1.5μ doubly-coated, and (c) 50μ doubly-coated particles. The scale bar is 100nm and 25nm for (a)/(b) and (c), respectively. While in (a), the metal/glass interface is clearly visible, diffusion of Ni atoms, probably from the heating process following the sol-gel process, blurred this boundary, as shown in (b) and (c). The TiO_2 coating thickness is seen to be in the range of 10nm to 30nm for (b) and 25nm to 60nm for (c). Note also the darker spots seen only in the TiO_2 coatings in (b) and (c). This leads us to believe that the TiO_2 coating was at least partially crystalline, the dark spots being regions with high scattering efficiency when the Bragg condition was satisfied when the electron micrograph was taken. With higher magnification, the apparently microcrystalline morphology of the TiO_2 layer can be seen even more clearly in (c). Reprinted with permission from Reference 8, Phys. Rev. Lett. 78, 1997, 2987, Tam et al.

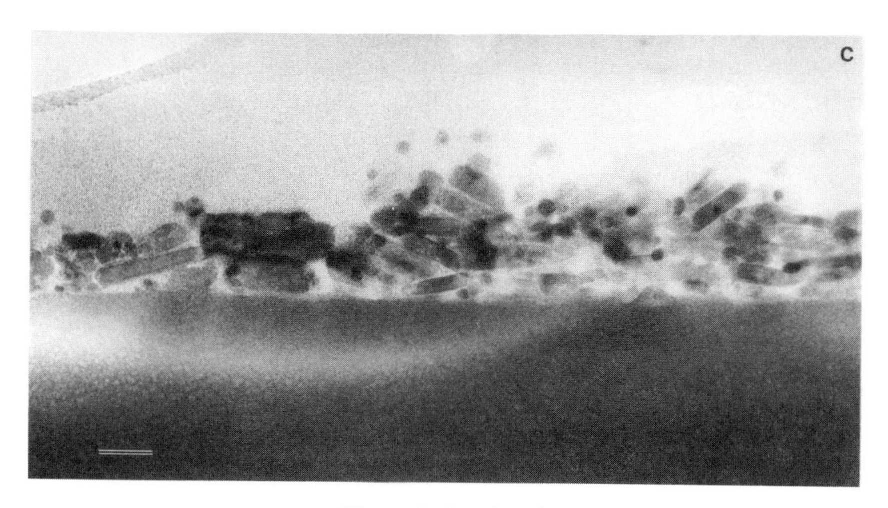

Figure 2. *Continued.*

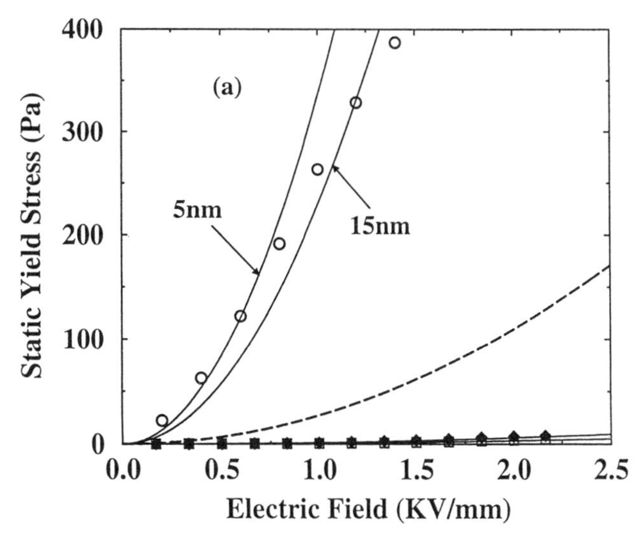

Figure 3. The measured (symbols) and calculated (lines) static yield stress of ER fluid using (a) 1.5μ doubly-coated particles, (b) 50μ doubly-coated particles, and (c) expanded scale of (a), uncoated particles (open squares) and singly coated with TiO_2 only (solid squares). As explained in the text, the solid line is yield stress obtained from first principles calculations using known dielectric constants and measured geometric information of the particles. Due to the variation of TiO_2 coating thickness as seen in Figures 1(b) and 1(c), the calculation was done for a range of TiO_2 thicknesses, as labeled in the figure. As a comparison, we have also calculated for the case of solid TiO_2particles of the same sizes (dashed line) showing that the doubly-coated particles have far superior static yield stress. Reprinted with permission from Reference 8, Phys. Rev. Lett. 78, 1997, 2987, Tam et al.

Continued next page

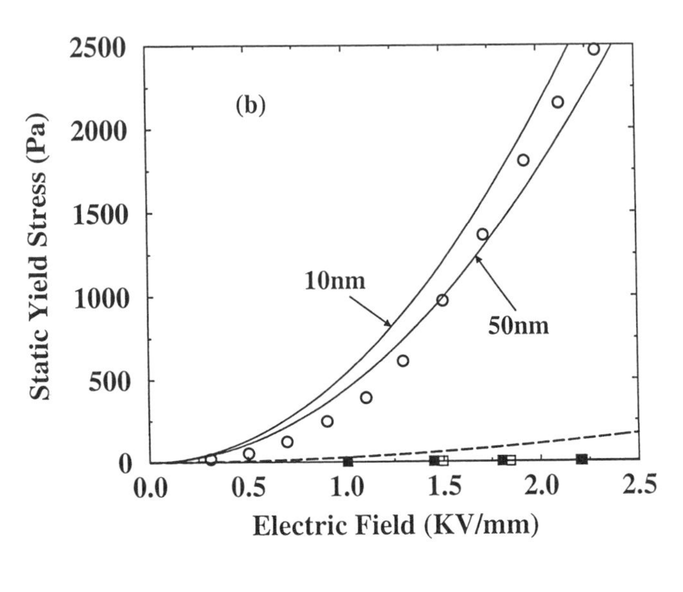

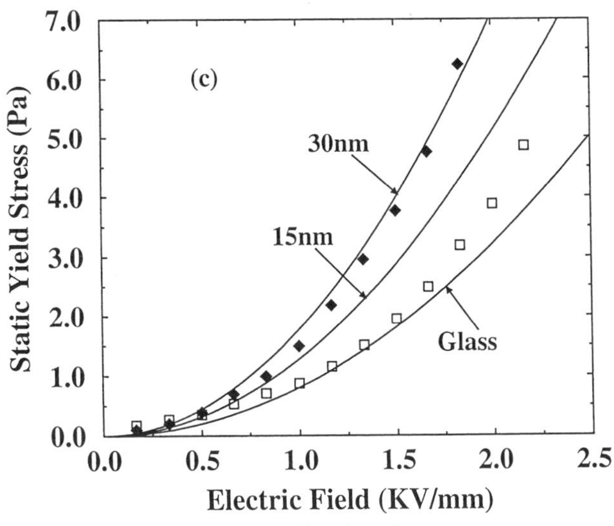

Figure 3. *Continued.*

particles. As a reference, we have also calculated the yield stress expected for TiO_2 particles of the same size, shown as dashed lines in Figures 3(a) and 3(b). Both the ability of the first principles calculations to account for the size and coating thickness dependencies of the static yield stress, and the efficacy of the doubly-coated DMP system, are clear from the figures.

In conclusion, we propose and experimentally demonstrated a new type of microstructured particles with improved ER performance. Parameters relevant to the ER static yield stress were identified through both simple physical arguments as well as first principles calculations, which are shown to account quantitatively all of our experimental findings. It is our believe that based on the composite particles idea, the coupling of materials choice, fabrication, characterization and physical understanding of the ER mechanism on a firm theoretical basis can ultimately lead to the goal of realizing ER fluids' applications potential.

This work is supported by HKUST Research Infrastructure Grant RI 93/94.SC09, and the William Mong Solid State Cluster Laboratory.

References

1. Tao, R. and Sun, J. M. *Phys. Rev. Lett.* **1991,** *67* , 398.
2. Ma, H.; Wen, W.; Tam, W. Y.; and Sheng, P. *Phys. Rev. Lett.* **1996,** *77,* 2499.
3. Clercx, H. and Bossis, G. *Phys. Rev. E* **1993,** *48,* 2721.
4. Friedberg, R. and Yu, Y. K. *Phys. Rev. B* **1992,** *46,* 6582.
5. Davis, L. C. *Phys. Rev. A* **1992** *46,* R719.
6. Korane, K. J. *Machine Design* **1991,** *63,* 52.
7. Goldstein, G. *Mech. Eng.* **1990,** *112,* 48.
8. Tam, W. Y.; Yi, G. H.; Wen, W.; Ma, H.; Loy, M. M. T.; and Sheng, P. *Phys. Rev. Lett.* **1997,** 78, 2987.
9. A calculation for a chain of metallic particles separated by finite spacings has been reported by Anderson, R. A. *Langmuir* **1994,** *10,* 2917.
10. Bergman, D. J. *Phys. Rep.* **1978,** *43,* 377.
11. Bergman, D. J. *Phys. Rev. B* **1979,** *19,* 2359.
12. Milton, G. W. *Appl. Phys. A* **1981,** *26,* 1207.
13. Milton, G. W. *J. Appl. Phys.* **1980,** *52,* 5286.
14. Heine, V. In *Solid State Physics*; Ehrenreich, H. and Turnbull, D.; Academic Press: New York, 1980, Vol. 35, p. 1.
15. Here the distortion is treated as symmetrical in the x and y directions since that is lower in free energy than any distortions that are anisotropic in x and y.
16. Severin, J. W.; Hokke, R.; Vanderwel, H.; and deWith, G. *J. Electrochem.. Soc.* **1993,** *140,* 682.
17. Briker, C. J. and Scherer, G. W. In *Sol-gel Science: The Physics and Chemistry of Sol-gel Processing*; Academic Press: New York, 1990.

Chapter 6

Composite Materials with Field-Induced Anisotropy

James E. Martin, Robert A. Anderson, and Chris P. Tigges

Sandia National Laboratories, Albuquerque, NM 87185

Composite materials consisting of a polymerizable liquid phase and a dispersed colloidal phase can be structured by a uniaxial or rotating electric or magnetic field and cured to form a material with anisotropic properties. We report the results of a computer simulation of the evolution of structure in these materials and characterize the structures that emerge through the anisotropy of their conductivity and permittivity. We find significant anisotropies are induced in these samples, in good agreement with experimental data we have collected. The anisotropies formed by a rotating field are found to be just the inverse of those produced by a uniaxial field.

When a fluid consisting of a colloidal phase dispersed in a liquid is exposed to a uniaxial electric or magnetic field, (say along the *z axis*) the particles will polarize and interact, provided a sufficient permittivity or permeability mismatch exists, leading to the evolution of complex anisotropic structures. At low concentrations the particles initially form chains and these then slowly coalesce into columns containing microcrystalline regions. At high concentrations the morphology is more complex, with branching between columns and a loss in anisotropy. When the same fluid is exposed to a field rotating at high frequency (10^2-10^4 Hz), (in the *x-y* plane) the particles form hexagonal sheets in the plane of the field, that then coalesce into plates. These plates branch by "staircase' structures.

In this paper we seek to develop an understanding of the evolution of structure in these fluids by large-scale, 3-D computer simulations of N spherical particles. We consider only very deep quenches where temperature does not play a role and find that structural coarsening proceeds without thermal fluctuations, due to the presence of defect structures that generate long range interactions between chains and sheets, respectively.

We characterize the structure and growth kinetics by the anisotropy of the conductivity and permittivity of these structures. The structures formed in a uniaxial field have a high conductivity and permittivity along the z axis, whereas the rotating field structures have high conductivity and permittivity in the *x-y* plane.

These simulation results are compared to experimental data collected for ceramic and metal particles field-cured in an epoxy resin.

Simulation Method

The first issue to address is what physics is essential to include in the simulation. We used a simple first-order approach that has enabled us to develop a statistical analysis of structure and properties in large scale systems. As a cross check we developed two codes that operate quite differently, yet give quantitatively similar results.

Equation of Motion. The equation of motion for the ith sphere in the system of N spheres is

$$m\mathbf{a}_i = F_h(\mathbf{v}_i) + \sum_{j\neq i} F_{hs}(r_{ij}) + \sum_{j\neq i} F_d(r_{ij},\theta_{ij}) \tag{1}$$

where $F_h(\mathbf{v}) = -6\pi\mu_0 a\mathbf{v}$ is the hydrodynamic Stokes force with μ_0 the solvent viscosity, F_{hs} is the hard sphere force, and F_d is the dipolar force. r_{ij} is the distance between the ball centers of mass and θ_{ij} is the angle the line of sphere centers makes to the applied field. The inertial term is set to zero. The spheres are modeled as nearly hard spheres, with a repulsive force dependent on the gap between spheres, $F_{hs}(r)=A/(r-cd)^{\alpha}$, where d is the sphere diameter and A and c are constants.

For the uniaxial field the dipolar potential is used, (*1*) which gives the interaction force between two spheres whose center of mass separation vector is of length r and inclined at an angle θ to the applied field E_0 (i.e. the z axis),

$$F_d(r,\theta) = -\frac{3\pi}{4}\varepsilon_0\varepsilon_c a^2\beta^2 E_0^2\left(\frac{d}{r}\right)^4\left[\left(3\cos^2\theta - 1\right)\hat{\mathbf{r}} + \sin 2\theta\hat{\theta}\right] \tag{2}$$

Here a is the sphere radius, ε_0=8.854x10^{-12} F/m is the vacuum permittivity, ε_c is the dielectric constant of the continuous (liquid) phase, and β is the dielectric contrast factor $(\varepsilon_p - \varepsilon_c)/(\varepsilon_p + 2\varepsilon_c)$, where ε_p is the particle dielectric constant. Note that the radial component of the dipolar force is attractive only when $\theta < 54.7°$. The potential of interaction for a rotating field in the x-y plane is just -1/2 times Eq. 2, where θ is the angle relative to the vector normal to the field plane (i.e. normal is z axis).

Timescale. In the absence of Brownian motion the strength of the dipolar interactions alters only the coarsening timescale, not the structural evolution. The dimensionless numerical equation of motion is thus of the form $\Delta u = \Delta s f(r,\theta)$, where the dimensionless length $\Delta u = \Delta r/2a$ and the dimensionless time $\Delta s = \Delta t\ \varepsilon_0\ \varepsilon_c\beta^2 E_0^2/16\mu_0$. For a viscosity of 1 cp, an applied field of 1.0 kV/mm, and $\varepsilon_c\beta^2 = 2$, $\Delta s \cong \Delta t \times 10^3$, so one dimensionless time unit is about a millisecond. We use this time conversion in all of our plots. Note that the timescale, and thus the coarsening kinetics, is independent of ball size.

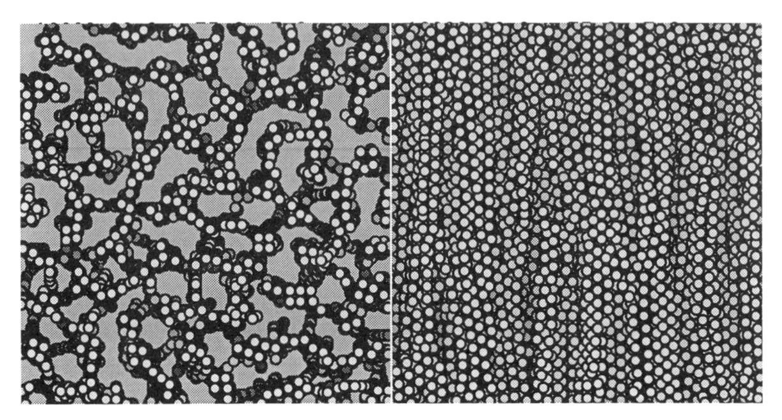

Figure 1. Views parallel (a) and perpendicular (b) to the uniaxial field (z) axis of a system of 10,000 balls at $\phi = 20$ vol. %, and at a time of 75 ms. Along the field axis the fluid structure is best described as a branched network. The pore size of the network increases slowly with time, and spatial correlations increasingly propagate down the z axis. Perpendicular to the field axis the aggregation of chains into columns can be seen.

Development of Composite Structure

Before giving the results of various numerical measures of structure, we first present a descriptive summary of the simulation results. The result of structure formation for 10,000 balls at a volume fraction of $\phi = 20$ vol. % in a uniaxial field is in Figures 1a&b, which nicely illustrates the pronounced anisotropy of these structures, with chains that form along the field axis aggregated into a kind of network. In fact, structure is exhibited on two scales; local crystalline order on length scales small compared to the domain size, and a mesomorphous order on length scales larger than the domain size and smaller than the correlation length. The crystalline order consists mostly of chains, hexagonal sheets, and body-centered tetragonal domains that Tao has shown minimize the electrostatic energy of the system. (*2*) At low concentrations and early times chains dominate; as time progresses or the concentration increases these chains form hexagonal sheets; and finally at large times and high concentrations bct structures predominate. The most pronounced anisotropy occurs at the lowest concentrations, and very little remains at 50 vol. %.

In a rotating field, Figures 2a&b, the attractive interactions are in the x-y plane, so the particles initially form sheets with a hexagonal packing in the field plane. These sheets then coalesce to form plate-like structures with a packing that is apparently random; dipolar lattice sums we have performed for the AB... packing of the hexagonal close pack (hcp) structure are only slightly less stable (by 0.0005%!) than the ABC... packing of the face-centered cubic structure, which we believe is the ground state. These plates are nearly uniformly spaced and have an extreme degree of structural anisotropy that is reflected in their properties.

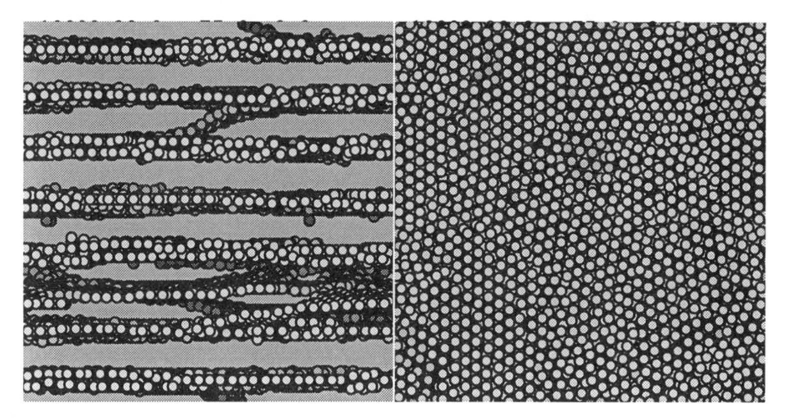

Figure 2. Views parallel (a) and perpendicular (b) to the rotating field (x-y) plane of a system of 10,000 balls at ϕ = 20 vol. %, and at a time of 75 ms. The aggregation of hexagonal sheets into plates is evident.

Anisotropic Properties

Conductivity. We computed the conductivity of the field-structured composites by making the simple approximation that the continuous phase does not conduct and that the contact resistance between conducting balls limits the conductivity. We further assumed that the contact resistance is a simple constant (in reality the contact resistance may depend logarithmically on the gap, but to some extent our gaps are determined by finite numerical jitter, so it is probably best to treat the contact resistance as a constant). Thus the balls are mapped onto a simple resistor network, where the ball centers are nodes.

Simulation Results. To find the node potentials we used a fast relaxation method to solve Kirkhoffs laws. The boundary conditions were formulated to insure that chains would indeed conduct: balls within 1 radius of the electrodes were fixed at the electrode potential, all other balls were at floating potentials. Cyclic boundary conditions were not applied to the faces orthogonal to the applied field, but were applied to the other two pairs of faces.

The current was computed at three locations in the simulation volume. First, we computed all the current flowing into the balls at fixed ground potential. Second, we computed the current flowing from the balls fixed at high potential, which should

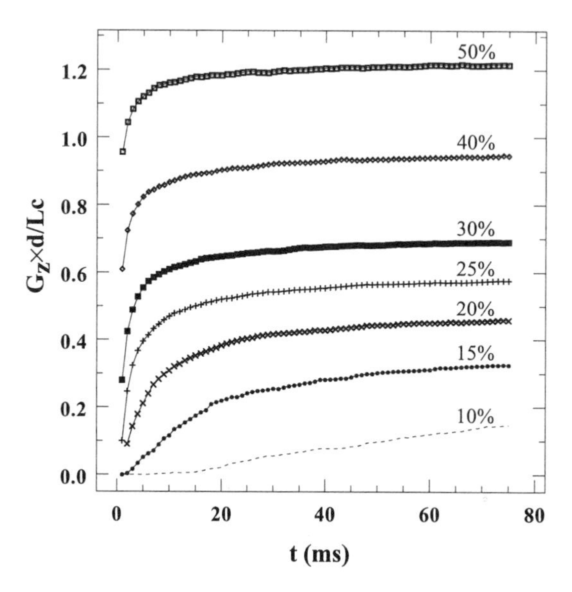

Figure 3. The *z*-axis conductivity of simulated structures in a uniaxial field increases rapidly in the first few milliseconds.

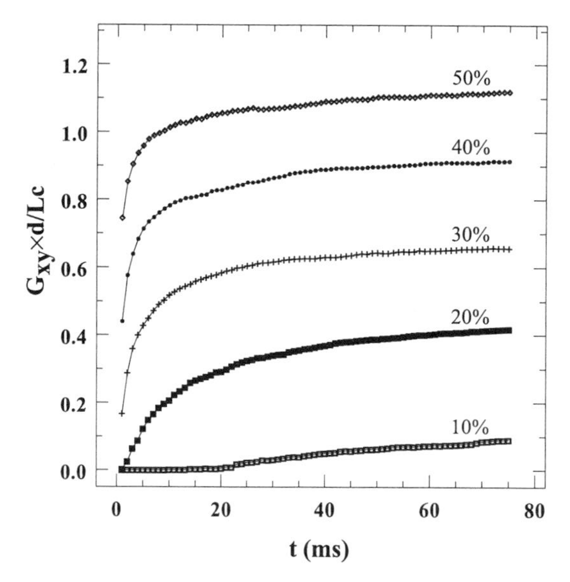

Figure 4. The *x*-*y*-plane conductivity of simulated structures in a rotating field.

be equal and opposite. Third, we computed the current flowing through a plane bisecting the sample and parallel to the electrode planes. All of these currents agreed to ~ 4-5 places, as indeed they should. The results of these studies are shown in Figures 3&4. The conductivity increases rapidly in the first few milliseconds, and then increases very slowly with time. Note that the conductivity anisotropy is inverted for the rotating field structures, relative to that produced by a uniaxial field.

The concentration dependence of the field-induced conductivities (at 75 ms) is shown in Figures 5&6. These are compared to that expected for the unstructured material which shows a classical percolation behavior, being zero until the percolation threshold ϕ_{perc}, then increasing more or less quadratically with $\phi-\phi_{perc}$. For the uniaxial structures the z axis conductivity increases almost linearly with ϕ, and the conductivity in the xy plane increases almost linearly with $\phi-\phi_c$, for $\phi > \phi_c$, where ϕ_c is the percolation threshold in the x-y plane for samples of 10,000 balls. For the rotating field samples the x-y plane conductivity increases almost linearly with volume fraction, and the z-axis conductivity is actually less than the zero field value.

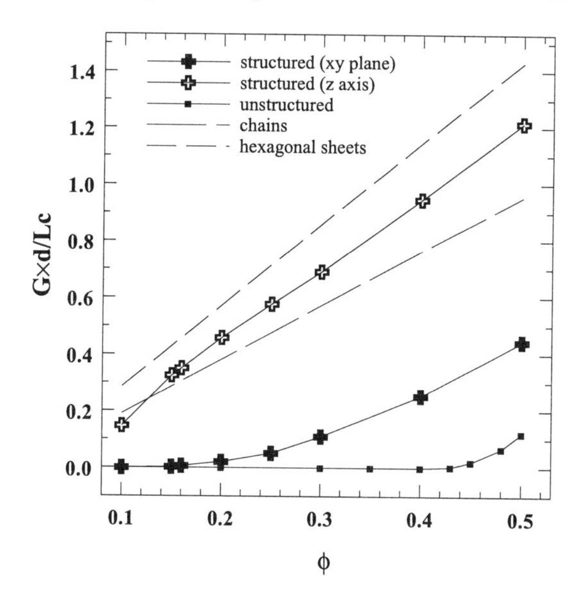

Figure 5. Simulated composites structured in a uniaxial field have conductivities (at 75 ms) much larger than those of the unstructured material, which shows percolation behavior. The z axis conductivity increases almost linearly with ϕ, and the conductivity in the xy plane increases almost linearly with $\phi-\phi_c$, for $\phi > \phi_c$, where ϕ_c is the percolation threshold in the xy plane for samples of 10,000 balls. Computed conductivities of chains and sheets are shown for comparison.

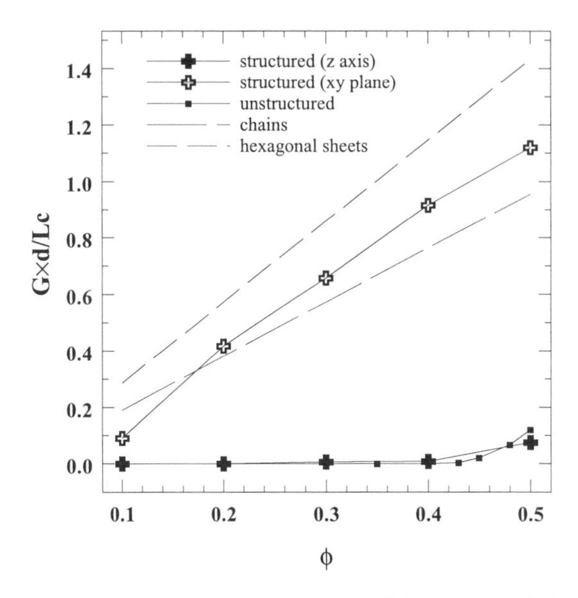

Figure 6. Composites structured in a rotating field have much larger anisotropy in the conductivity, with the *z* axis values actually lower than the percolation values.

Conductivity Anisotropy. For the uniaxial field structures the anisotropy in the conductivity, Figure 7, is strongly dependent on volume fraction and sample size; the data for our system is described by the expression $G_z/G_{xy} \sim |\phi - \phi_c|^{-1.0}$, where ϕ_c = 0.137. As the system size increases we would expect ϕ_c to decrease, eventually to zero.

Analytic Results. The nearly linear increase of the conductivity along the *z* axis with volume fraction can be rationalized surprisingly well by comparison to a system of chains or sheets. A chain of spheres of diameter *d* that spans a volume *L* on a side will contain $n = L/d$ spheres. The conductance of this chain will be $g_{chain} = c/n$, where *c* is the inverse contact resistance. The number of such chains in the volume of *N* spheres will be $m = N/n$, so the conductance of the sample will be $g_z = g_{chain}m$. The conductivity is then $G_z = g_z/L = c\,(6/\pi)\phi/d$, or $d\times G_z/c \cong 1.91\ \phi$. Our simulated currents are very close to this, with $d\times G_z/c \cong 2.2\ \phi$. Of course, this simple chain calculation gives zero conductivity in the *xy* plane.

Hexagonal sheets and other structures give additional currents paths. A hexagonal sheet formed of chains aligned along the *z* axis will have additional currents paths that zig-zag back and forth between adjacent chains, which are out of registry. Each zig-zag path will have half the conductivity of a chain path, and for a sheet there is just one zig-zag per chain, so a system of hexagonal sheets will have a conductivity of $d{\times}G_z/c = (9/\pi)\ \phi$. Likewise, a system consisting of large bct domains will have two zig-zag paths per chain, so $d{\times}G_z/c = (12/\pi)\ \phi$. That the conductivity we observe in either the uniaxial or rotating field cases is greater than the chain limit is due to the presence of these additional current paths.

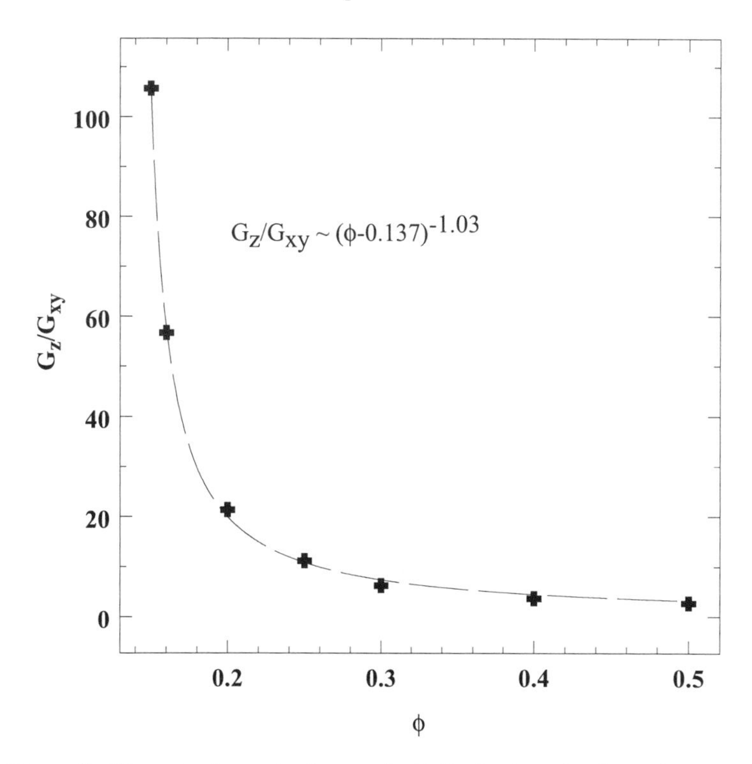

Figure 7. The anisotropy in the conductivity is strongly dependent on volume fraction and sample size; the data for our simulated uniaxial composite is well described by a critical point expression, although as the system size increases we would expect ϕ_c to decrease, eventually to zero.

The conductivity in the *xy* plane for the uniaxial samples is not so easily understood. The conductivity of a bct lattice in the *xy* plane is just 3/4 of the conductivity along the *z* axis, clearly too large to explain the small *xy* plane conductivities we observe. Even at 50 vol. % the *xy* conductivity is only 1/3 of the *z* axis conductivity. In fact, by plotting the current carrying paths in the *xy* plane we see that these paths are very indirect, consisting of ill-connected domains.

Alternative Formulation. During the course of these studies a fourth method of measuring the current became apparent, and this clarifies the connection of the conductance problem to the permittivity problem. In a volume L on a side containing N spheres the total charge displacement along the w axis after a time δt is

$$Q\delta w = \sum_{k=1}^{\zeta N} c\Delta V_k d \cos\theta_k \delta t \quad (3)$$

where Q is the charge, ζ is the average coordination number of the lattice, ΔV_k is the voltage drop across the kth resistor, and θ_k is the angle this resistor makes to the w axis. The macroscopic current is the charge density times the velocity times the area, $I = (Q/L^3)(\delta w / \delta t)L^2$, and the conductivity is $G_w = (I / \Delta V) / L$, where ΔV is the voltage applied across the sample. Combining these we obtain the useful expression

$$G_w = \frac{cd}{L^2} \sum_{k=1}^{\zeta N} \frac{\Delta V_k}{\Delta V} \cos\theta_k \quad (4)$$

We shall see that this formulation of the conductivity allows us to directly map the conductivity problem onto the permittivity problem.

Experimental Results. The predicted conductivities can be compared to experimental results we obtained for 10 wt. % stainless steel *fibers* dispersed with the surfactant Triton X-100 in an epoxy resin structured with a magnetic field of 700 Gauss and cured at 70° C. In conductivity units where the unstructured sample is 1, the z-axis conductivity measured 38.1 and the *x-y* plane conductivity measured 15.2. Fibers are not an ideal comparison, but samples made with iron particles unfortunately showed capacitive effects (frequency dependent conductivities) due to oxide formation at the particle surface, so this work is still in progress.

Permittivity. Our formulation of the conductivity is based on the extreme contrast one can in practice obtain between the conductivities of the continuous and discrete phases, for example, in the magnetic alignment of conducting particles of stainless steel in an epoxy resin. (*3*) This contrast far exceeds that which can be obtained in dielectric systems, but it is useful at least to point out the analogy between these two.

Relation to Conductivity. Let us assume that the capacitance due to the gap between two high dielectric constant spheres dominates all other contributions to the capacitance. This capacitance is a function of the log of the gap. (*4*) If we assign a fixed gap capacitance to each contacting pair then the material capacitance is just the capacitance of the capacitor network. Like conductors, the total capacitance of capacitors in parallel add, and the inverse of the capacitances add when connected in series. Also, from Kirkhoffs law the sum of the charges on the capacitor plates connected to each floating point ball is zero, so the node voltages are exactly those determined for the resistor network. The w-axis permittivity, $\varepsilon_w\varepsilon_0$, of the material is then just the capacitance times the material thickness divided by the area, and so is given by our final expression for the conductivity,

$$\varepsilon_w\varepsilon_0 = \frac{cd}{L^2}\sum_{k=1}^{\zeta N}\frac{\Delta V_k}{\Delta V}\cos\theta_k \tag{5}$$

where c is now the gap capacitance. Thus, the behavior illustrated in the figures for the conductivity directly apply to the permittivity. Because this limit is not easily achieved, we will now discuss a more sophisticated approach to the dielectric problem.

Self-Consistent Point-Dipole Moment Approach. In principle, the permittivity of simulated structures could be calculated to any desired precision by multipole-expansion techniques. This would be a daunting task, however, when a disordered three-dimensional arrangement of thousands of particles must be accommodated. As an alternative, we have developed practical algorithms for approximating the dielectric constants of the simulated structures. The results obtained are sufficiently accurate for elucidation of the concentration and temporal dependencies, as well as for comparison with experimental results.

The local field at each particle is affected by the fields from other nearby polarized particles, which thereby make a major contribution to the macroscopic permittivity. To account for this many-body effect, code was written to self-consistently calculate the local field at the center of each of the N particles. Surface effects were eliminated by utilizing the cyclic boundary conditions to surround each dipole with a volume of nearby dipoles the size of the primary cubic simulation volume and centered on the dipole under consideration. Fields from the induced dipole moments of the N-1 surrounding particles are summed, then augmented by the applied field, $\mathbf{E}_0$, and the Lorentz-cavity field due to the boundaries of the volume. For the purposes of this calculation, the coarse structures that develop within the simulation volume are ignored in the application of the Lorentz cavity correction; the cavity wall charge density is simply the average dipole-moment density. Each particle is treated as a point dipole with polarizability equivalent to that of a conductive sphere (β =1 limit, $\mathbf{p}=4\pi\varepsilon_c\varepsilon_0 a^3\mathbf{E}_{loc}$) and its dipole moment is updated according to the calculated local field at its site. This procedure is iterated to convergence.

It should be noted that our algorithm employs all three cartesian components of the individual dipole moments, although only the component parallel to the permittivity direction directly contributes to the desired diagonal element of the permittivity tensor. The transverse induced moments, however, also give rise to fields parallel to the applied field which then influence the parallel moments. We found that ignoring the transverse components can lower the calculated permittivities by as much as 10%.

For convenience we treat the continuous phase as having the permittivity of free space. The effective dielectric constant is simply $1+P/(\varepsilon_0 E_0)$, where the macroscopic polarization, P, is the density of the discrete-phase induced dipole moment. The actual permittivity of the continuous phase can then be accounted for by recognizing that the results so obtained, for $\beta = 1$, correspond to the normalized dielectric constant, $\varepsilon_f/\varepsilon_c$.

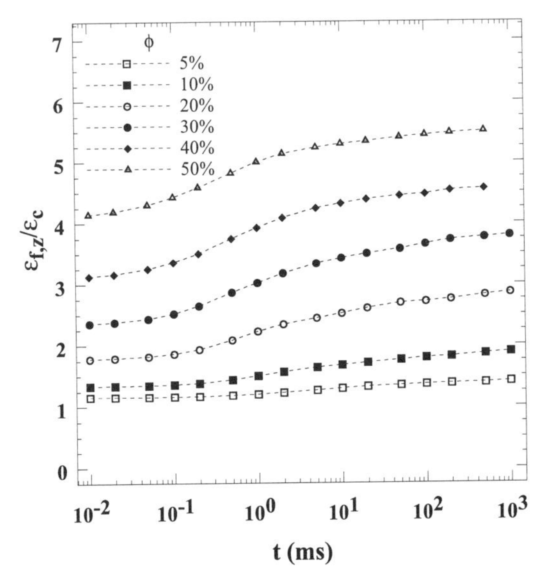

Figure 8. The time dependence of the dielectric constant along the z axis, computed by the self-consistent point-dipole moment method for uniaxial field simulation files of 1000 balls.

The time dependence of dielectric constants, calculated according to the above procedure, are shown in Figures 8&9. The permittivity parallel to the field direction in the dynamic simulations (z axis) increases with time, but decreases slightly in transverse directions (along x and y axes). The concentration dependence of the dielectric constant is shown in Figures 10&11. A notable resemblance to the conductivity data is observed, with the permittivity along the z axis increasing nearly linearly with the particle concentration. The dielectric constants for the unstructured samples, obtained from the randomized zero time files for the dynamic simulation, are compared to the familiar Clausius-Mossotti prediction for random spheres (Maxwell's relationship), $\varepsilon_{f,r}/\varepsilon_c = (1+2\phi)/(1-\phi)$, evaluated at $\beta = 1$, and agreements within a few percent are found at all volume fractions investigated (5% - 50%).

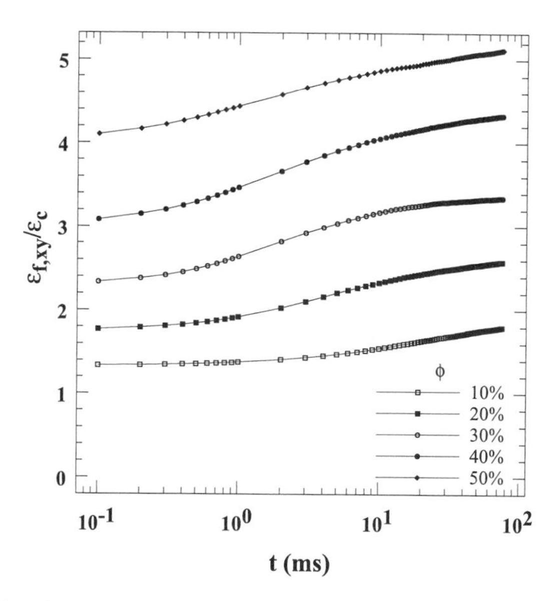

Figure 9. The dielectric constant in the *x-y* plane for rotating field simulation files, computed by a self-consistent point-dipole method.

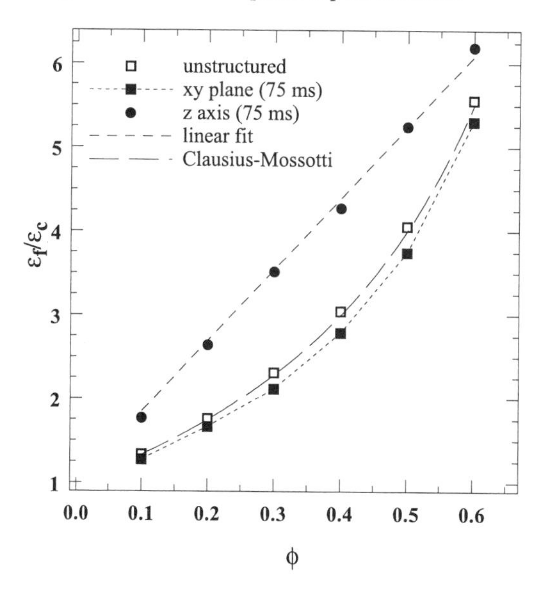

Figure 10. The dependence of the dielectric constant on particle volume fraction is shown for simulated samples structured for 75 ms. Along the field cure axis the dielectric constant increases nearly linearly with particle volume fraction; the unstructured samples are well described by the Clausius-Mossotti equation.

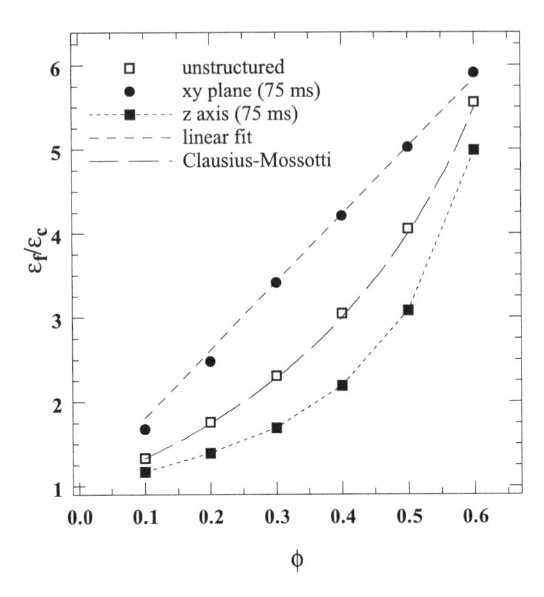

Figure 11. The rotating field samples show behavior that is inverted from that shown in Figure 10.

Ball-Contact Moments Method. An independent numerical method has been developed that captures the higher-multipole influence on the permittivity of tightly aggregated structures and is therefore closely related to Eq. 5, but makes an additional assumption that particle potentials vary uniformly along the applied field direction. This postulate is reasonable when a permittivity measuring field is applied parallel to the axes of long chains or sheets of balls having invariant cross sections.

This method approximates the permittivity from a summation of the localized dipole moments induced at each of the contact points between particles. For the purposes of this calculation, a pair of balls is considered to be contacting if the center-to-center distance, $r = 2a\,(1+\delta)$, does not exceed the ball diameter by more than 5%. If contacting particles are viewed as conductive spheres, the localized dipole moment depends linearly on the potential difference and logarithmically on the separation gap δ. (*4*) The gaps we observe in our computer simulations are sensitive to the specific interaction potential, discrete time step etc., so we have arbitrarily chosen $\delta = 0.004$ in calculating the contact moments of particle pairs selected according to the contact criterion. Under the uniform-field assumption, each moment contains a $\cos(\theta)$ factor, where θ is the angle of the vector between a pair of balls, relative to the applied field direction. An additional $\cos(\theta)$ factor expresses the component of the contact moment along the applied field direction. From (*4*) the parallel component of each contact moment is then

$$m_{cont} = 4\pi\varepsilon_c\varepsilon_0 \cos^2(\theta)a^3(1+\delta)\left[(1+\delta)\ln\left(1+\delta^{-1}\right)-1\right] \quad (6)$$

where the permittivity of the continuous phase is taken as that of free space. As previously, the normalized dielectric constant is then found by using the β=1 formula, $\varepsilon_f/\varepsilon_c = 1+P/(\varepsilon_0 E_0)$.

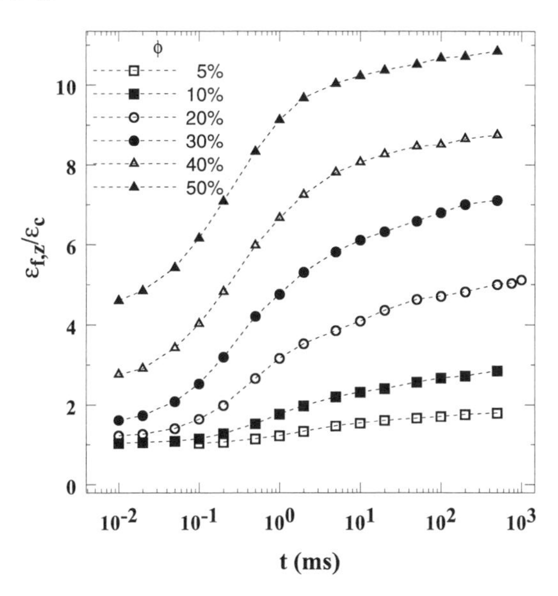

Figure 12. The concentration dependence of the dielectric constant of simulated uniaxial field-structured composites, using the contact moment method.

Dielectric constants of $19.9\varepsilon_c$ and $15.2\varepsilon_c$, parallel and normal to the c axis, are found when this numerical procedure is applied to a perfect, ϕ=69%, bct lattice, for an anisotropy of 1.31. Calculations by Clercx and Bossis (*5*) find corresponding values of 16.84 and 13.71, for an anisotropy of 1.23 with a bct lattice of the same density. This fair agreement indicates that our numerical procedure is sufficiently accurate for analysis of the dynamic simulation results, Figure 12. These dielectric constants are nearly twice those of the self-consistent-local-field calculation, but the trends are similar.

Permittivity Anisotropy. The anisotropy of the permittivity in mature, field-structured materials can be predicted from an analysis of the dynamic simulation results. As described in the foregoing, we use the contact-moment and local-field analysis methods, respectively, to approximate the dielectric constants parallel and normal to the alignment axis. We find that a maximum ratio of approximately 3.3

occurs at volume fractions near 30%, and that this ratio remains near 3.0 from 20% to 50%.

Experimental Results. We prepared dispersions of $BaTiO_3$, a ferroelectric, in an epoxy resin, using the surfactant Triton X-100 as a dispersant. We capacitively coupled the electrodes to the sample using thin sheets of Mylar. The measured dielectric constants, Figure 13, show a marked resemblance to the simulation data, both in the concentration dependence and in the observed anisotropy.

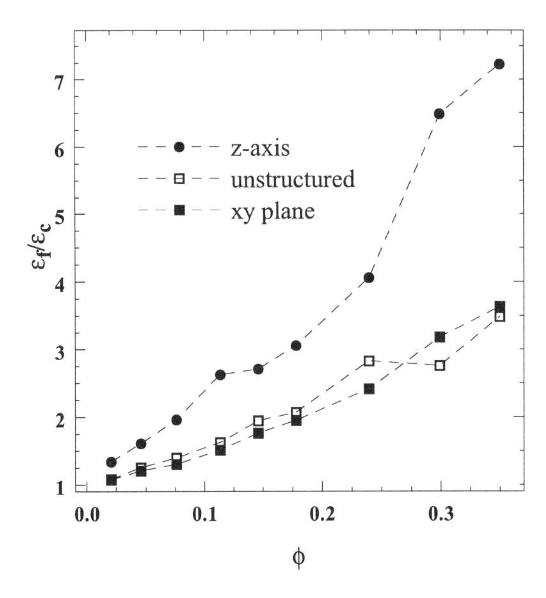

Figure 13. The concentration dependence of the dielectric constant of $BaTiO_3$ dispersed in an epoxy resin with a dielectric constant $\varepsilon_f \cong 4.5$.

Conclusions

These simulation studies of the formation and properties of anisotropic, uniaxial field-structured composites have demonstrated that materials with significant anisotropy in structure and materials properties can be produced. The anisotropy is a monotonic function, increasing as the concentration decreases. This anisotropy is strongly reflected in the thermal and electrical conductivity, and for the uniaxial case the ratio of the conductivity along the *z* axis to that in the *x-y* plane diverges to infinity at a critical concentration that decreases to zero as the system size increases to infinity. The rotating field case is just the opposite, with the ratio of the conductivity in the *x-y* plane to that along the z axis diverging to infinity. The permittivity reflects this anisotropy less strongly; but permittivity anisotropies of roughly 3 can be achieved for high dielectric particles dispersed in a much lower dielectric continuous phase.

Literature Cited

1. Gast, A. P.; Zukoski, C. F. *Adv. Coll. Interface Sci.* **1988,** *30*, 153.
2. Tao, R.; Sun, J. M. *Phys. Rev. Lett.* **1991,** *67*, 398.
3. Martin, J. E.; Odinek, J. ; Anderson, R. A.; Tigges, C. P. *to be published.*
4. Anderson, R. A. *Langmuir* **1994,** *10*, 2917.
5. Clercx , H. J. H.; Bossis, G. *Phys. Rev. E* **1993,** *48*, 2721.

Chapter 7

The Impact of Dielectric Confinement on Electron Dynamics and Optical Nonlinearities in Metal Nanocrystals in Glasses

Fabrice Vallée, Natalia Del Fatti, and Christos Flytzanis[1]

Laboratoire d'Optique Quantique du Centre National de la Recherche Scientifique, Ecole Polytechnique, 91128 Palaiseau, France

We review the impact of electron confinement on the optical nonlinearities and electron dynamics in metal nanocrystals in glasses. The main contribution to the optical Kerr nonlinearity results from the photoinduced redistribution of electron population close to the Fermi level, a hot electron contribution, enchanced by the morphological surface plasmon resonance. The electron dynamics in these nanocrystals are also analysed in the light of recent studies using time resolved nonlinear optical techniques in the femtosecond time scale and it is shown that the dielectric confinement and the Fermi statistics have a major impact on the interplay of electron-electron and electron-phonon interactions.

In line with the present surge in the study (1-3) of the optical and electronic properties of quantum confined semiconductor nanostructures attention has been also directed to those of metal nanostructures. The study of the later (4-6) actually started much ealier, already in the turn of this century and has been pursued since then without interruption although in a quite different context. These two classes of artificial nanostructures at the outset share certain commun features which however show up in a strikingly different manner. Anticipating the main conclusion of the subsequent discussion we may already now state that when electron statistics are properly incorporated the main optical properties of the metallic particles can be accounted for within a classical picture of electron dynamics in confined space while those of the semiconductor particles require quantum mechanical considerations.

One feature that underlies the present interest on all these nanostructured materials with delocalized electrons, metals or semiconductors, is the modification of the electronic density of states distribution that the confinement imparts and by the same token that of the optical oscillator strength distribution and relaxation

[1]Corresponding author.

processes ; both play an essential role in shaping the linear and nonlinear optical properties (7) of these materials and consequently determine their potential use (3) in microoptoelectronics and other areas. Another feature of much interest, recurrent to all composite materials obtained by interfacing materials of different chemical constitutions and more or less mobile electrons, is the interfacial charge transfer (7) and similar processes that may take place there and can be exploited to produce new functionality (3) artificial materials. The later aspect clearly involves quite sophisticated chemical as well as physical considerations but the former, namely the confinement, can be addressed and its main consequences followed up to a large extent by physical considerations regarding the electron behavior in confined geometries.

Here we shall review the impact of the confinement on the electron dynamics and optical nonlinearities of metal nanocrystals embedded in transparent dielectrics, mainly glasses and also compare it with that observed (9) in semiconductor nanocrystals also embedded in glasses. Several approaches (4-6) exist for preparing such mesoscopic systems, metal or semiconductor nanocrystals uniformly embedded in a liquid or solid matrix. To most purposes they can be considered as spherical and their lattice structure has been found to be the same as in bulk. There are several indications that these nanocrystals are defect free and in particular exempt of impurities. The average crystallite size can cover a wide range of values, from several micrometers down to a few nanometers, with more or less narrow size distribution depending on the preparation technique and the quality of the interface with the surrounding dielectric that can substantially differ from case to case. The later has much relevance on the "surface" features but usually for sufficiently large nanocrystals containing thousand or more atoms corresponding to particle diameters larger than a few nanometers we may simply assume that the interface with the surrounding dielectric is sharp and of infinite potential height. The volume concentration can also vary over a wide range down to concentrations as low as 10^{-6} or up as high as several percent which allows one to study single particle features as well as features that arise from interparticle interactions. In certain cases ordered arrays (10) of such particles have been achieved by physicochemical or artificial techniques, for instance self-organization on semiconductor or polymeric substrates or artificial nanofabrication techniques involving particle or photon beams ; these are still in an experimental stage and mostly concern two dimensional arrays.

Here we shall be concerned with metal particles uniformly and randomly dispersed in a glassy or other transparent dielectric with concentrations low enough that interparticle excitation transfer can be neglected and only the electron confinement within such a single particle embedded in the glass has to be considered.

Electron confinement

All composite materials formed (6,7) by uniformly and randomly dispersed metal or semiconductor crystallites in a transparent dielectric share in common two important features. First, in the metal or semiconductor nanocrystals the otherwise delocalized valence electrons in the bulk find themselves confined in regions much smaller than their natural delocalization length ; this drastically modifies their quantum motion as probed by optical beams but also their interaction with other degrees of freedom. Second, because the size of the crystallites is much smaller than the wavelength and their dielectric constant is very different from that of the surrounding transparent dielectric, the distribution of the electric field that acts on and polarizes the charges inside these crystallites can be vastly different from the macroscopic Maxwell field in the composite.

These two effects, the first quantum mechanical and the second classical, go under the names of quantum and dielectric confinements respectively and are particularly conspicuous, in the optical frequency range. The first can be treated within the effective mass approximation (11,12) if the particles are large enough so that surface effects can be neglected with respect to volume effects ; the second can be treated within the effective medium approach (13-15) if the particle size and volume concentration are small enough. The most conspicuous signature of both confinements however is the appearance of morphological resonances related to quantum confined dipolar transitions and collective dipolar modes for the quantum and dielectric confinement respectively. They have a drastic influence on the electron dynamics and optical nonlinearities in these systems.

Dielectric confinement. For very low volume fraction p of spherical crystallites in the transparent dielectric, or $p << 1$, and particle size $d << \lambda$, where $d = 2a$ is the spherical particle diameter and a its radius, λ is the optical wavelength which is also larger than the interparticle distance, one can introduce an effective dielectric constant for the composite which within the effective medium (13,14) approach is written

$$\tilde{\varepsilon}(\omega) = \varepsilon_d + 3p\,\varepsilon_d \frac{\varepsilon_m(\omega) - \varepsilon_d}{\varepsilon_m(\omega) + 2\varepsilon_d} \qquad (1)$$

with ε_d and $\varepsilon_m(\omega)$ being the dielectric constants of the transparent dielectric and the embedded crystallite respectively. This is a straightforward consequence of the Clausius-Mosotti approximation (14) for the dipole induced in a spherical polarizable particle immerged in a dielectric and can also be related to the Mie theory of light scattering (16) from a diluted gas of spherical particles by imposing the vanishing of the forward scattering amplitude and neglecting all terms of higher order than the dipolar one.

Taking into account the dielectric polarization effect since $a << \lambda$ the field E_{in} inside the particles is related to the Maxwell field in the composite by the relation $E_{in}(\omega) = f_\ell(\omega)E(\omega)$ where $f_\ell(\omega)$ is approximately given by

$$f_\ell(\omega) = \frac{3\varepsilon_d}{\varepsilon_m(\omega) + 2\varepsilon_d} \qquad (2)$$

To the extent that ε_d is frequency independent and real while $\varepsilon_m(\omega)$ is frequency dependent and complex, $\varepsilon_m(\omega) = \varepsilon'_m(\omega) + i\varepsilon''_m(\omega)$, the absorption coefficient of the composite is easily obtained from (1) and (2) as

$$\tilde{\alpha} = \frac{\omega}{nc} p \left| f_\ell(\omega) \right|^2 \varepsilon''_m(\omega) \quad (3)$$

where $n \approx \varepsilon_d^{1/2}$ is the refractive index of the composite.

It is evident from (1) and also (2) and (3) that these quantities are resonantly enhanced close to a frequency ω_s such that

$$\varepsilon'_m(\omega_s) + 2\varepsilon_d = 0 \quad (4)$$

which is the condition for the surface plasmon resonance (15,16). It arises from the restoring force of the build-up of surface charge distribution that leads to resonant density fluctuations of the conduction electrons and accordingly (4) can also be viewed as the condition for a collective dipolar mode which introduces an antishielding. Its width which also measures the coherent damping of the mode is determined by the imaginary part $\varepsilon''_m(\omega)$; the value of ω_s clearly depends on the values of ε'_m and ε_d which can be artificially modified.

An intense light beam can modify the dielectric constant through the optical Kerr nonlinearity (6,17) by an amount

$$\delta\tilde{\varepsilon}(\omega) = 12\pi\tilde{\chi}^{(3)}(\omega, -\omega, \omega)E(\omega)E^*(\omega) \quad (5)$$

where by extending (18,19) the previous approach to the nonlinear regime one finds

$$\tilde{\chi}^{(3)}(\omega, -\omega, \omega) = p \left| f_\ell(\omega) \right|^2 f_\ell^2(\omega)\, \chi_m^{(3)}(\omega, -\omega, \omega) \quad (6)$$

for the effective third order susceptibility of the composite related to the optical Kerr effect, when the contribution of the surrounding dielectric is neglected with respect to that of the nanoparticles which is resonantly enhanced, $\chi_m^{(3)}(\omega, -\omega, \omega)$ being the corresponding third order susceptibility of the later. The resonances are either the surface plasmon one or the quantum confined ones and inspection of (5) and of the quantum mechanical expression of $\chi_m^{(3)}(\omega, -\omega, \omega)$ clearly shows (6,7) that in the case of the optical Kerr effect because of the frequency degeneracy the resonant enhancement is multiple and the neglect of the nonresonant Kerr nonlinearity of the surrounding medium is justified.

In the previous discussion we have tacitly assumed that one can replace the summation over the size distribution by an average value for the optical coefficients of the nanoparticles. In a more rigorous approach (6) such aspects can be properly and simply incorporated and in fact lead to minor deviations from the above sketched derivation.

Quantum confinement. For metal and semiconductor nanocrystals containing a few thousand or more atoms the effective mass approximation (EMA) provides (11,12) a convenient framework to discuss the modification of the electron states that results from the confinement. This approach is concerned (20-22) with the behavior of the delocalized electrons in a crystal perturbed by an aperiodic potential and was initially devised for impurity and electron-hole pair states in crystals and subsequently extended to the case of quantum confined nanostructures (11,12).

In the one-electron picture of a perfect crystal the electrons occupy band states which form energy band continua $\varepsilon_n(\underline{k})$ separated by forbidden energy regions ; their wavefunctions are of the form

$$\psi_{nk}(\underline{r}) = e^{ikr}\, u_{nk}(r) \tag{7}$$

where $\hbar\underline{k}$ is the pseudo-momentum, n labels the band, $u_{n\underline{k}}$ has the periodicity of the lattice and is closely connected with the atomic wave-functions that form the basis set of the bands and exp(ikr) is the enveloppe wave function. The band states are filled up to an energy level E_F, the Fermi level. Within this picture the electronic transitions are viewed as the promotion of an electron from an allowed occupied state below E_F to an unoccupied allowed state above E_F leaving behind a positively charged hole the two interacting predominantly through Coulomb forces within the sea of all carriers. The main distinction between a metal and a semiconductor is related to the Fermi level E_F being an allowed state situated within a band, a half filled band in the case of a metal, or being a forbidden state situated within a gap between a filled valence and empty conduction band in the case of a semiconductor or equivalently that the electron-hole spectrum extends down to zero or to a finite energy gap E_g respectively.

As a consequence of the zero energy gap and large density of states at E_F in a metal the wave vector dielectric constant $\varepsilon(\underline{k})$ is infinite for $\underline{k} = 0$ resulting in a complete screening (23-25) of the electron-hole interaction potential within a distance $r_F \approx 1/k_F$, the inverse of the Fermi wavevector which is of the order of a few Angströms or roughly equal to the lattice constant. Thus the electrons and holes on either side of E_F can behave and move as free noninteracting particles over any distance in a perfect metal occasionally interacting and being scattered by temporal and spatial disorders ; such interruptions of the coherent free electron motion can be expressed in terms of a free mean path for the electrons (26) which in an ideal metal we will assume to be limited by the unklapp (27) electron-electron and electron-phonon scattering processes lumped together. This behavior underlies the success of the essentially classical Drude modelling (28) of the dielectric constant and can also be rigorously justified within the general Landau theory of Fermi liquids (23) : one finds there that

$$\varepsilon_m(\omega) = 1 + \delta\varepsilon_{inter} - \frac{\omega_p^2}{\omega(\omega + i/\tau_0)} \tag{8}$$

for the dielectric constant of a metal were ω_p is the plasma frequency, $\omega_p^2 = 4\pi e^2 n/m$, n is the electron density, $\delta\varepsilon_{inter}$ is the interband contribution and τ_0 is a scattering time for the electrons which for an ideal defect free metal we assume to be related to the compound effect of electron-electron and to electron-phonon scattering.

The situation is drastically different in semiconductors (12,29) where because of the finite energy gap the dielectric constant is finite and the sceening is only partial leading to substantial electron-hole interactions and even to the occurrence of bound states, such as excitonic ones and others types of bound states which play a key role in electron dynamics and optical nonlinearities.

Within this context in a metal nanocrystal the main modification then will result from an additional restriction of the free electron motion (15,30) because of the presence of the interface with the surrounding dielectric which to a good approximation can be visualized as a spherical potential well of infinite height. In addition the electron-phonon coupling has to be modified because of the modification of the vibrational spectrum (31) in the metal nanocrystal with respect to the bulk. Assuming the two processes uncorrelated this only amounts (15,30) to replacing $1/\tau_0$ in the denominator in (8) by

$$\frac{1}{\tau} = \frac{1}{\tau_0'} + \frac{v_F}{2a} \tag{9}$$

where v_F is the electron velocity on the Fermi level and $1/\tau_0'$ now is the electron-phonon relaxation rate in the crystallite.

This classical picture which however properly incorporates the Fermi statistics can be also justified within a quantum mechanical approach (15) in terms of the confined electronic states in the spherical potential well set up by the interface with the surrounding dielectric. These are now discrete states which incorporate the boundary conditions because of the requirement that the wave functions must vanish at the interface ; more realistic models introduce only minor changes (32) and need not be taken into account here. The periodicity being destroyed in the nanocrystal the wavefunctions are wavepackets of band states and written in the form

$$\psi_n(\underline{r}) = F_n(r)\, u_n(r) \tag{10}$$

for each band where n is the set of quantum numbers that label the states, $F_n(\underline{r})$ is the enveloppe of the wavepacket that vanishes at the interface or $F_n(a) = 0$, while $u_n(r)$ is assumed to be the same as in the bulk metal and actually can be approximated $u_n(\underline{r}) \approx 1$. $F_n(\underline{r})$ satisfies then the Schrödinger equation

$$\left[-\frac{\hbar^2}{2m}\nabla^2 + W(r)\right] F_n(\underline{r}) = E_n\, F_n(\underline{r}) \tag{11}$$

where the spherical well potential $W = 0$ for $r < a$ and $W = 0$ for $r = a$ which implies $F_n(a) = 0$ as previously stated. This is the "free electron in a box" model (33) precursor to the models of quantum confinement in semiconductor nanostructures.

Although the enveloppe equation (11) together with the boundary condition lead to a discrete spectrum with energy spacing $\Delta E \sim E_c = \hbar^2/2ma^2$ as we move up in the energy spectrum this becomes close spaced and in fact one may again introduce a density of states

$$\gamma(E) = \frac{V}{2\pi^2}\left(\frac{2m}{\hbar^2}\right)^{3/2} E^{1/2} = \frac{2}{3}\frac{E^{1/2}}{E_c^{3/2}} \quad (12)$$

where V is the spherical particle volume, and a Fermi level

$$E_F = \frac{\hbar^2}{2m}\left(3\pi^2 n\right)^{2/3} \quad (13)$$

both independent on the nanocrystal size and equal to their corresponding bulk values. Calculating the transition dipole moment between such states and introducing a common coherence relaxation time T_2 related to electron-lattice interactions for all allowed dipole transitions one may proceed to calculate the optical properties. Taking into account also transitions from the filled d-band (35) which is unaffected by the quantum confinement one obtains (15,30,34) the remarkable result that the dielectric constant takes the form

$$\varepsilon_m(\omega) = 1 + \delta\varepsilon_{inter} - \frac{\omega_p^2}{\omega(\omega + i/\tau)} \quad (14)$$

namely the same as for the bulk with the damping time τ in the Drude term given by expression (9) previously obtained by a classical argument, while the imaginary part of the interband term is given by

$$\delta\varepsilon''_{inter} = \frac{4\pi^2 e^2}{3m^2\omega^2}\left|P\right|^2 J(\omega) \quad (15)$$

where P is the transition matrix element of the momentum operator between the d- and the conduction s-p bands, and $J(\omega)$ is the joint density of states for the two bands (35). Thus one recovers for the metal particles an expression similar to that of the bulk metal (8) with the the damping replaced by expression (9) ; this also justifies the use of the bulk expression for $\varepsilon_m(\omega)$ in équation (1) as anticipated. With this expression for $\varepsilon_m(\omega)$ we obtain

$$\omega_s^2 = \omega_p^2/(\varepsilon_b + 2\varepsilon_d) \quad (16)$$

for the expression of the surface plasmon resonance in terms of the dielectric constants ε_d and $\varepsilon_b = 1 + \delta\varepsilon_{inter}$.

In contrast the spectrum of a semiconductor nanocrystal (6,9,11,12) is drastically different from that of the bulk and dominated by the appearance of well resolved discrete states with spacing $\Delta E \sim \hbar^2/2ma^2$ and well developed quantum confinement in the optical frequency range with the absorption threshold being blue-shifted to $E_g + \hbar^2/2ma^2$ from its value E_g in the bulk. In addition because the screening in semiconductors is only partial and not as complete as in metals residual electron-hole interactions and interfacial charge trapping are present and have substantial impact on the electron dynamics, energy spectrum and optical nonlinearities of the semiconductor nanocrystals. The broadening of the transitions between the discrete quantum confined states is also of quite different character (9) : it results from the combined effect of an intrinsic broadening, essentially the electron-phonon interaction (36) and an extrinsic broadening, essentially because of the size distribution of the nanocrystals in the glass, leading to inhomogeneously broadened lines as has been evidenced (6,9) by hole burning and other nonlinear spectroscopic techniques and also implied by the temperature dependence of the line widths.

Optical nonlinearities

The optical Kerr nonlinearity of noble metal nanocrystals in glasses has been extensively studied both experimentally and theoretically as has been also the case for the semiconductor nanocrystals in glasses. In both cases an understanding of their main aspects has emerged (6,9). The difference between the two cases are striking and to a large extent can be traced back to the differences in the electron dynamics in these confined nanocrystals and in particular in the energy spectrum as previously pointed out.

The optical Kerr effect results from the photoinduced modification of the complex refractive index of the medium or $\delta\tilde{n} = n_2 I$, where I is the light beam intensity. The optical Kerr coefficient n_2 is simply related (7,17) to the third order susceptibility $\chi^{(3)}(\omega, -\omega,\omega)$ or $\chi^{(3)}(\omega, -\omega',\omega')$ which can be resonantly enhanced whenever ω and/or ω' are close to a resonance of the medium. In the case of the metal or semiconductor nanocrystals in glasses the relevant resonances are the morphological ones resulting from the dielectric and quantum confinements. A simple inspection of the expressions (6,7) of the third order optical Kerr susceptibilities shows that one has multiple resonant behavior there leading to strong enhancement close to these resonances and also strong dependence on the features of these resonances which also reflect those of the confinement.

Broadly speaking, in such a resonant regime, the physical mechanisms contributing to $\chi^{(3)}$ can be distinguished in two classes. The first type involves real population of excited states or excitation of charge carriers there resulting from real transitions ; it is accompanied by linear absorption and the response speed is limited by the population relaxation time. The second type of mechanisms contributing to $\chi^{(3)}$ effectively results from light induced shifts on the electronic levels, direct or indirect ones, and can be connected with virtual transitions. The most conspicuous direct contribution here is the one related to two-photon resonances while indirect ones may result either from interpair electron-hole (biexcitonic) interactions which can be viewed as an optical Stark shift mutually induced between photoinduced e-h pairs, or through interaction with the field set up by an interface trapped photocarrier which can be viewed as a static Stark effect ; the three mechanisms can be distinguish through their different frequency dependence and different response times they imply. In semiconductor nanocrystals all these mechanisms have been evidenced (6,9) by appropriate time and frequency resolved nonlinear techniques.

In the case of metal nanocrystal on the other hand because of the complete screening that is operative there all contributions of the second class to the Kerr nonlinearity mentioned above are drastically reduced and only those of the first class contribute there. This has also been confirmed (37) experimentally. On the other hand in metal nanocrystals the effective third order Kerr susceptibility is fourfold resonantly (19) enhanced close to the surface plasmon resonance which is not the case in semiconductor nanocrystals where the dielectric confinement has only a minor effect (6,7,9) and does not lead to any sizeable consequences either on the optical nonlinearities or the optical spectrum.

The optical nonlinearities of metal nanocrystals in transparent dielectrics have been studied to a certain extent (6) over the last ten years and a good understanding of these nonlinearities and their relation to the electron dynamics has emerged that confirm the main aspects stated above. Reliable predictions regarding their behavior in noble metals are indeed feasible. Thus the fourfold resonant enhancement of $\tilde{\chi}^{(3)}$ of the composite material close to the surface plasmon resonance (16) as predicted by (5) has been strikingly demonstrated (19) in gold and silver colloids first and subsequently confirmed (37-39) also in the case where the gold and silver particles are embedded in glass. Similar behavior has also been evidenced (39) in copper nanocrystals in glass where the surface plasmon resonance strongly overlaps with the interband transitions. Using the experimental values of $\varepsilon'_m(\omega)$ and $\varepsilon''_m(\omega)$ and relation (4) one easily finds that $|f_e| \sim 3\varepsilon_d/\varepsilon''_m(\omega)$ has values in the range 10-20 for noble metals resulting in enhancement factors in the range 10^3-10^5 for $\tilde{\chi}^{(3)}$ over the surface plasmon resonance width (9) ; in contrast in semiconductor doped glasses f_ℓ hardly reaches unity and the corresponding factor in $\chi^{(3)}$ is less than unity. On the other hand the optical nonlinearities of semiconductor nanocrystals are strongly enhanced by the quantum confined resonances and the concomitant oscillator strength concentration there while this is not the case in the metal nanocrystals as will be discussed below.

The impact of the quantum confinement is related to the behavior of $\chi^{(3)}_m(\omega,-\omega,\omega)$ which can be extracted from (5) after the factor $f_\ell(\omega)$ has been accounted for in terms of the dielectric confinement as stated above using expression (2) and condition (4). The detailed experimental study of ref. 37 using the degenerate four wave interaction technique or optical phase conjugation where several parameters were varied, such as the size of the nanocrystals, the temperature or the polarization state of the input beam and in addition the phase of $\chi^{(3)}$ was determined, allowed to analyse its behavior in the light of a detailed theoretical modelling and single out the origin of the nonlinearity ; it confirmed on the one hand the minor role played by the quantum confinement there and on the other the key role played by the photoinduced population rearrangement close the Fermi level and its dynamics.

Careful analysis of the quantum mechanical expression of $\chi^{(3)}_m(\omega,-\omega,\omega)$ indeed shows (37) that the second class of contributions previously termed as photoinduced level shifts are irrelevant in metal nanocrystals and only the first class of contributions, termed photoinduced population changes are responsible for the optical nonlinearities there. These can be additionally split into three contributions :

an intraband contribution involving only transitions between two quantum confined states within the conduction s-p conduction band which can be written

$$\chi^{(3)}_{intra} = -\,i\,\kappa_1\,T_1\,T_2\,\frac{E_F^4}{a^3\omega^7}\,(1 - a/a_0) \tag{17}$$

where κ_1 is a constant factor, T_1 et T_2 are respectively the energy lifetime and dephasing time and $a_0 \approx T_2(2E_F/m)^{1/2}$ which is roughly 100-200Å in metal nanocrystals ; thus $\chi^{(3)}_{intra}$ is imaginary with Im $\chi^{(3)}_{intra} < 0$ and strong size dependence characteristic of quantum confinement,

an interband contribution involving transitions between a d-band state and a quantum confined s-p conduction band state which can be written

$$\chi^{(3)}_{inter} = -\,i\,\kappa'\,T_1'\,T_2'\,J(\omega)\,(\underline{P}/\omega)^4 \tag{18}$$

where κ' is a constant, T_1' and T_2' are the energy lifetime and dephasing time for the two-level system and P and $J(\omega)$ were previously defined in equ. (15) ; thus $\chi^{(3)}_{inter}$ is imaginary with $\mathrm{Im}\chi^{(3)}_{inter} < 0$ and size independent,

a hot electron contribution which can be viewed as a hybrid term of the previous two and results from the photoinduced modification of the occupation of the conduction band states above and below the Fermi level accessible to the d-electrons. This modification is provoked by the absorption of photons, close to the surface plasmon resonance frequency, by the conduction electrons which are promoted to initially unoccupied states above the Fermi level and the same time liberate states below it. Assuming this leads to a quasithermal distribution in a subpicosecond to picosecond time scale at an electronic temperature T_e (see Fig. 1) much higher than the lattice temperature T_ℓ one gets

$$\chi^{(3)}_{he} = \frac{i}{24\pi^2 C}\,\frac{\omega_p^2}{\omega^2}\,\frac{\tau_0}{\tau_{eff}}\,\frac{\partial\varepsilon_L''}{\partial T} \tag{19}$$

for this contribution where C is the conduction electrons heat capacity, τ_0/τ_{eff} measures the number of all collisions the electrons suffer before they thermalize, ε_L'' is the imaginary part of interband term (35) at the point L of the Brillouin zone which makes the dominant contribution close to the surface plasmon resonance frequency ; thus $\chi^{(3)}_{he}$ is imaginary with $\mathrm{Im}\chi^{(3)}_{he} > 0$ and size independent.

The detailed experimental study in ref. 37 showed that in the size range 30-100 Å the hot electron contribution is the dominant one in $\chi^{(3)}_{xxxx}$ being by one and two orders of magnitude larger than the inter and intraband ones respectively. In $\chi^{(3)}_{xyxy}$ this hot electron contribution being incoherent does not contribute while the interband one does and this is confirmed by the measured value and sign on the ratio $\chi^{(3)}_{xxxx}/\chi^{(3)}_{xyxy}$. Similar conclusions were also reached in the detailled study (39) of the optical nonlinearities in copper nanoparticles in glasses where the interband transitions strongly overlap with the intraband ones involved in the surface plasmon resonance.

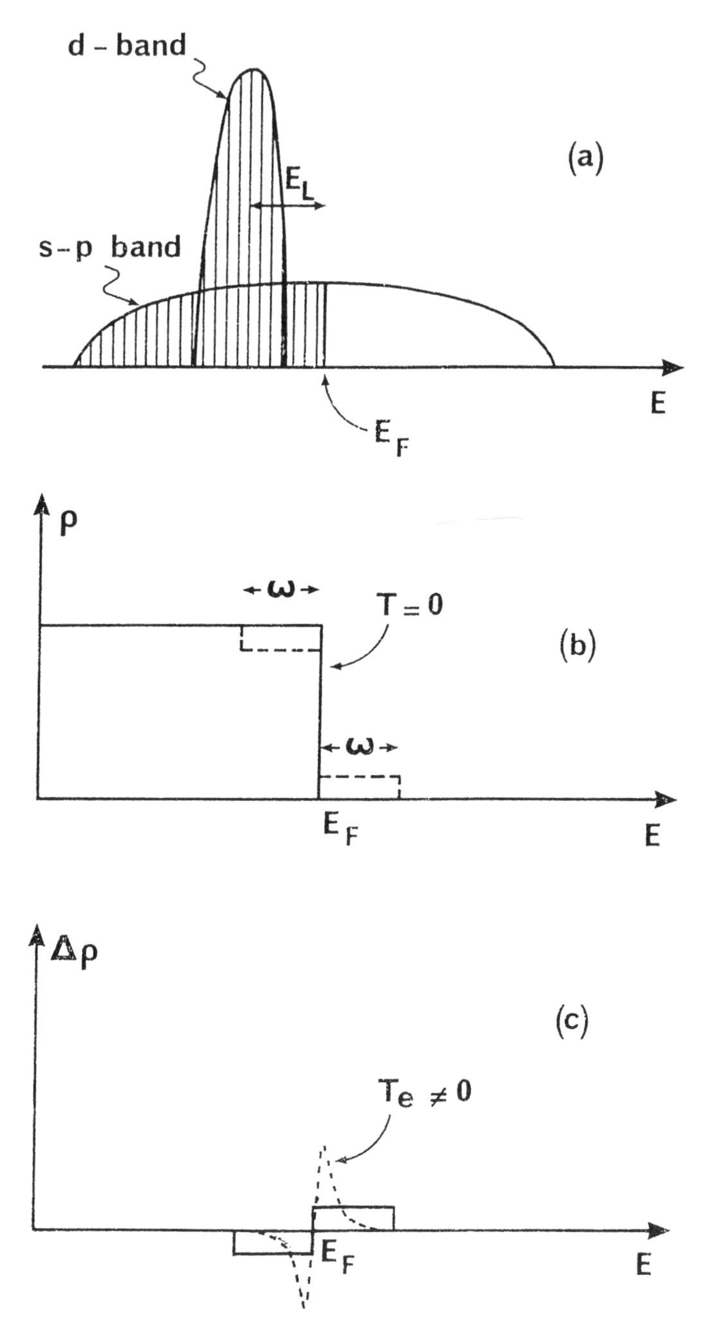

Fig. 1: Schematic representation of the density of states in noble metals (a) and of the photoinduced modification $\Delta\rho$ of the Fermi distribution ρ initially assumed at temperature $T = 0$ (b) and (c). We indicated by E_L the energy separation between d- and s-p-band states at point L of the Brillouin zone which is relevant for the interband transitions $E_L \approx \hbar\omega \approx \hbar\omega_s$. In (c) we also indicated a thermal electronic distribution (T_e) of the photoexcited electrons.

Electron dynamics

The study of the optical nonlinearities in metal nanocrystals brought into focus the important role played there by the electron dynamics in a confined space. The study of such processes was initiated quite early in electron transport (40,26) in connection with the electrical conductivity in metallic wires and plates or thin films and it was associated there to the modification of the mean free path of the electrons because of their collisions with the walls. The Landau theory of Fermi liquids provides (23-25) a rigorous albeit complicated framework to study such aspects in terms of the interaction of quasi particles (electrons and holes) among themselves or with other excitations and with the walls. Much insight however can also be gained with a simpler description and by proper accounting for the Fermi statistics.

In the past these experimental and theoretical studies were predominantly conducted in the frequency domain (23,24) under the assumption that the different interactions that cause damping were uncorrelated and the long time regime prevails. Deviations from such a behavior however were clearly noticeable but difficult to analyze in frequency domain. The recent advances in ultrashort laser pulse techniques (41-43) opened the way to address these problems directly in the time domain, in particular the short time behavior, and assess the interplay of the different processes there. The most commonly used time resolved technique is the pump and time delayed probe technique and its variants : in essence in all these techniques the spatiotemporal evolution of a quickly photoinduced modification of an optical property is interrogated by a short probe pulse with variable delay and related to the dynamics of the relevant electronic transitions.

Electronic transitions. As previously indicated the optical properties in the visible are connected (35,25) with two classes of electronic dipole transitions in the metal nanocrystals : intraband ones between the quantum confined states of the wide half-filled s-p band and interband ones between such states and the states of the filled narrow d-band which in the case of noble metals (Cu,Ag,Au,Pt) overlaps the s-p band below the Fermi energy E_F (see Fig. 1). The d-states are not affected by the confinement while the s-p are affected but their density of states resembles that of the bulk as one approaches the Fermi level which is the same as in the bulk metal.

When such oscillating transition dipoles are induced in the metal nanoparticles of sufficiently small diameter, a surface charge is built up that sets up a restoring oscillating force that resonantly enhances the electron density oscillations inside the nanoparticle at a suitable frequency, namely that of the surface plasmon mode ω_s, as defined by (4) and (14) and produces a dipole oscillation which interacts with an external electromagnetic field ; the surface plasmon oscillation can thus be viewed as a coherent envelope of a packet of such dipolar transitions between states bound in the finite volume of the sphere : It is essential to stress here that the interface does not act as a scatterer but actually determines the eigenstates of the system and is integrated there through the boundary condition. Otherwise stated an external transverse electromagnetic field through the existence of the interface directly couples with the collective dipole of the surface plasmon mode whose damping proceeds via the damping of the individual constituant electronic transitions and can be related (15) to the correlation of the different random forces acting on the electrons involved in these individual dipolar transitions.

Damping processes. In a bulk metal (23-25) the damping processes are related to electron-electron, electron-phonon and electron-impurity collisions and scattering which limit the conductivity of the electrons. These electrons or the quasiparticles in the Landau theory of Fermi liquids (23), are nearly free in a homogeneous environment and occasionally scattered by electrons, phonons or impurities and one may define a mean free path ℓ and relate it to the electrical conductivity (26) ; alternatively one introduces a scattering time or a damping rate for each such scattering process and if one assumes that these processes are uncorrelated one may add their contributions to obtain the total damping. If we restrict ourselves to perfect metals and disregard the impurities which are extrinsic to the electron and phonon system a key role in the scattering processes and the establishment of a stationary equilibrium regime is played by the unklapp processes (27,21) both in the electrons and phonons which conserve the quasiparticle quasimomentum $\hbar$k only up to an arbitrary vector $\hbar$K of the reciprocal space ; this is a consequence of the space homogeneity and infinite periodic extension of the metal consistent with the boundary conditions.

Momentum non conservation. When we consider the damping processes in the metal nanocrystals the previous picture cannot be transposed as such for different reasons but most and foremost because the electrons are now bound in the finite volume of a sphere. This in particular breaks down the space homogeneity and periodicity and the concept of the mean free path cannot be used as such ; as a matter of fact the very notion of a quasi particle in the sense of Landau's Fermi liquid has to be reconsidered in such small metal particles. As previously stated the surface of the nanocrystal is not a scatterer but is built into the electron states which now become discrete through the boundary condition ; incidentally the same is true for the phonons (31). The transitions now involve such states for which the momentum is not conserved. Otherwise stated the role of the momentum non-conserving unklapp processes in the electron-electron collisions in the bulk metal is replaced by the momentum non-conserving collisions of the quasiparticles with the wall set up by the interface. This is also the fundamental reason of the identical result for the size dependent term in the damping of the surface plasmon in (9) being reached either proceeding classically or quantum mechanically with the assumption that the random e-e and e-p interactions are uncorrelated.

This modification of the electron-electron (e-e) interactions introduced into the damping because of the surface of the nanoparticle is also accompanied by a concomitant modification of the electron-phonon (e-p) interaction since not only the electrons but also the phonons are bound in the finite volume of the spherical metal nanoparticle. This modifies both the coupling strength of the electron-phonon interaction as well as the phonon density of states. The impact of the confinement on the vibrational spectrum of metal nanocrystals has already been addressed in a few cases (2,31) in particular regarding its impact on the heat capacity of such particles and more recently an approach has been sketched (44) to calculated the electron-phonon coupling along lines similar to the ones used in the case of the semiconductor nanocrystals (36). From these approximate approaches one can infer a size dependence of the damping processes related to the electron-phonon interaction as well which can be included in (9) in a straightforward manner if we assume that the random interactions are uncorrelated so that their damping rates just add.

Interference of damping processes. Even with such provisions for the vibrational spectrum and electron-phonon interaction in the metal nanocrystal the previous analysis flaws (41-43) in a very crucial aspect by assuming that the e-e and e-p collisions are uncorrelated and the corresponding rates add. Although experimental studies in frequency domain roughly confirm the additivity of the damping rates as implied by (9) this picture is indeed naive if not incorrect since it overlooks a very subtle interplay of the e-e and e-p collisions in the evolution of the density distribution of the electrons in metals in general and irrespective of the presence of any confinement as we will shortly discuss. Before doing so we point out that concomitant to the previous assumption of the e-e and e-p damping processes being uncorrelated is also the assumption that when the electron distribution in metals is brought off equilibrium by a photo-excitation or other sudden external perturbation first the e-e collisions bring the electron system in internal thermal equilibrium characterized by a temperature T_e and subsequently the e-p collisions bring it to external thermal equilibrium with the lattice characterized by a temperature T_ℓ. There is now first hand evidence (41-43) for bulk metals and also for thin metallic films obtained by different time resolved techniques in the subpicosecond time scale that this two temperature thermal model does not properly describes the evolution and certainly fails in the early stages following the photo excitation.

The proper way to proceed in the analysis of off equilibrium situations is to set up the Boltzmann equation describing (45,41,42) the evolution of the electron density distribution n_ε in time and containing the e-e and e-p collision integrals $C_{ee}(n_\varepsilon)$ and $C_{ep}(n_\varepsilon)$ respectively

$$\frac{dn_\varepsilon}{dt} = C_{ee}(n_\varepsilon) + C_{ep}(n_\varepsilon) \tag{20}$$

where the expressions of the collisions integrals C_{ee} and C_{ep} can be found in the literature. Assuming that the number of photoexcited quasiparticles is very low so that these only scatter off the Fermi sea one easily sees (43) that after each such scattering event three new quasiparticles appear which can each lose energy at a rate $\dot{q}$ by the e-p interaction. Thus the e-e interaction and the concomitant change of the electron distribution leads to an acceleration of the energy rate to the lattice causing in its turn a substantial change in the electron distribution as well.

The extension of the Boltzmann equation to the case of the electrons in confined metal nanoparticles encounters certain difficulties because of the inherent space inhomogeneity introduced through the interface which implies that n_ε is not only momentum dependent but also space dependent even under isothermal regime (46) which complicates the detailed analysis but the previous interplay of e-e and e-p collisions remains intact even in the metal nanocrystals with the only provision now that the momentum non-conserving unklapp electron-electron collisions in the bulk are replaced by those of the electron-interface collisions and similarly for the electron-phonon collisions. Thus the interplay of two random processes, the electron-surface and electron-bound phonon interactions, introduce a correlation and an interference in the damping rates which will be shown up in an acceleration of the overall damping rate in the initial stage with respect to what one expects when the two processes are assumed uncorrelated ; deviations from an exponential decay could be expected. Similar effects are expected in bulk metals as well.

Time resolved dynamics. The recent application of powerful time resolved nonlinear optical techniques in the femtosecond time scale opened (41-43) the way to study the early stages of the thermalization of photoexcited electron populations in metals and show the inadequacy of the two temperature thermal model to account for this regime. Indeed the observed behavior confirms the importance of the interplay of the e-e and e-p collisions in the process of thermalization as this is introduced by the Boltzmann equation (20). Numerical solutions (41-43) with certain simplifying approximations fully confirm this effect.

The extension of these techniques to the case of metal nanocrystals embedded in glass was only recently attempted (47,48,49) but additional studies are needed to ascertain the impact of the interplay of electron-surface and electron-phonon collisions there. The case of copper metal particles where some of these studies were performed presents the inconvenience that the interband transitions in copper strongly overlap with the intraband ones of the surface plasmon resonance. The case of the gold or silver particles embedded in glass should be exempt of such complications.

Recently studies were performed (50) in silver nanoparticles embedded in a glass matrix ; here the plasmon resonance and interband transitions are well separated permitting a selective probing of the surface plasmon resonance. The preliminary results obtained with 100 fs resolution indicate that thermalization does not set in before several hundred femtoseconds. The study yielded evidence of a time dependent red shift $\Delta\omega_s$ and broadening of the surface plasmon resonance which set in instantaneously and follow the thermalization process. The frequency shift can be related to the photoinduced modification of ε_b in (16) because of the optical Kerr effect (19) while the broadening is attributed to the increment of the collision rate of the electrons with the surface as their average velocity increases. Additional studies with shorter time resolution are needed to unveil the athermal behavior.

Extensions

The study of the optical nonlinearities and electron dynamics in metal nanocrystals embedded in glass matrix is connected with some fundamental aspects of the behavior of electrons in metals in general and in confined geometries in particular. In this respect two aspects should retain our attention : one is the crucial role played by the dielectric confinement to enhance the nonlinear optical coefficients close to the surface plasmon resonance and other morphological resonances ; the other is the restrictions imposed by the Pauli principle and Fermi statistics on the available phase space for the electrons leading to interference of the scattering processes and eventually to the breakdown of the addivity assumption of the damping rates and that of the thermal evolution of the photoexcited electronic distribution. Indeed the impact of these two aspects is such that any specifically quantum mechanical features are effectively washed out. This is in striking contrast with the situation in the semiconductor nanocrystals where specific quantum mechanical features prevail in the optical spectra and electron dynamics.

At the same time the recognition of the important role played by the dielectric confinement and the Fermi statistics shows the way to some interesting future studies. Thus similar studies in transition metal nanoparticles where the d-band is also half filled or noble metal nanoparticles containing magnetic impurities should show some new features regarding electron dynamics and optical nonlinearities because of the drastic modification of the electron density distribution at the Fermi level that occur there. Another line of study is that of much higher volume concentrations for the metal nanoparticles where the Maxwell-Garnett effective medium approach breaks down and must be replaced by more sophisticated ones like the Bruggemann approach (51) or more complex ones where the topology and fractality are properly introduced. Such studies have already been initiated uncovering some new aspects (52). Finally there is certainly interest in the study of glasses co-doped with metal nanocrystals and rare earth ions (53) where the dielectric confinement through the surface plasmon mode it sets up inside the nanocrystal may also show its effect on the rare earth ions outside the metal nanocrystals. In these and other cases the radiative damping (15) which was althogether diregarded in the previous discussion may play a certain role for sufficiently large nanocrystals since its contribution grows as the cube of the particle size and must now be included. In this respect we wish to point out that the absorption losses (54) in the far infrared of the composites formed by dispersing metal nanocrystals in dielectrics can not be accounted for even when all the refinements of the present theory are included. These and other studies will be pursued both for their fundamental interest as well as for their impact in designing (3) artificial composite materials with specifies functions.

Literature Cited

1 - Esaki L., Tsu R., IBM J. Res. Develop. Jan 1970, p.61 ; Appl. Phys. Lett. 1973, 22, 562
2 - See for instance *Physics and Applications of Quantum Wells and Superlattices*, Mendez E.E., von Klitzing K., NATO ASI Series, Plenum Press, N.Y., 1987, Vol 170 ; *Nanostructures and Mesoscopic Systems*, Reed M., Kick W. (Eds), Academic, San Diego, 1991 ; *Nanostructured Systems, in Semiconductors and Semimetals*, Willardson R.K., Beer A.C., Weber E.R. (Eds), Academic, San Diego, 1992, Vol. 35
3 - *New Functionality Materials*, Tsuruta T., Doyama M., Seno M., (Eds), Elsevier, Amsterdam, 1993
4 - Perenboom J.A.A., Wyder P., Maier F., Phys. Reports 1981, 78, 173
5 - Halperin W.P., Rev. Mod. Phys. 1986, 58, 533

6 - Flytzanis C., Hache F., Klein M.C., Ricard D., Roussignol Ph., in Progress in Optics, Wolf. E (Ed.), North Holland, Amsterdam, 1991, Vol. XXIX, p. 321
7 - Flytzanis C., Hutter J., in Contemporary Nonlinear Optics, Agrawal G.P., Boyd R.W. (Eds), Academic Press, Orlando, 1992, Ch. 8.
8 - Miller R.D., Mc Lennon G.L., Nozik A.J., Schmickler W., Willig F., Surface Electron Transfer Processes, VCH, N.Y., 1995 ; Murray C.B., Norris D.J., Bawendi M.G., J. Am. Chem. Soc. 1993, 115, 8706 ; Bowen Katari J.E., Calvin V.L., Alivisatos A.P., J. Phys. Chem. 1984, 98, 4109 ; Belloni J., Current Opinion in Colloid and Interface Science, 1996, 1, 184
9 - Schanne-Klein M.C., Ricard D, Flytzanis C. in Quantum Dot Materials for Nonlinear Optics Applications, Haglund R.F. (Ed), NATO/ASI Series, Kluwer Publ., Amsterdam, 1997
10 - Thin Films, Growth Techniques for Low-Dimensional Structures, Farrow R.F.C., Parkin S.S.P., Dobson P.J., Neave J.H., Arrott A.S. (Eds), NATO/ASI, Plenum Press, N.Y., 1987, Vol. 163
11 - Efros Al.L., Efros. A.L., Sov. Phys. Semic. 1982, 16, 772
12 - Banyai L., Koch S.W., Semiconductor Quantum Dots, World Scientific, Singapour, 1993 ; Bastard G., Wave Mechanics Applied to Semiconductor Heterostructures, Eds de Physique, Les Ulis, 1988
13 - Maxwell-Garnett J.C., Philos. Trans. R. Soc. (London) 1904, 203, 385 ; ibid 1906, 205, 237
14 - Böttcher C.J., Theory of Electric Polarization, Elsevier, Amsterdam 1973 ; Kreibig U., Wollmer M., Optical Properties of Metal Clusters, Springer-Verlag, Berlin, 1995
15 - Kawabata A., Kubo R.J., Phys. Soc. Jap. 1966, 21, 1765 ; Kubo R., Kawabata A., Kobayashi S., Ann. Rev. Mater. Sci., 1984, 14, 49
16 - Mie G., Am. Phys. (Leipzig) 1908, 25, 377 ; Van der Hulst H.C., Light Scattering by Small Particles, Dover Publ. Inc., N.Y., 1981
17 - Shen Y.R., Principles of Nonlinear Optics, John Wiley, N.Y., 1984
18 - Rustagi K.C., Flytzanis C., Opt. Lett. 1984, 9, 344
19 - Ricard D., Roussignol P., Flytzanis C., Opt. Lett. 1985, 10, 511
20 - Ashcroft N.W., Mermin N.D., Solid State Physics, Holt Rinehart and Winston, Philadelphia, 1976
21 - Ziman J.M., Electrons and Phonons, Oxford University Press, Oxford, 1960
22 - Weinreich G., Solids : Elementary Theory for Advanced Students, John Wiley, N.Y., 1965
23 - Pines D., Norzieres P., The Theory of Quantum Liquids, W.A. Benjamin Inc. N.Y., 1996, Vol. 1
24 - Abrikosov A.A., Lectures on the theory of Normal Metals, Hindustan Publ. Corp. Delhi, 1968
25 - Wilson A.H., The theory of Metals, Cambridge Un. Press, Cambridge, 1965
26 - Sondheimer E.H., Adv. Phys., 1, 1, 1952
27 - Peierls R.E., Am. Phys. 1932, 12, 154 ; also Quantum Theory of Solids, Oxford Univ. Press, Oxford, 1955

28 - Drude P., Am. Phys. (Leipzig) 1900, _1_, 566 ; ibid _3_, 369 ; Sommerfeld A., Phys. 1928, _47_, 1

29 - Smith R.A., _Wave mechanics of Crystalline Solids_, Chapman and Hall, London, 1969

30 - Genzel L., Martin T.P. and Kreibig U., Z. Phys. 1975, _1321_, 339 ; Rupin R., Yatom H., Phys. Stat. Sol. (b), 1975, _74_, 14 ; Kreibig U., Genzel L., Surf. Sci. 1985, _156_, 678

31 - Nishiguchi N., Sakuma T., Sol. St. Comm. 1981, _38_, 1073 ; Tamura A., Higeta K., and Ichinokawa T., J. Phys. C, 1582, _15_, 4975

32 - de Heer W.A., Knight W.D., Chore M.Y., Cohen M.L., in _Solid State Physics_, Ehrenreich H., Seitz F., Turnbull D. (Eds), Academic Press, N.Y., 1987, Vol. 40 p. 94

33 - Kuhn H., J. Chem. Phys. 1948, _16_, 840

34 - Hache F., _Thèse_, Université de Paris, Orsay 1988

35 - Rosei R., Phys. Rev. 1974, _B10_, 474 ; Rosei R., Culp C.H., Weaver J.H., ibid 1974, _1310_, 484

36 - Takagahara T., J. Luminescence, 1996, _70_, 129

37 - Hache F., Ricard D., Flytzanis C., Kreibing U., Appl. Phys. 1998, _A47_, 347 ; Hache F., Ricard D., Flytzanis C., J. Opt Soc. Am. B, 1986, _3_, 1647

38 - Bloemer M.J., Haus J.W., Ashley P.R., J. Opt. Soc. Am. 1990, _B7_, 790

39 - Uchida K., Kaneko S., Ondi S., Hata C., Tanji H., Asahara Y., Ikushima A.J., Tokizaki T., Nakamura A., J. Opt. Soc. Am., 1994, _B11_, 1236

40 - Fuchs K., Proc. Cambridge Phil. Soc. 1938, _34_, 100

41 - Sun C.K., Vallée F., Acioli L.H., Ippen E.P., Fujimoto J.G., Phys. Rev. 1994, _B50_, 15337

42 - Groenveld R.H.M., Sprik R., Lagendjik A., Phys. Rev. 1995, _B51_, 11433

43 - Tas G., Maris H.J., Phys. Rev. 1994, _B49_, 15046 ; Ritchie R.H., J. App. Phys. 1966, _37_, 2276

44 - Belotskii E.D., Tomchuk P.M., Int. J. Electr. 1992, _73_, 955

45 - Allen P.B., Phys. Rev. Lett. 1987, _59_, 1460. Kaganov M.I., Lifshitz I.M., Tanatarov L.V., Zh. Ehs. Teor. Fiz. 1956, _31_, 232 (Sov. Phys. JETP 1957, _4_, 173) ; Grimvall, G., _Electron-phonon interaction in metals_, North-Holland, Amserterdam, 1981

46 - Lifshitz I.M., Pitaevski L.P., _Physical Kinetics_, Pergamon Press, Oxford, 1981

47 - Tomizaki T., Nakamura A., Kaveko S., Uchida K., Omi S., Tanji H., Asahara Y., Appl. Phys. Lett. 1994, _65_, 941

48 - Roberti T.W., Smith B.A., Zhang J.Z., J. Chem. Phys. 1995, _102_, 3860

49 - Bigot J.Y., Merle J.C., Cregut O., Dannois A., Phys. Rev. Lett. 1995, _75_, 4702

50 - Vallée F., Del Fatti N., Flytzanis C. to be published

51 - Bruggemam D.A.G., Ann. Phys. (Leipzig) 1935, _24_, 636

52 - Shalaev U.M., Phys. Reports, 1996, _272_, 61 ; Shalaev V.M., Moskovits M., Rev. Mod. Phys., 1985, _57_, 783

53 - Malta O.L., Santa Cruz P.A., de Sa G.F. and Auzel F., J. Sol. St. Chem. 1987, _68_, 314

54 - Genzel L., Kreibig U., Z. Phys. 1980, _1337_, 93 ; Sen P.N., Tanner D.B., Phys. Rev. 1982, _B26_, 3582

Chapter 8

Fractal-Surface-Enhanced Optical Nonlinearities

V. M. Shalaev[1], V. P. Safonov[2], E. Y. Poliakov[1], V. A. Markel[1], and A. K. Sarychev[3]

[1]Department of Physics, New Mexico State University, Las Cruces, NM 88003
[2]Institute of Automation and Electrometry, Novosibirsk 630090, Russia
[3]Center for Applied Problems of Electrodynamics, Moscow 127412, Russia

Optical excitations of nanomaterials with fractal structure result in highly localized areas of large fields leading to strong enhancements of optical nonlinearities. The localized modes of fractals cover a broad spectral range, from the visible to the far-infrared.

Optical phenomena experience giant enhancements in metal nanocomposites and rough thin films consisting of small nanometer-sized particles and roughness features, respectively. The enhancement is associated with excitation of surface plasmons which are collective modes and strongly depend on the morphology (geometrical structure) of the material. Fractal structures are prevalent in composites and rough thin films. The emergence of fractal geometry was a significant breakthrough in the description of irregularity. Fractal objects do not possess translational invariance and, therefore, cannot transmit running waves. Accordingly, collective excitations, such as surface plasmons, tend to be localized in fractals (*1,2*). Formally, this is a consequence of the fact that plane running waves are not eigenfunctions of the operator of dilation symmetry characterizing fractals.

In fractals, collective plasmon oscillations are strongly affected by the fractal morphology, leading to the existence of "hot" and "cold" spots (i.e., areas of high and low local fields). It is interesting to note that in many cases local enhancements in the "hot spots" exceed the average surface enhancement by many orders of magnitude; this is because the local peaks of the enhancement are spatially separated by distances much larger than the peak sizes. Also, the spatial distribution of these high-field regions are very sensitive to the frequency and polarization of the applied field (*2-6*). The positions of the "hot spots" change chaotically but reproducibly with frequency or/and polarization. This is similar to speckle created by laser light scattered from a rough surface with the important difference that the scale-size for fractal plasmons in the hot spots is in the nanometer range rather than in the micrometer range for photons.

Two varieties of surface plasmons are commonly recognized: localized surface plasmons (LSP) and surface plasmon waves (SPW). [SPW are also called surface plasmon polaritons (SPP) - coherent mixture of plasmons and photons.] SPW propagate laterally along a metal surface whereas LSP are confined to metal particles that are much smaller in size than the wavelength of the incident light. However, in fractal media plasmon oscillations in different features strongly interact with each other via dipolar or, more generally, multipolar forces. Thus, plasmon oscillations on a self-affine surface and in a fractal aggregate are neither conventional SPW nor independent LSP. Rather, they should be treated theoretically as collective eigenmodes arising from multipolar interactions in a fractal object.

With the aid of well-established chemical and depositional methods, fractal nanostructured materials may be fabricated. For example, colloidal clusters may be grown possessing a measured fractal dimension, 1.78 (*7*). Alternatively, fractals may be grown by particle-cluster aggregation (termed Witten-Sander aggregation or WSA (*7*)), having the fractal dimension, 2.5. Also, by controlling conditions of atomic beam deposition and substrate temperature, self-affine thin films may be grown with various fractal dimensions (*8*). Finally, metal-dielectric films (called also semicontinuous metal films) produced by metal sputtering onto an insulating substrate also include fractal clusters of metal granules near the percolation threshold (*7,8*).

The fractal plasmon, as any wave, is scattered from fluctuations of density (or, in other words, fluctuations of polarization). The strongest scattering occurs from inhomogeneities of the same scale as the wavelength. In this case, interference in the process of multiple scattering results in Anderson localization. Anderson localization corresponds typically to uncorrelated disorder. A fractal structure is in some sense disordered, but it is also correlated for all length scales, from the size of constituent particles, in the lower limit, to the total size of the fractal, in the upper limit. Thus, what is unique for fractals is that because of their scale-invariance, there is no characteristic size of inhomogeneity; inhomogeneities of all sizes are present from the lower to the upper limit. Therefore, whatever the plasmon wavelength, there are always fluctuations in a fractal with similar sizes, so that the plasmon is always strongly scattered and, consequently, localized (*2,9*).

Surface-Enhanced Optical Responses

Below we consider enhancement of optical responses on different fractal surfaces, such as aggregates of colloidal particles, self-affine thin films, and semicontinuous metal films in a vicinity of the percolation threshold. We assume that each site of the surface possesses a required nonlinear polarizability, in addition to the linear one. The local fields associated with the light-induced eigenmodes of a fractal surface can significantly exceed the applied macroscopic field, $\mathbf{E}^{(0)}$. For metal surfaces, this enhancement increases toward the infrared part of the spectrum where resonance quality-factors are significantly larger, in accordance with the Drude model of metal (*2,4*).

We begin our consideration with the Kerr-type nonlinearity, $\chi^{(3)}(-\omega;\omega,\omega,-\omega)$, that is responsible for nonlinear corrections to absorption and refraction. This type of optical nonlinearities can be in particular used for optical switches and optical limiters. The local nonlinear dipole, in this case, is proportional to

$|\mathbf{E}(\mathbf{r})|^2\mathbf{E}(\mathbf{r})$, where $\mathbf{E}(\mathbf{r})$ is the local field at the site $\mathbf{r}$. For the resonant eigenmodes, the local fields exceed the macroscopic (average) field by a quality-factor, q.

The fields generated by the nonlinear dipoles can also excite resonant eigenmodes of a fractal surface resulting in the additional "secondary" enhancement $\propto \mathbf{E}(\mathbf{r})/\mathbf{E}^{(0)}$. Accordingly, the surface-enhanced Kerr-susceptibility, $\bar{\chi}^{(3)}$, can be represented as (the angular brackets in the following formulas denote an ensemble-average) (*2,4,10*)

$$\bar{\chi}^{(3)}/p\chi^{(3)} = G_K = \frac{\left\langle |\mathbf{E}(\mathbf{r})|^2 \left[\mathbf{E}(\mathbf{r})\right]^2 \right\rangle}{\left[\mathbf{E}^{(0)}\right]^4}. \tag{1}$$

Here $\chi^{(3)}$ is the initial "seed" susceptibility; it can be associated with some adsorbed molecules (then, $\bar{\chi}^{(3)}$ represent the nonlinear susceptibility of the composite material consisting of the adsorbed nonlinear molecules and a surface providing the enhancement). The seed $\chi^{(3)}$ can be also associated with an isolated colloidal particle; then, G_K represents the enhancement due to clustering initially isolated particles into aggregates, with the average volume fraction of metal given by p. The applied field with the frequency in the visible, near IR or IR parts of the spectrum is typically off resonance for an isolated colloidal particle (e.g., silver) but it does efficiently excite the eigenmodes of fractal aggregates of the particles; the fractal eigenmodes cover large frequency interval including the visible and infrared parts of the spectrum (*2,4*).

Note that definition (1) for G_K, as well as the definitions of other enhancement factors introduced in this paper, are generic, i.e., applicable to any surface or 3-dimensional cluster, irrespectively of their fractal nature. However, the fractality is important because it results in especially large values of these enhancement factors as we will see below.

For simplicity, we assume that $\mathbf{E}^{(0)}$ in (1) is linearly polarized and therefore can be chosen real. In Refs. (*4,6*), the above formula was proven from rigorous first-principle considerations. Note also that G_K depends on the local-field phases and it contains both real and imaginary parts.

Because of a random character of fractal surfaces the high local fields associated with the localized eigenmodes look like strong spatial fluctuations. Since a nonlinear optical process is proportional to the local fields raised to some high power the resultant enhancement associated with the fluctuation area ("hot" spot) can be extremely large. In a sense, one can say that enhancement of optical nonlinearities is especially large in fractals because of very strong field fluctuations.

Four-wave mixing (FWM) is determined by the nonlinear susceptibility similar to (1) $\chi^{(3)}_{\alpha\beta\gamma\delta}(-\omega_s;\omega_1,\omega_1,-\omega_2)$, where $\omega_s = 2\omega_1 - \omega_2$ is the generated frequency, and ω_1 and ω_2 are the frequencies of the applied waves. Coherent anti-Stokes Raman scattering (CARS) is an example of FWM. In one elementary CARS process, two ω_1 photons are transformed into the ω_2 and ω_s photons. Another example is degenerate FWM (DFWM); this process is used for optical phase conjugation (OPC) that can result in complete removal of optical aberrations (*11*). In DFWM, all waves have the same frequency ($\omega_s = \omega_1 = \omega_2$) and differ only by their propagation directions and, in general, by polarizations. In a typical OPC experiment, two oppositely directed pump beams, with field amplitudes $\mathbf{E}^{(1)}$ and $\mathbf{E}'^{(1)}$, and a probe beam, with amplitude $\mathbf{E}^{(2)}$ (and propagating at a small angle to the pump beams), result in a OPC beam which propagates against the probe beam. Because of the interaction geometry, the wave vectors of the beams satisfy to the relation: $\mathbf{k}_1 + \mathbf{k}'_1 = \mathbf{k}_2 + \mathbf{k}_s = 0$. Clearly, for the two pairs of oppositely directed beams, that have the same frequency ω, the phase-matching conditions are automatically fulfilled (*11*).

The nonlinear susceptibility, $\chi^{(3)}$, that results in DFWM, also leads to the considered above nonlinear refraction and absorption that are associated with the Kerr optical nonlinearity. Note also that as above the nonlinear susceptibility, $\chi^{(3)}$, can be associated with either the fractal surface itself or molecules adsorbed on the surface.

For coherent effects, including the ones discussed in this section, averaging is performed for the generated field amplitude (rather than intensity) or, equally, for the nonlinear polarization of a medium. The average polarization, $P^{(3)}(\omega)$, is proportional to the nonlinear susceptibility, $P^{(3)}(\omega) \propto \bar{\chi}^{(3)} = \chi^{(3)} G_K$. The measured signal for coherent processes is proportional to $|\bar{\chi}^{(3)}|^2$. Thus we conclude that the resultant enhancement for degenerate (or near degenerate) four-wave mixing can be expressed in terms of the enhancement for the Kerr susceptibility as follows (*4*)

$$G_{FWM} = \left|G_K\right|^2 = \left|\frac{\left\langle\left|\mathbf{E}(\mathbf{r})\right|^2\left[\mathbf{E}(\mathbf{r})\right]^2\right\rangle}{\left[\mathbf{E}^{(0)}\right]^4}\right|^2. \tag{2}$$

Note that one can equally describe a medium optical response in terms of the nonlinear currents rather nonlinear polarizations; these two approaches are completely equivalent (*6*).

Raman scattering is a linear optical process. For small Stokes shifts, however, the surface-enhanced Raman scattering is proportional to the average of fourth power of the local fields (*4,12*), so that the corresponding enhancement factor is (*4*)

$$G_{RS} = \frac{\left\langle\left|\mathbf{E}(\mathbf{r})\right|^4\right\rangle}{\left[\mathbf{E}^{(0)}\right]^4}. \tag{3}$$

Note that in contrast to the enhanced Kerr nonlinearity considered above, G_{RS} is real and the local enhancement is phase insensitive, so that there is no destructive interference of signals from different points of a surface. As a result, typically, $G_{RS} \gg G_K$.

Under some simplifying conditions, the enhancement for the second harmonic generation can be written as (*2*)

$$G_{SHG} = \left|\left\langle\left[\frac{E_\omega(\mathbf{r})}{E_\omega^{(0)}}\right]^2\left[\frac{E_{2\omega}(\mathbf{r})}{E_{2\omega}^{(0)}}\right]\right\rangle\right|^2, \tag{4}$$

where $E_{2\omega}^{(0)}$ and $E_{2\omega}(\mathbf{r})$ are the macroscopic and local *linear* fields at frequency 2ω. If there is no enhancement at frequency 2ω, then $E_{2\omega}(\mathbf{r}) = E_{2\omega}^{(0)}$.

The above formula can be easily generalized for the n-th harmonic generation:

$$G_{nHG} = \left| \left\langle \left[\frac{E_{\omega}(\mathbf{r})}{E_{\omega}^{(0)}} \right]^{n} \left[\frac{E_{2\omega}(\mathbf{r})}{E_{2\omega}^{(0)}} \right] \right\rangle \right|^{2}. \qquad (5)$$

Note that the above formulas are valid for arbitrary surface fractal or non-fractal. In fractals, however, because of the extremely large field fluctuations the ensemble average enhancements are typically much larger than for non-fractal surfaces. In addition, the fractal modes provide enhancements in a very large spectral range including the infrared part, where the enhancement is particularly large because of the high quality-factor for metal surfaces in this spectral range (*2*). We will also show below that the local enhancements in the "hot" zones (associated with the localized eigenmodes) can exceed the ensemble-average enhancement by many orders of magnitude.

Fractal Aggregates of Colloidal Particles. As well known, there is only one dipolar mode that can be excited by a homogeneous field in a spherical object. For a 3-dimensional (non-fractal) collection of small particles, such as randomly close-packed hard spheres of particles (CPHS) or a random gas of particles (RGP), absorption spectra are peaked near the relatively narrow resonance of the individual particles, i.e., all eigenmodes of the collection of particles are located within a small spectral interval (*4*).

In contrast to such conventional (non-fractal) 3-dimensional systems, dipolar interactions in low-dimensional fractal clusters are not long range; this results in the spatial localization of the eigenmodes at various random locations in the cluster (*2-6,9*). The spectrum of these eigenmodes exhibits strong inhomogeneous broadening as a result of spatial variability of the local environment. It is also important to note that, despite the asymptotically zero mean density of particles in a fractal cluster, a high probability always exists of finding a number of particles in close proximity to a given particle (stated more precisely, in fractals, the pair correlation $g \propto r^{D-d}$ ($D < d$), where D is the fractal dimension and d is the dimension of the embedding space; accordingly, g becomes large at small r). Thus, objects with fractal morphology possess an unusual combination of properties; namely, despite the fact that the volume fraction filled by N particles in a fractal is very small, $\propto N^{1-d/D} \rightarrow 0$, strong interactions nevertheless exist between neighboring particles (*2*). These strong interactions between neighboring particles, which are highly variable because of variability of local particle configurations in the cluster, lead to the formation of inhomogeneously broadened eigenmodes covering a broad spectral range (*4*).

Localization of eigenmodes in fractals leads to a patchwork-like distribution of local fields associated with "hot" and "cold" zones (*2,3,9*). This, in turn, results in large spatial fluctuations of local fields in fractal composites and in the giant enhancement of various optical effects (*2-6,9*).

For the special case of fractals formed by metal particles, the dipole eigenmodes span the visible and infra-red regions of the spectrum; since the mode quality-factors increase with wavelength, local fields are especially large in the long wavelength part of the spectrum (*2,4*).

In Figure 1, we show an electron microscope picture of a typical aggregate of silver colloidal particles. The fractal dimension for these aggregates is $D \approx 1.78$. Using the well-known model of cluster-cluster aggregation, colloidal aggregates can be readily simulated numerically (*7*). Note that voids are presented in all scales from the minimum one (about the size of a single particle) to the maximum one (about the size of the whole cluster); this is indication of the statistical self-similarity of a fractal cluster. The size of an individual particle is $\sim$ 10 nm, whereas the size of the whole cluster is $\sim 1\mu$m.

In Figures 2a and 2b, we show results of our calculations for enhancement of the real, G'_K, and imaginary, G''_K (see formula(1)), parts of the enhancement of the Kerr-nonlinearity, and, in Figure 2c, the enhancement for the four-wave mixing, $G_{FWM} = |G_K|^2$, in silver colloidal aggregates. We use the experimental data for the optical constants of bulk silver from (*13*). The enhancements increase toward the infrared part of the spectrum where the resonance quality-factors are larger and the localization of the eigenmodes is stronger (*2,4*).

The nonlinear susceptibility, $\bar{\chi}^{(3)}$ of the composite material consisting of fractal aggregates of colloidal particles in solution is given by $\bar{\chi}^{(3)} = p \cdot G_K \chi_m^{(3)}$, where $\chi_m^{(3)}$ is the susceptibility of the non-aggregated colloidal particles and p is the volume fraction filled by metal. The experimentally measured value of $\chi_m^{(3)}$ re-estimated for $\lambda = 532$ nm is $\chi_m^{(3)} \approx 10^{-8}$ esu (*14*). As follows from Figure 2, $|G_K| \sim 10^4$ for $\lambda = 532$ nm. With the indicated value of $\chi_m^{(3)}$ and $p \approx 5 \cdot 10^{-6}$ used in our experiments, we obtain $|\bar{\chi}^{(3)}| \approx 5 \cdot 10^{-10}$ esu. This is in good agreement with the experimentally (using the degenerate four-wave mixing) obtained value, $|\bar{\chi}^{(3)}| = 5.7 \cdot 10^{-10}$ esu (*15*).

Note that p is a variable quantity and can be increased. We can assign the value 10^{-4} esu to the nonlinear susceptibility, $\chi^{(3f)}$, of fractal aggregates, i.e., $\bar{\chi}^{(3)} = p \cdot \chi^{(3f)}$, with $\chi^{(3f)} \sim 10^{-4}$ esu. This is a very large value for a third-order nonlinear susceptibility.

The above high nonlinearities were obtained at the pump intensity $I_L \leq 1.5$ MW/cm^2. At higher intensities a light-induced modification (selective in frequency and polarization) occurs resulting in local re-structuring of resonant domains (hot spots) in the irradiated cluster; this in turn leads to a dip in the aggregate absorption spectrum in the vicinity of laser wavelength (*15,16*).

In the other series of experiments, using Z-scan technique, both nonlinear refraction and absorption were measured (*16*). It has been found that for $\lambda = 540nm$ and $p = 5 \cdot 10^{-6}$, the aggregation of silver colloidal particles into fractal clusters is accompanied by the increase of the nonlinear correction, α_2, to the absorption, $\alpha(\omega) = \alpha_0 + \alpha_2 I$, from $\alpha_2 = -9 \cdot 10^{-10}$ cm/W to $\alpha_2 = -5 \cdot 10^{-7}$ cm/W, i.e., the enhancement factor is 560 (*16*). The measured nonlinear refraction at $\lambda = 540nm$ for fractal aggregates of silver colloidal particles is $n(\omega) = n_0 + n_2 I$, with $n_2 = 2.3 \cdot 10^{-12}$ cm^2/W. Similar measurments at $\lambda = 1079$ nm give the following values $n_2 = -0.8 \cdot 10^{-12}$ cm^2/W and $\alpha_2 = -0.7 \cdot 10^{-7}$ cm/W.

The measured n_2 and α_2 allows one to find the real and imaginary parts of the Kerr susceptibility; they are $Re[\bar{\chi}^{(3)}] = 1 \cdot 10^{-10}$ esu, $Im[\bar{\chi}^{(3)}] = -0.8 \cdot 10^{-10}$

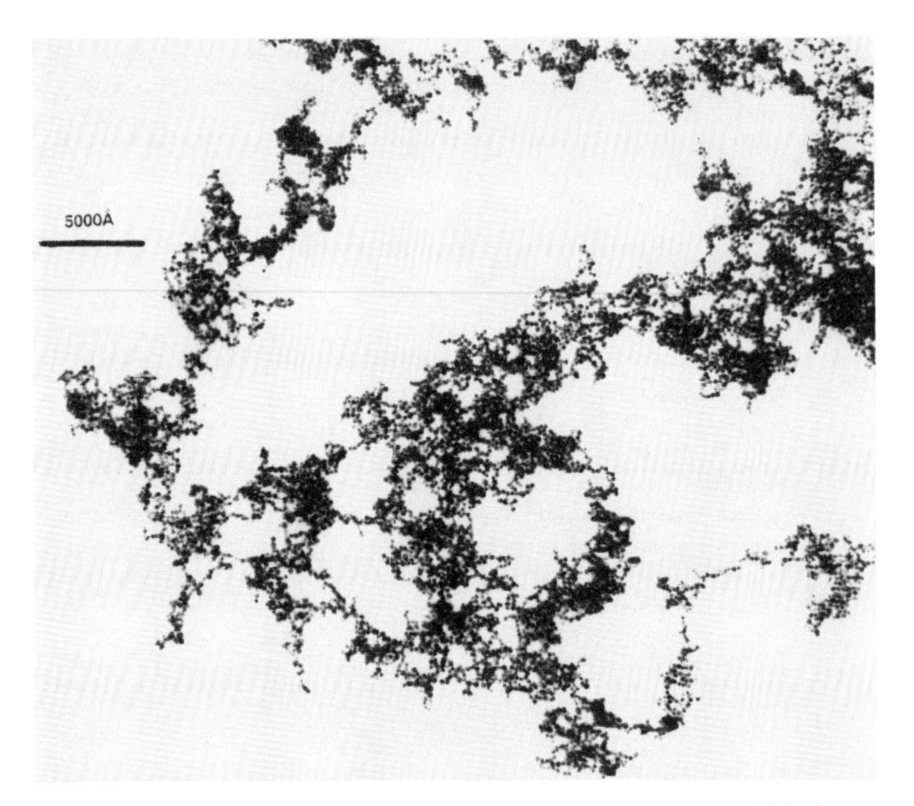

Figure 1. Electron micrograph of a silver cluster-cluster aggregate (CCA).

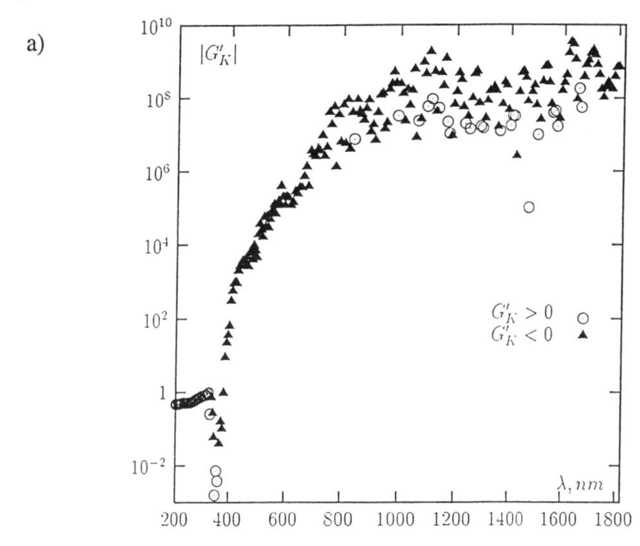

Figure 2. Enhancements on silver CCAs for the real, G'_K, (a) and imaginary, G''_K, (b) parts of the Kerr-type optical nonlinearity and for four-wave mixing, G_{FWM}, process (c).

Continued next page

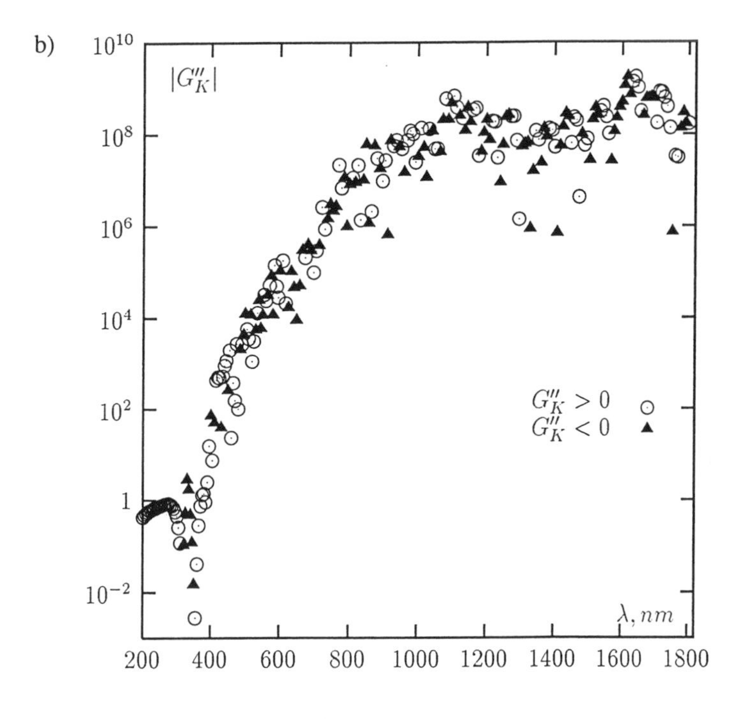

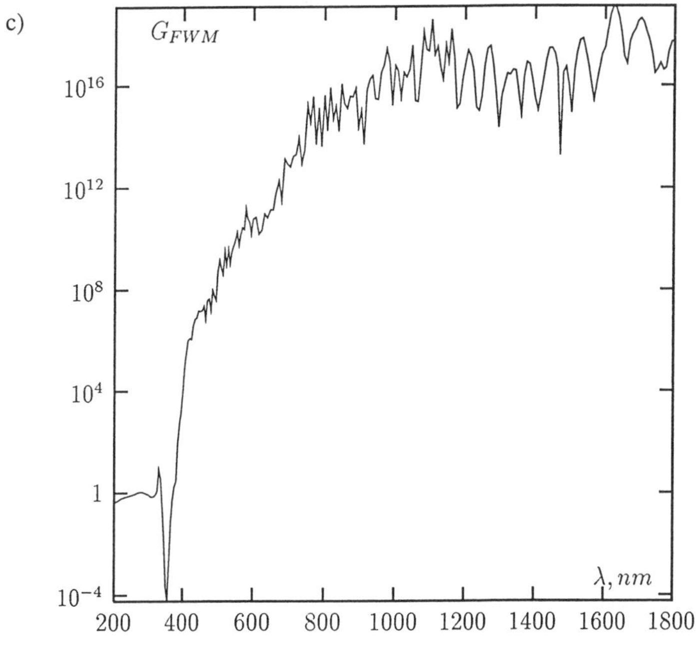

Figure 2. *Continued.*

esu, for $\lambda = 540nm$, and $Re[\bar{\chi}^{(3)}] = -3.5 \cdot 10^{-11}$ esu, $Im[\bar{\chi}^{(3)}] = -2.7 \cdot 10^{-11}$, for $\lambda = 1079$ nm. This means that the saturation of absorption and the nonlinear refraction provide comparable contributions to the nonlinearity. Note that the real part changes its sign with the wavelength.

Using a different technique based on a dispersion interferometer, the nonlinear correction to the refractive index, n_2, was also measured for $\lambda = 1064$ nm (at $p = 5 \cdot 10^{-6}$ and the intensity below the photomodification threshold); the obtained value is $n_2 = -1.5 \cdot 10^{-11}$ cm^2/W that corresponds to $Re[\bar{\chi}^{(3)}] \approx -7 \cdot 10^{-10}$ esu (*16*).

Note that fractal aggregates of colloidal particles can be placed into polymer or gel matrices. Thin films can then be prepared with fractal aggregates in such matrices. The volume fraction filled by metal fractal aggregates in such thin films is typically larger than in the case of colloidal solution and therefore the nonlinearities are significantly higher.

The laser pulse duration used in the above experiments was $\sim 10ns$. The Kerr type third-order nonlinearity was also detected with the use of 30 ps laser pulses; however, the observed optical nonlinearities were in this case significantly smaller than in the experiments with ns laser pulses. Our studies indicate that there are probably two different types of the optical nonlinearities, $\bar{\chi}^{(3)}$. A smaller one has the time of nonlinear response in the picosecond scale and a larger one in the nanosecond scale. The first one is associated with polarization of metal particles in an aggregate, whereas the second one probably involves absorption-related effects resulting in a holographic grating in colloidal clusters.

We also mention that very high optical susceptibilities, $|\bar{\chi}^{(3)}| \approx 10^{-6}$ esu with a subpicosecond response time was recently obtained for films containing J-aggregates of a dye (PIC) in a PVP matrix when fractal clusters of gold or silver are added to it (*17*).

We mention that rather strong enhancements of optical nonlinearities were obtained in non-fractal composite materials as well (*18*).

To conclude this section, we also show in Figure 3 results of our theoretical calculations and experimental studies of SERS in fractal aggregates of silver colloidal particles (*4,12*). As seen from the figure, the theory successfully explains the experimental observations. Note that the enhancement increases toward the red part of the spectrum, where the local fields associated with the localized fractal modes are significantly larger.

Self-Affine Thin Films. We also performed studies of linear and nonlinear optical properties of self-affine thin films. Rough thin films, formed when an atomic beam condenses onto a low-temperature substrate, are typically self-affine fractal structures (*8*). Contrary to the case of "usual" roughness, there is no correlation length for self-affine surfaces, which implies that inhomogeneities of all sizes are present (within a certain size interval) according to a power-law distribution. Self-affine surfaces obtained in the process of film growth belong to the Kardar-Parisi-Zhang universality class. Unlike statistically self-similar structures, to reveal scale-invariance, a self-affine surface (SAS) requires different scaling factors in the (x, y)-plane and in the normal direction, z.

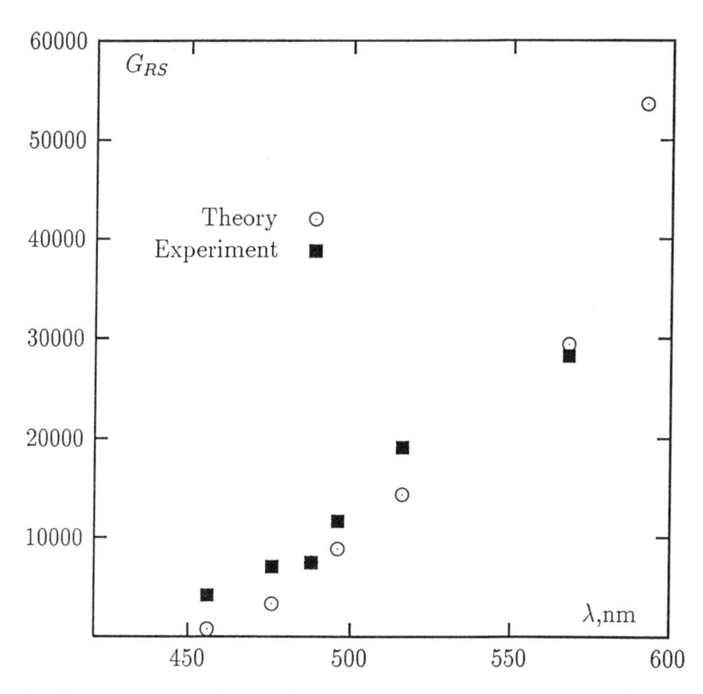

Figure 3. Theoretical and experimental enhancement factors, G_{RS}, for silver colloid aggregates as a function of the wavelength.

To simulate a self-affine film, we used the restricted solid-on-solid (RSS) model (for details see (*5*) and references therein). In this model, a particle is incorporated into the growing aggregate only if the newly created interface does not have steps which are higher than one lattice unit, a. The surface structure of such deposits is relatively simple, because there are no overhangs. In this way strong corrections to scaling effects are eliminated and the true scaling behavior appears clearly, even for small dimensions. In the long-time regime, the height-height correlation function for a self-affine surface has the form

$$\langle [h(\mathbf{R}) - h(0)]^2 \rangle \sim R^{2H},$$

where $\mathbf{R}$ is the radius-vector in the plane normal to the growth direction, z, and the scaling exponent (co-dimension), H, is related to the fractal dimension, D, through the formula $H = 3 - D$. For the RSS model, $D = 2.6$ and the scaling formula above is valid for large values of the average height, $\bar{h}$, (which is proportional to the deposition time), such that $\bar{h} \gg l^{\zeta}$, where $\zeta = 2(d+1)/(d+2) = 2 - H$ (l is the linear size of a system and d is the dimension of the embedding space. Our simulations satisfied this condition, and the above scaling relation was well pronounced.

Our analysis revealed that the eigenmodes of a self-affine film possess strongly inhomogeneous spatial distributions characterized by various degrees of localization (*5*). On a metal self-affine film, the intensities in areas of high local fields ("hot" zones) exceed the applied field intensity by approximately three orders of magnitude in the visible part of the spectrum. In common with the mode-structure of fractal aggregate media, the spatial locations of "hot" zones on a self-affine thin film are very strong functions of the frequency and polarization of the incident light.

In Figure 4a, we show a computer-generated self-affine film obtained within the restricted solid-on-solid model ($D = 2.6$); the model provides an accurate description for *Ag* cold-deposited films having approximately the same fractal dimension (*8*).

The importance of field localization in accounting for the optical properties of self-affine surfaces have recently been demonstrated experimentally by the University of Toronto research team led by Prof. M. Moskovits (*19*). Self-affine fractal surfaces were prepared by gravitationally depositing fractal aggregates of colloidal silver particles out of solution onto pyrex microscope cover-slides. Thick deposits were prepared in order to ensure a self-affine structure. Our numerical simulation of this process showed that the surface resulting from deposition of many clusters on top of each other is self-affine with the fractal dimension $D \approx 2.5$.

Note that a similar strong dependence on frequency and polarization of the applied field was also obtained in previous experiments on near-field scanning optical microscopy of localized optical modes in *Ag* fractal aggregates deposited onto a surface (*3*).

In Figure 4b, we show results of our theoretical calculations for the average enhancement of Raman scattering for both small and large Stokes shifts on self-affine films generated in the RSS model. $G_{RS,\parallel}$ and $G_{RS,\perp}$ describe enhancements

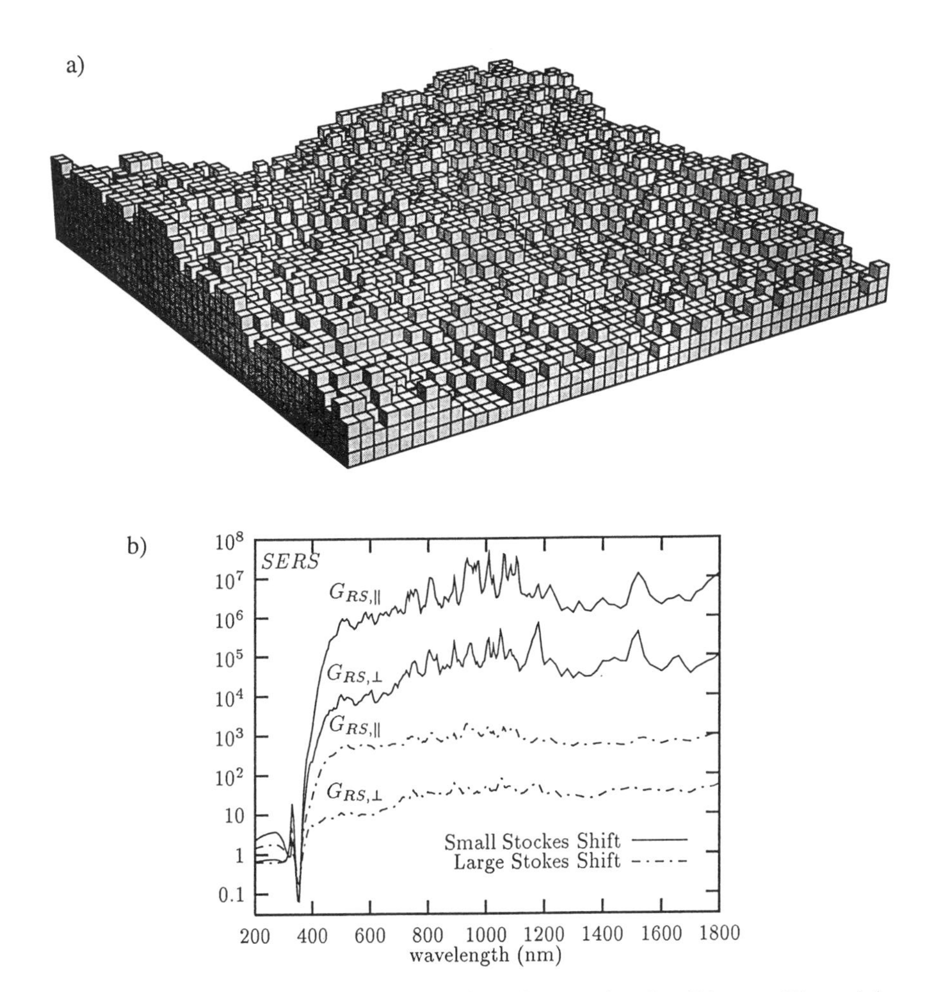

Figure 4. (a) Self-affine film obtained in the restricted solid-on-solid model. (b) The enhancement of Raman scattering on silver self-affine films for the applied field polarized in the plane of the film ($G_{RS,\parallel}$) and normal to the plane ($G_{RS,\perp}$) for small and large Stokes shifts.

for the applied field polarized in the plane of the film and perpendicular to it, respectively. As seen in the figure, the enhancement increases toward the long-wavelength part of the spectrum and reaches very large values, $\sim 10^7$; this agrees well with the experimental observations of SERS on cold-deposited thin films (*12*).

In Figures 5 a-c, the field spatial distributions at the fundamental and Stokes frequencies are shown. As seen in the figure, the distributions contain "hot" spots, where the fields are very high. Although the Stokes signal is proportional to the local field at the fundamental frequency, ω, the generated Stokes field, with frequency ω_s, excites, in general, other eigenmodes. Hence the field spatial distributions produced by the applied field and by the Raman signal can be different, as clearly seen in the figure.

This pattern is expected to be typical for various optical processes in strongly disordered fractal systems, such as self-affine thin films. Specifically, hot spots associated with fields at different frequencies and polarizations can be localized in spatially separated nm-sized areas. These novel nano-optical effects can be probed with NSOM providing sub-wavelength resolution (*3,19*).

If molecules possess the nonlinear susceptibility, $\chi^{(2)}$, then second harmonic generation (SHG) can be strongly enhanced when adsorbing the molecules on a metal self-affine surface. In Figure 6, we plot the calculated enhancement for SHG from molecules on a silver self-affine surface (for the applied field polarized parallel and perpendicular to the surface, $G_{SHG,\|}$ and $G_{SHG,\perp}$, respectively). To calculate the enhancements we used formula (4). As seen in the figure, the enhancement is very large and increases toward larger wavelengths.

Semicontinuous Metal Films. Self-affine thin films are typically produced by condensing atomic beams onto a *low-temperature* substrate. When placed into the air at room temperatures these films are annealed and change their morphology, yet remaining random. Our preliminary calculations showed that such annealed films produced originally at low-temperature still provide large enhancement of local fields, however, the spectral range where the enhancement occurs is somewhat narrowed.

For applications, it is important to have fractal films that retain their fractal morphology at room temperatures, such as two-dimensional semicontinuous metal films near the percolation threshold. In the vicinity of this threshold, fractal clusters are formed from metal granules (produced by thermal evaporation or sputtering of metal onto an insulator substrate). In contrast to self-affine films that are essentially 3-dimensional, the 2-dimensional semicontinuous films remain stable at room temperatures.

In the growing process of metal semicontinuous films, small metallic grains are first formed on the substrate. As the film grows, the metal filling factor increases and coalescences occur, so that irregularly shaped clusters are formed on the substrate eventually resulting in $2d$ fractal structures. The sizes of these structures diverge in the vicinity of the percolation threshold. A percolating cluster of metal is eventually formed, when a continuous conducting path appears between the ends of a sample. The metal-insulator transition (in other words, the conductivity threshold) is very close to this point.

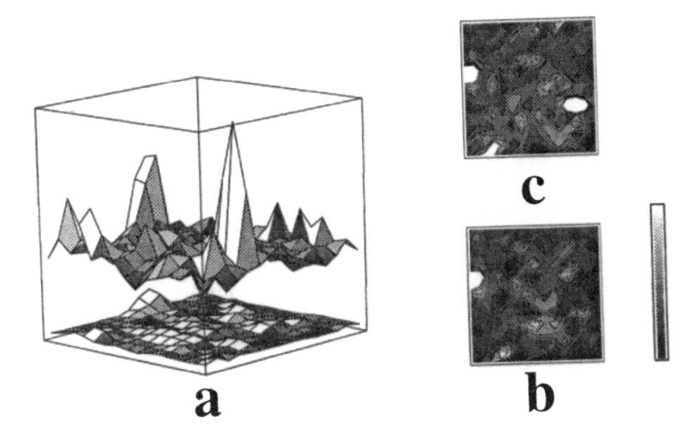

Figure 5. a): The spatial distributions for the local fields at the fundamental frequency, $\lambda = 550nm$, (bottom; the field distribution is magnified by 3) and for the Stokes fields, $\lambda_s = 600nm$, (top). [The applied field is linearly polarized in the plane of the film.] b) and c) : The contour-plots for the field distributions shown on a).

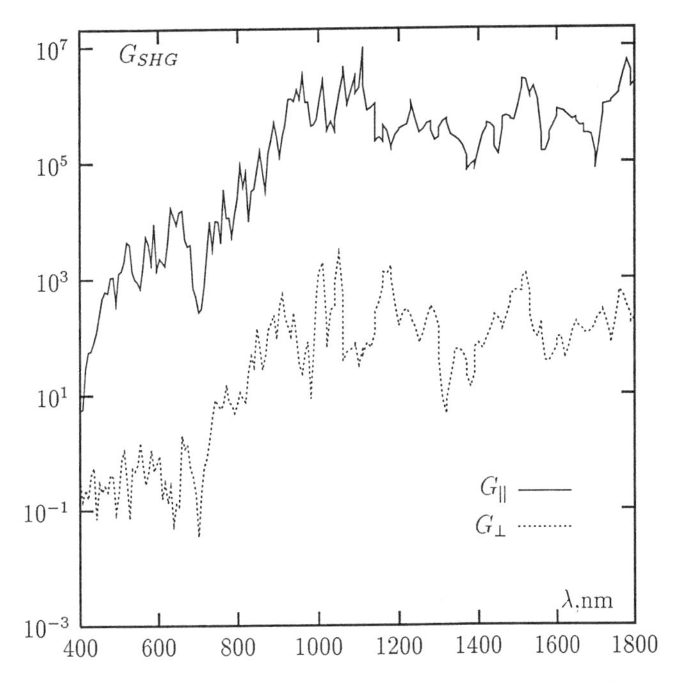

Figure 6. The enhancement for second harmonic generation, G_{SHG}, on the self-affine silver surface for the applied field polarized in the plane, $G_{\|}$, and normal to the plane, $G_{\perp}$.

We showed that SERS from Raman active molecules adsorbed onto a semicontinuous films in the vicinity of the percolation threshold is given by (*6*)

$$G_{RS} = \frac{\left\langle \left|\varepsilon(\mathbf{r})\right|^2 \left|\mathbf{E}(\mathbf{r})\right|^4 \right\rangle}{\varepsilon_d^2 \left[\mathbf{E}^{(0)}\right]^4} \approx 9.5 \frac{|\varepsilon'_m|^3}{\varepsilon_d \varepsilon''_m}, \tag{6}$$

where ε is the dielectric constant of the film consisting of a dielectric substrate, with $\varepsilon = \varepsilon_d$, and metal metal grains, with $\varepsilon = \varepsilon_m = \varepsilon'_m + i\varepsilon''_m$. The second approximate equality in (6) was obtained using the scaling analysis (*6*). Note that formula (6) differs from (3) by factor $\left|\varepsilon(\mathbf{r})/\varepsilon_d\right|^2$; the difference results from the fact that in (3) we assumed that both the Raman and linear polarizabilities are associated with the same site on a fractal surface, whereas to obtained (6) we supposed that the linear polarizability is due to a metal grain on the film but the Raman polarizability is due to a molecule adsorbed on the grain.

In Figure 7 we show results of our Monte Carlo simulations for Raman scattering from a silver semicontinuous film at the percolation threshold (based on the first equality in (6)) and results of the calculations based on the approximate scaling formula (second equality in (6)). One can see that the scaling formula works well for almost all frequencies, except the small frequencies that are comparable or smaller than the relaxation constant.

In Figure 8 we show distributions of the local Raman signals on a silver semicontinuous film at $p = p_c$, for different wavelengths, $\lambda = 1.5\mu$m and $\lambda = 10\mu$m. As seen in the figure, the Stokes field intensities have the forms of sharp peaks sparsely distributed on the film, with the magnitudes increasing toward the long-wavelength part of the spectrum. The local enhancement in peaks achieve 10^9, for $\lambda = 1.5\mu$m, and 10^{12}, for $\lambda = 10\mu$m. [The Raman signal distributions were calculated using the first formula in (6) but with no averaging over samples (so that the sign $< .. >$ should be omitted in this case).] The average enhancement, G_{RS}, is much less, of the order of 10^6, for both wavelengths (see Figure 7). As seen in Figure 8, the peak positions strongly depend on the frequency. This non-trivial pattern for the local SERS distribution can be probed by means of near-field optical microscopy. If the density of Raman-active molecules is small so that it approximately equals to the surface concentration of the metal grains on the film, then each peak in Figure 8 is due to Raman scattering from a *single* molecule. Thus the presented picture of the SERS distribution opens an unique opportunity to perform Raman spectroscopy of a single molecule on a semicontinuous metal film.

Let molecules, possessing the nonlinear susceptibility $\chi^{(2)}(-2\omega;\omega,\omega)$, be adsorbed on a dielectric substrate; then the adding metal grains on the film results in enhancement of the second-harmonic generation by factor:

$$G_{SHG} = \left| \left\langle \frac{\varepsilon_{2\omega}(\mathbf{r})}{\varepsilon_{d,2\omega}} \left[\frac{E_\omega(\mathbf{r})}{E_\omega^{(0)}}\right]^2 \left[\frac{E_{2\omega}(\mathbf{r})}{E_{2\omega}^{(0)}}\right] \right\rangle \right|^2, \tag{7}$$

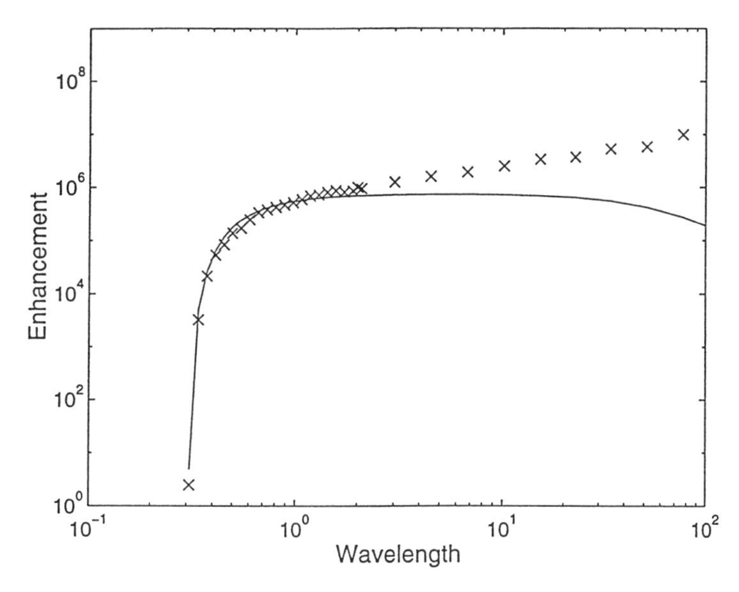

Figure 7. The enhancement factor, G_{RS}, for Raman scattering from a silver semicontinuous film at the percolation threshold (wavelength is in μm). The points are results of the numerical simulations based on the exact formula (6); the solid line represents results of calculations based on the scaling formula (second approximate equality in (6)).

where $E_{2\omega}^{(0)}$ and $E_{2\omega}(\mathbf{r})$ are the macroscopic and local *linear* fields at frequency 2ω. The difference between (7) and (4) is due to the same reason as the difference between (6) and (3) (see the discussion following formula (6) above).

The average enhancement, G_{SHG}, on a silver semicontinuous film as a function of the wavelength at $p = p_c$ is shown in Figure 9a. The enhancement increases toward larger wavelengths where the local fields are higher.

We also calculated the local distributions of the nonlinearly scattered light, with frequency 2ω found from the formula

$$g_{SHG}(\mathbf{r}) = \frac{\varepsilon_{2\omega}(\mathbf{r})}{\varepsilon_{d,2\omega}} \left[\frac{E_\omega(\mathbf{r})}{E_\omega^{(0)}} \right]^2 . \qquad (8)$$

[For simplicity, we assumed in (8) that there is no "secondary" enhancement for the generated field at frequency 2ω.] The distribution of the local SHG enhancements, $g_{SHG}(\mathbf{r})$, are shown in Figure 9b for $\lambda = 1.5\mu$m and $\lambda = 20\mu$m at $p = p_c$. Similar to the case of the SERS local distributions, the peaks in the SHG become much larger with the increase of the wavelength; the spatial separations between them, however, also increase with λ. For the "hot spots" the local enhancement can be giant, reaching values up to 10^{15}, whereas the ensemble-average enhancement is relatively small, as seen from comparison Figs. 9a and 9b.

Huge local enhancements for Raman scattering, SHG, and other nonlinear optical scattering, that significantly exceed the ensemble average enhancements, opens a fascinating possibility to perform Raman and nonlinear spectroscopy for a

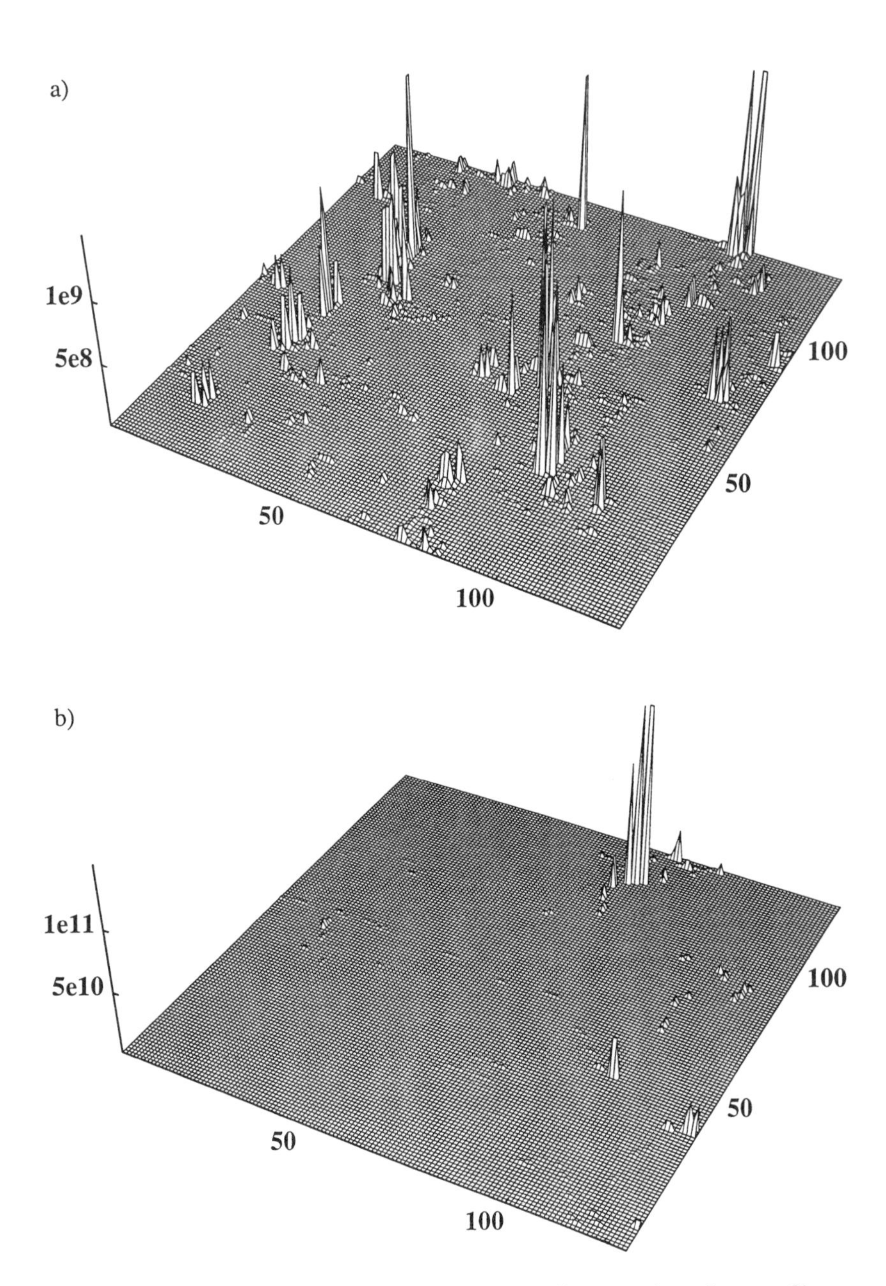

Figure 8. The local SERS distribution on a silver semicontinuous film at the percolation threshold for different wavelengths: a) $\lambda = 1.5 \mu m$, and b) $\lambda = 10 \mu m$.

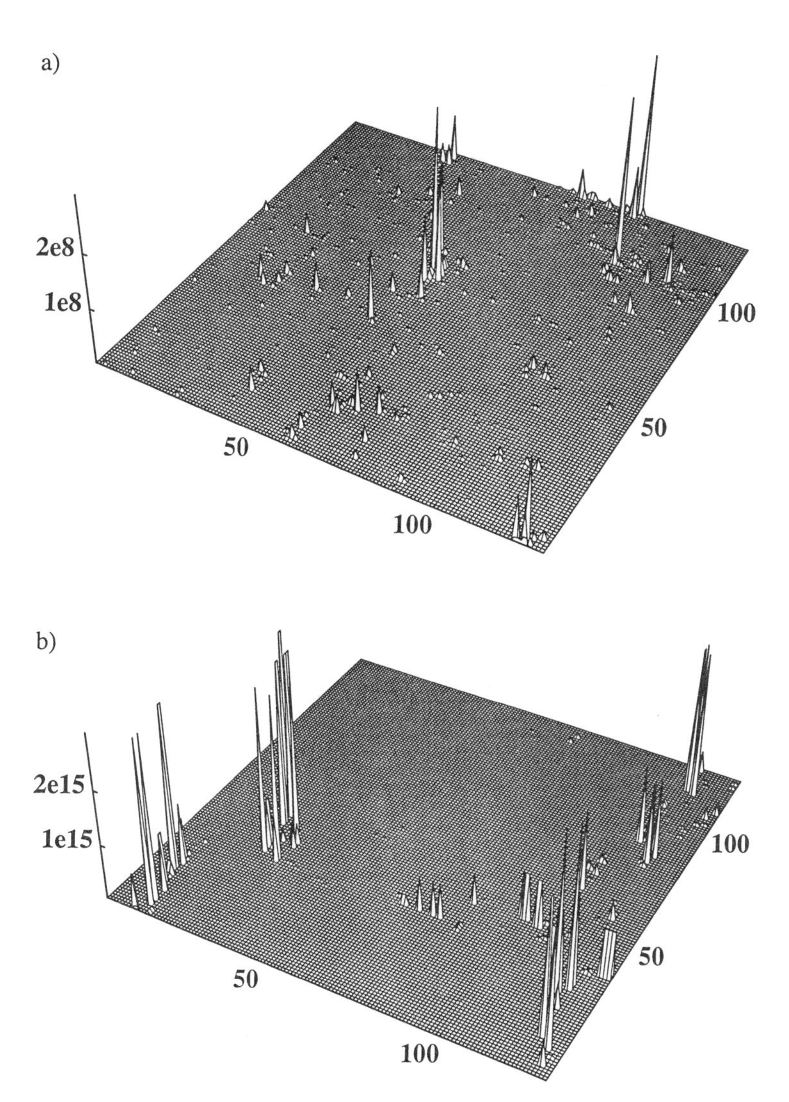

Figure 9. a): The average SHG enhancement, G_{SHG}, on a silver semicontinuous film as a function wavelength at $p = p_c$. The film sizes are 512x512. Distributions of the local SHG enhancements, $g_{SHG}(\mathbf{r})$, for $\lambda = 0.5\mu m$ (b) and $\lambda = 20\mu m$ (c) at $p = p_c$.

single molecule. We note that SERS from a single molecule on a colloidal particle and a fractal aggregate of colloidal particles was recently reported in (*20*).

Conclusions

To conclude, light excitation of objects with fractal morphology, such as aggregates of colloidal particles, self-affine surfaces, and semicontinuous thin films, results in extremely high fields localized in nm-sized "hot spots". This provides huge enhancements for a number of optical phenomena, including Raman scattering, nonlinear refraction and absorption, and second-harmonic generation. The local enhancements exceed the ensemble average by many orders of magnitude.

Acknowledgments

Work was supported by NSF under grant 9500258 and by RFBR under grant 96-02-19331.

Literature Cited

1. *Fractals and Disorder*; Bunde, A., Ed.; North-Holland: 1992; Havlin, S.; Bunde, A. In *Ibid*; p. 97; Alexander, S.; Orbach, R. *J. Phys. (Paris) Lett.* **1982,** *43*, L1625; Rammal, R.; Toulouse, G. *J. Phys. (Paris) Lett.* **1983,** *44*, L13; Sapoval, B.; Gobron, Th.; Margolina, A. *Phys. Rev. Lett.* **1991,** *67*, 2974.
2. Shalaev, V. M. *Phys. Reports* **1996,** *272*, 61.
3. Tsai, D. P.; Kovacs, J.; Wang, Z.; Moskovits, M.; Shalaev, V. M.; Suh, J.; Botet, R. *Phys. Rev. Lett.* **1994,** *72*, 4149; Shalaev, V. M.; Moskovits, M. *Phys. Rev. Lett.* **1995,** *75*, 2451.
4. Markel, V. A.; Shalaev, V. M.; Stechel, E. B.; Kim, W.; Armstrong, R. L.; *Phys. Rev. B* **1996,** *53*, 2425; Shalaev, V. M.; Poliakov, E. Y.; Markel, V. A. *Phys. Rev. B* **1996,** *53*, 2437.
5. Shalaev, V. M.; Botet, R.; Mercer, J.; Stechel, E. B. *Phys. Rev. B* **1996,** *54*, 8235; Poliakov, E. Y.; Shalaev, V. M.; Markel, V. A.; Botet, R. *Opt. Lett.* **1996,** *21*, 1628.
6. Brouers, F.; Blacher, S.; Lagarkov, A.; Sarychev, A. K.; Gadenne, P.; Shalaev, V. M. *Phys. Rev. B* **1997,** *55*, 13234; Gadenne, P.; Brouers, F.; Shalaev, V. M.; Sarychev, A. K. *J. Opt. Soc. Amer. B*, submitted.
7. Jullien R.; Botet, R. *Aggregation and Fractal Aggregates*; World Scientific: Singapore, 1987; Feder, J. *Fractals*; Plenum Press: New York, 1988.
8. Chiarello, R.; Panella, V.; Krim, J.; Thompson, C. *Phys. Rev. Lett.* **1991,** *67*, 3408; Douektis, C.; Wang, Z.; Haslett, T. L.; Moskovits, M. *Phys. Rev. B* **1995,** *51*, 11022; Barbasi, A.-L.; Stanley, H. E. *Fractal Concepts in Surface Growth*; Cambridge U. Press: Cambridge, 1995.
9. Markel, V. A.; Muratov, L. S.; Stockman, M. I., George, T. F. *Phys. Rev. B* **1991,** *43*, 8183; Stockman, M. I.; Pandey, L. N.; Muratov, L. S.; George, T. F. *Phys. Rev. Lett.* **1994,** *72*, 2486; Stockman, M. I.; Pandey, L. N.; Muratov,

L. S.; George, T. F. *Phys. Rev. B* **1995** *51*, 185; Stockman, M. I.; Pandey, L. N.; George, T. F. *Phys. Rev. B* **1996,** *53*, 2183.

10. Stroud, D.; Hui, P. M. *Phys. Rev. B* **1988,** 37, 8719.
11. Boyd, R. W. *Nonlinear Optics*; Academic Press: New York, 1992.
12. Moskovits, M. *Rev. Mod. Phys.* **1985,** *57* , 783; Otto, A.; Mrozek, I.; Grabhorn, H.; Akemann, W. *J. Phys.: Condens. Matter* **1992,** *4*, 1143; *Surface Enhance Raman Scattering*, Chang R. K.; Furtak, T. E., Eds.; Plenum Press: New York, 1982; Stockman, M. I.; Shalaev, V. M.; Moskovits, M.; Botet, R.; George, T. F. *Phys. Rev. B* **1992,** *46*, 2821.
13. Johnson, P. B.; Christy, R. W. *Phys. Rev. B* **1975,** *6*, 4370.
14. Flytzanis, C. *Prog. Opt.* **1992,** *29*, 2539; Ricard. D.; Roussignol, Ph.; Flytzanis, C. *Optics Letters* **1985,** *10*, 511; Hache, F.; Ricard, D.; Flytzanis, C.; Kreibig, U. *Applied Physics A* **1988,** *47*, 347.
15. Rautian, S. G.; Safonov, V. P.; Chubakov, P. A.; Shalaev, V. M.; Stockman, M. I. *JETP Lett.* **1988,** *47*, 243; Butenko, A. V.; *et al.*, *Z. Phys. D* **1990, 17,** 283.
16. Danilova, Yu. E.; Rautian, S. G.; Safonov, V. P. *Bulletin of the Russian Academy of Sciences, Physics* **1996,** *60*, 374; Danilova, Yu. E.; Drachev, V. P.; Perminov, S. V.; Safonov, V. P. *Bulletin of the Russian Academy of Sciences, Physics* **1996,** *60*, 342; Danilova, Yu. E., Safonov, V. P. in: *Fractals Reviews in the Natural and Applied Sciences*; Novak, M. M., Ed.; Chapman & Hall: London, 1995; p. 101; Danilova, Yu. E., Lepeshkin, N. N.; Rautian, S. G.;, Safonov, V. P. *Physica A* **1996,** in press.
17. Zhuravlev, F. A.; Orlova, N. A.; Shelkovnikov, V. V.; Plehanov, A. I.; Rautian, S. G.; Safonov, V. P. *JETP Lett.* **1992,** *56*, 260.
18. Sipe, J. E.; Boyd, R. W. *Phys. Rev. B* (1992) *46*, 1614 (1992); Gehr, R. J.; Fisher, G. L.; Boyd, R. W.; Sipe, J. E. *Phys. Rev. A* **1996,** *53*, 2792; Fisher, G. L.; Boyd, R. W.; Gehr, R. J.; Jenekhe, S. A.; Osaheni, J. A.; Sipe, J. E.; Weller-Brophy, L. A. *Phys. Rev. Lett.* **1995,** *74*, 1871; Boyd, R. W.; Gehr, R. J.; Fisher, G. L.; Sipe, J. E. *Pure Appl. Opt.* **1996,** *5*, 505.
19. Zhang, P.; *et al. Phys. Rev. B*, submitted.
20. Kneipp, K.; *et al. Phys. Rev. Lett.* **1997,** *78*, 1667; Nie, S.; Emory, S. R. *Science* **1997,** *275*, 1102.

Chapter 9

Third-Order Nonlinear Optical Properties of Selected Composites

George L. Fischer[1] and Robert W. Boyd[2]

[1]**Air Force Research Laboratory, Hanscom Air Force Base, MA 01731–2909**
[2]**Institute of Optics, University of Rochester, Rochester, NY 14627**

We describe several experimental studies of the optical properties of composite materials. The enhanced nonlinearities of each system support local field enhancement theories developed for layered and interdispersed composites. This enhancement results from differences in the linear refractive indices between the two constituents. After demonstrating the validity of the technique, composites are next used to eliminate nonlinearities and finally in an optical limiter. The size scale in the composites ranges from Ångstrom in the interdispersed composites, to nanometer in the layered and colloidal composites, to sub millimeter in the nonlinear Christiansen filter.

In electronics nonlinearities play a large role in the function of various components such as transistors and diodes. In harmonic generation, rectification, heterodyning and many other electronic processes, nonlinearity is the key characteristic of the process. Transferring these technologies to their optical counterparts has proven to be very difficult, however, owing to the small nonlinear optical coefficients of materials. For linear optical materials $P = \chi^{(1)}E$, where P is the electric polarization or dipole moment per unit volume, E is the electric field strength, and $\chi^{(1)}$ is the polarizability or linear susceptibility. $\chi^{(1)}$ is related to the index of refraction n and to the dielectric constant $\varepsilon^{(1)}$ by $n^2 = \varepsilon^{(1)} = 1 + 4\pi\chi^{(1)}$. For most optical materials P is very well described by these simple equations. When more terms are needed P is often expanded in a power series

$$P = \chi^{(1)}E + \chi^{(2)}E^2 + \chi^{(3)}E^3 + \ldots \tag{1}$$

For optical computers, and many other complex devices where the output from one component is the input to the next, it is necessary for the signals to remain at the same wavelength. The third-order contribution to the polarization $\chi^{(3)}E^3$ is the first that has a contribution at the starting wavelength. This can be seen by assuming a monochromatic applied field given by $\tilde{E}(t) = \mathcal{E}\cos\omega t$. Since $\cos^3\omega t = \frac{1}{4}\cos 3\omega t + \frac{3}{4}\cos\omega t$, the nonlinear polarization can be expressed as

$$\tilde{P}^{(3)}\omega t = \chi^{(3)}\tilde{E}(t)^3 = \tfrac{1}{4}\chi^{(3)}\mathcal{E}^3\cos 3\omega t + \tfrac{3}{4}\chi^{(3)}\mathcal{E}^3\cos\omega t. \quad (2)$$

Here the last term describes the nonlinear contribution to the polarization at the starting frequency, the intensity-dependent refractive index. Nonlinear optical switching devices for use in photonics and real-time coherent optical signal processors are examples of applications of nonlinear optics that would benefit from the development of fast, low-loss materials with large values of the third-order nonlinear susceptibility $\chi^{(3)}$ which is directly related to this last term (*1-3*).

The most common approach to the development of new nonlinear optical materials involves the synthesis of materials in which the constituent molecules possess an inherently large nonlinear response (*4-9*). An alternative approach, on which this chapter concentrates, entails combining components with known properties into a composite material. In forming a composite one hopes to create a material possessing a desirable feature, or set of features not found in any one starting component. Composites can be designed that display some average of the characteristics of the constituents, as in a Faraday rotator rod which combines the polarization rotation of a paramagnetic ion in a magnetic field with the stability and good optical qualities of glass. More interesting is the composite which possesses enhanced or even totally new properties. In this chapter four examples of this second type of composite material are described.

Although the use of composites in nonlinear optics is quite new, composites have been selected to solve problems in classical optics for hundreds of years. Three such composite devices, chosen from three centuries are the achromatic lens, the Christiansen filter, and the gradient index lens. The last experimental section of this work is devoted to the nonlinear version of the Christiansen filter. The four composite geometries included in this chapter are depicted in Figure 1. A composite is made from two or more starting materials that remain distinct in the final product on some distance scale. Three such scales are used to describe the component size of the composites presented here. In the layered composites of Figure 1a, the two materials are intermixed on a distance scale of the order of 50 nm, which is much larger than an atomic dimension but much smaller than the wavelength of light used in the experiment. Consequently, the structural properties of each constituent material are essentially the same as those of a bulk sample of that material, but the propagation of light through the composite can be described by effective values of the linear and nonlinear optical susceptibilities that are obtained by performing suitable volume averages. The Maxwell Garnett composite (Figure 1b) contains gold particles, whose diameters are of the same order as the layer thicknesses of the layered composites. The average pore diameter in the Bruggeman composites (Figure 1c) is 40 Å. The average particle size of the nonlinear Christiansen filter (Figure 1d) lies at the other end of the distance scale. In this optical power limiter the glass particles are

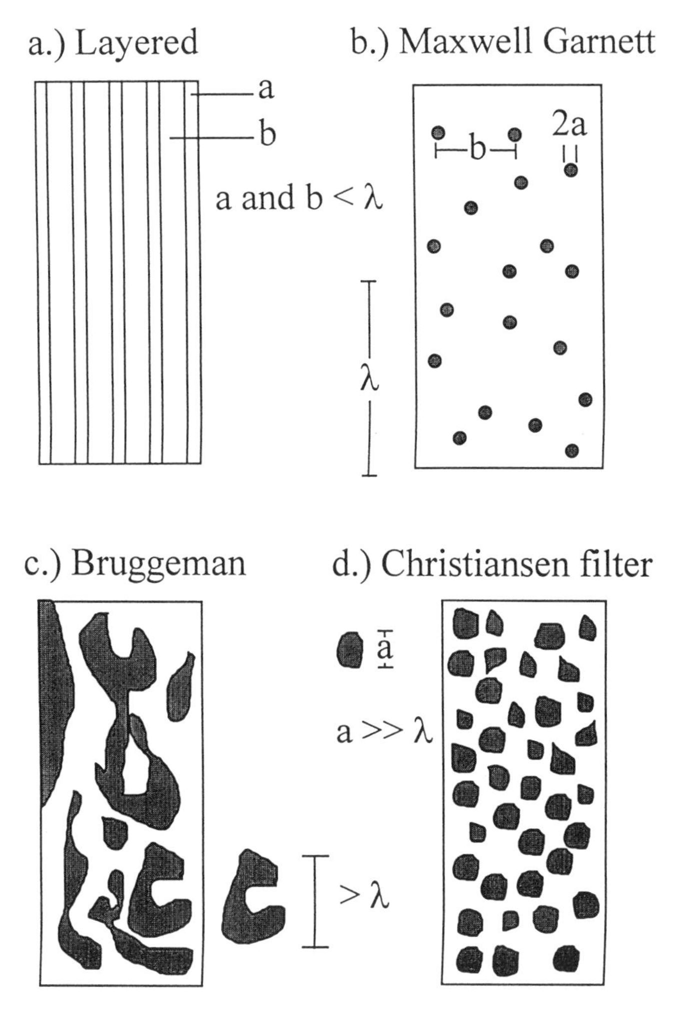

Figure 1. Composite geometries used.

much larger than a wavelength of light, thus ensuring that scattering can take place at the many interfaces.

The first theoretical descriptions of the linear optical properties of composite materials are also very much older than their nonlinear counterparts. In 1892 Lord Rayleigh considered a regular array of first cylindrical then spherical "obstacles" in a material and calculated the conductivity, fluid flow, and refractive index of the composite (*10*). In a paper by Maxwell Garnett, published in 1904, a very successful effective medium theory assuming spherical inclusions in a host was developed (*11,12*). This theory was used to accurately predict the color of precious metal doped glass and certain other samples of stained glass. Two weeks after the first of Maxwell Garnett's papers Wiener's derivation for the effective index of refraction of a layered composite was published (*13*). In 1935 Bruggeman published his predictions on interdispersed composites (*14*). The first nonlinear optical version of effective medium theory appeared in 1985 (*15*). It was modeled after the linear theory of Maxwell Garnett and determined the third-order susceptibility of metal colloids. Several experimental and theoretical advancements followed with this same type of composite (*16-18*), culminating in the full tensor description of Sipe and Boyd (*19*). This formulation also allows for nonlinearities in either or both components. Boyd and Sipe have also derived nonlinear effective medium theories for various nonlinear optical processes in layered composites (*20*). In the theories for both the Maxwell Garnett and the layered composite, enhancement of $\chi^{(3)}$ is predicted under certain conditions. To see how the formation of a composite can lead to an enhanced nonlinearity, the local field effect in homogeneous materials is briefly visited.

Local Field Enhancement

Three of the composites described below display enhanced $\chi^{(3)}$ due to local-field effects. The local electric field E_{loc} is the field experienced within a small region of a sample. It depends on the polarization of the surrounding material as well as on the applied field. E_{loc} can be related to the applied field through the local-field enhancement factor. For homogeneous materials, this enhancement factor is always greater than unity, but it is prescribed by the dielectric constant of the material. The local field enhancement factor experienced by one component in a given two-component composite structure is also fixed, however in this case the enhancement factor is prescribed not by the dielectric constant of the component, but by the ratio of dielectric constants of both materials and by their fill fractions. These two extra degrees of freedom were first capitalized on by forming a composite of a nonlinear optical material with a relatively low dielectric constant and a material having a relatively higher value of the dielectric constant (*21*). In such a system the two constituent materials can be optically lossless, and the response time of the composite is essentially the same as that of the nonlinear constituent. Since this method can be used for a broad class of composites, information learned from one composite is useful in the construction of many other composites as well. This research is also totally compatible with the molecular synthesis efforts to increase $\chi^{(3)}$ taking place around the world; the nonlinearities of any new material can be enhanced by combining it with a material that has a high index of refraction.

The essence of this theory is that the electric field amplitude of an incident laser beam becomes nonuniformly distributed between the two constituents of the composite, and under suitable conditions the electric field strength within the more nonlinear constituent will exceed the spatially averaged field strength. When this is the case, the effective third-order susceptibility $\chi_{eff}^{(3)}$ of the composite can exceed that of either of its constituents. The importance of local fields will first be demonstrated for the case of layered composites.

Layered Composites

In this section, the successful effort to demonstrate enhanced nonlinear optical response of layered composite materials is presented. Enhancement takes place when the dielectric constant of the more nonlinear material is lower than that of the less nonlinear material. The main difference between this analysis and those for other geometries is that enhancement of layered composites is very polarization dependent. As long as each layer is much thinner than a wavelength, the enhancement does not depend on the spacing of the layers. The uniformity of the thickness of each layer is also unimportant, as long as the thickness doesn't vary appreciably over lengths of the order of a wavelength. The two components of the layered composite were carefully chosen not only for their linear and nonlinear optical properties, but also for their chemical compatibility and for simplicity in the construction of the composite. Construction of the samples involved spin coating five layer pairs of the two components. The first component was a highly nonlinear organic polymer, poly(p-phenylene benzobisthiazole) PBZT solubilized in a complex coordinated solution. The second component, titanium dioxide, served as the high linear refractive index material. It was deposited from a sol-gel precursor. A modified form of the z-scan technique (*22*) was developed (*23*) for many of the experimental measurements described in this chapter. The new formulation of this well established technique was employed to measure the relative $\chi^{(3)}$ values of the samples using a Nd:YAG laser (where YAG denotes yttrium aluminum garnet).

Theoretical Response. The nonlinear susceptibility of a layered composite is given by

$$\chi_{eff}^{(3)}(\omega = \omega + -\omega) = \left|\frac{\varepsilon_{eff}(\omega)}{\varepsilon_a(\omega)}\right|^2 \left(\frac{\varepsilon_{eff}(\omega)}{\varepsilon_a(\omega)}\right)^2 f_a \chi_a^{(3)} + \left|\frac{\varepsilon_{eff}(\omega)}{\varepsilon_b(\omega)}\right|^2 \left(\frac{\varepsilon_{eff}(\omega)}{\varepsilon_b(\omega)}\right)^2 f_b \chi_b^{(3)}. \quad (3)$$

Note that the third-order effective susceptibility shows a fourth-order dependence on the local-field enhancement factor. The enhancement predicted by equation 3 can be appreciable. The predictions of this equation are presented in Figure 2 with the assumption that only component *a* possesses a third-order nonlinear optical response. Special cases that are shown include the curve labeled $\varepsilon_b/\varepsilon_a = 4$, which corresponds to materials that differ by a factor of 2 in linear refractive index, and the curve labeled $\varepsilon_b/\varepsilon_a = 1.77$, which corresponds to the materials used in our experimental investigation. In the experimental case the maximum predicted enhancement is 35%.

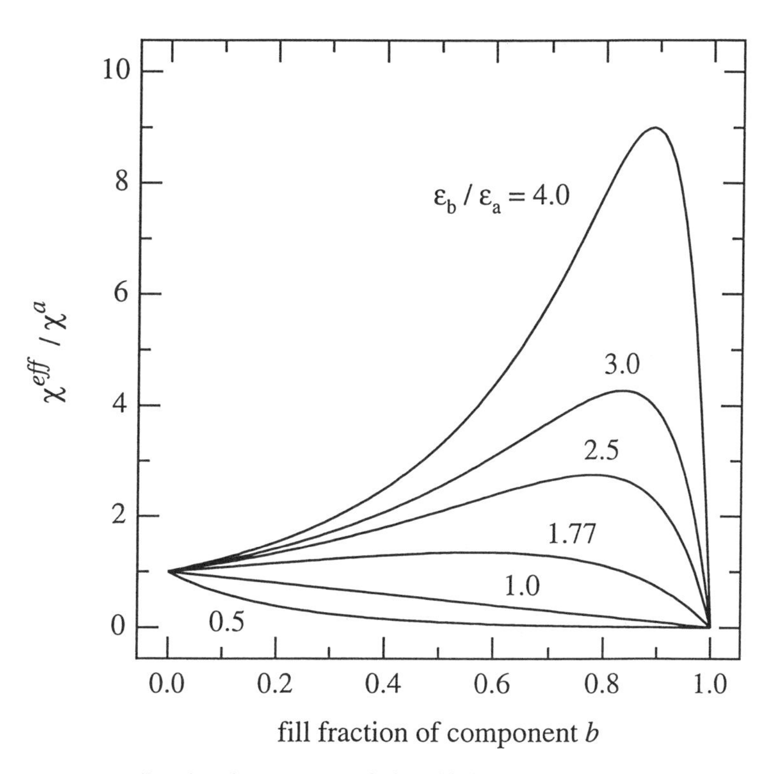

Figure 2. Predicted enhancement of the third-order nonlinear susceptibility plotted as a function of the volume fill fraction of linear component *b* for several values of the ratio of the optical frequency dielectric constants of the two components. (Reproduced with permission from ref. 20. *J. Opt. Soc. Am. B* **1994**, 11, 297-303, Boyd R. W.; Sipe, J. E. Copyright 1994 The Optical Society of America. All rights reserved.)

Note that the maximum enhancement in the top curve is nearly an order of magnitude. This maximum occurs at a fill fraction f_a of approximately 0.1. The size and location of this maximum illustrate two attractive features of local-field enhancement; using only ten percent of the nonlinear material the nonlinearity could be increased tenfold. Nonlinear materials often have higher absorption coefficients than TiO_2, so decreasing their thicknesses would lead to lower attenuation. By capitalizing on local-field enhancement, a material with higher nonlinearity and lower absorption can be constructed.

As can be seen in Figure 2 the level of predicted enhancement depends on the relative fill fractions of the two materials in the composite. This is true for all ratios $\varepsilon_b/\varepsilon_a$, but is more pronounced for large ratios. For the curve labeled $\varepsilon_b/\varepsilon_a = 1.77$, which corresponds to the TiO_2 and PBZT used in our experiment, the value of $\chi^{(3)}_{eff}$ appears nearly constant for $f_b < 0.7$. Even this shallow curve has a maximum, however, and to give ourselves every possible advantage in measuring the nonlinearities we calculated the fill fraction ratio that would give us the largest value of $\chi^{(3)}_{eff}$. Starting with equation 4 we assume that the TiO_2 has only a linear response. Setting the derivative of the effective nonlinear susceptibility with respect to f_a to zero to find the maximum yields

$$f_{a\,\mathrm{MAX}} = \frac{\varepsilon_a}{3(\varepsilon_b - \varepsilon_a)}, \tag{4}$$

where $f_{a\,\mathrm{MAX}}$ denotes the fill fraction of nonlinear material a that gives the maximum enhancement of $\chi^{(3)}_{eff}$. Equation 4 is general and can be used to find the ideal fill fraction for layered composites of the type discussed in this chapter with any ratio $\varepsilon_b/\varepsilon_a > 1$, providing that material a is the one nonlinear component. If $\varepsilon_b/\varepsilon_a < 1$ there is no enhancement possible and the ideal fill fraction of material a is unity.

Layered Composite Experiment. For these components $f_{a\,\mathrm{MAX}} = 0.43$. By varying the details of the spin-coating process, a series of TiO_2 films, each with a different thickness was made. In this manner a procedure was developed which would repeatably yield densified TiO_2 layers with the desired thickness of 50 nm. When combined with 40 nm PBZT layers the composite $f_a = 0.44$. These layer thickness are much smaller than a wavelength, but large enough to display properties of the bulk components. They are also thick enough that the nonlinearity of the composite can be accurately measured. Thicknesses were also measured at various stages in the fabrication process.

The full enhancement occurs when the electric field vector is perpendicular to the layers. Unfortunately waveguiding could not be achieved for our samples. Instead the samples were angled so that a component of the vector would be perpendicular to the layers. As a control the samples were also angled in the plane of the electric field vector where no enhancement is predicted. The predictions of local field enhancement for this experiment are shown in Figure 3 along with the experimental data. The derivation of the curves shown in Figure 3 is lengthy (*24*).

Since the local field enhancement factor is raised to the fourth power, large increases in $\chi^{(3)}$ should be possible with suitable starting materials. We have used local field effects to raise the nonlinearity of the layered composites that we constructed by 35 %. The predictions of our effective medium theory are in excellent agreement with our experimental results.

Bruggeman Composites

A second type of composite which was studied approximated the Bruggeman geometry; here it is assumed that the constituents are interdispersed. Several expressions have been derived for the linear and nonlinear properties of such composites (*14, 25, 26*). The linear dielectric constant can be shown to be

$$0 = f_a \frac{\varepsilon_a - \varepsilon_{eff}}{\varepsilon_a + 2\varepsilon_{eff}} + f_b \frac{\varepsilon_b - \varepsilon_{eff}}{\varepsilon_b + 2\varepsilon_{eff}}. \tag{5}$$

By assuming that the electric field within each grain is uniform and that the nonlinearity only perturbs the refractive index, the expression for the effective dielectric constant may be expanded in a Taylor series. The expression for the effective third-order susceptibility is then

$$\chi_{eff}^{(3)} = \frac{1}{f_a}\left|\frac{\partial\varepsilon_{eff}(\omega)}{\partial\varepsilon_a(\omega)}\right|\left(\frac{\partial\varepsilon_{eff}(\omega)}{\partial\varepsilon_a(\omega)}\right)\chi_a^{(3)} + \frac{1}{f_b}\left|\frac{\partial\varepsilon_{eff}(\omega)}{\partial\varepsilon_b(\omega)}\right|\left(\frac{\partial\varepsilon_{eff}(\omega)}{\partial\varepsilon_b(\omega)}\right)\chi_b^{(3)}, \tag{6}$$

where for a Bruggeman composite the partial derivative with respect to ε_a is given by

$$\frac{\partial\varepsilon_{eff}}{\partial\varepsilon_a} = \frac{3f_a\varepsilon_{eff}}{3f_a\varepsilon_a + f_a\left(2\varepsilon_b + \varepsilon_{eff}\right)\left(\varepsilon_a + 2\varepsilon_{eff}\right)^2 / \left(\varepsilon_b + 2\varepsilon_{eff}\right)^2}. \tag{7}$$

Just as was the case with the layered composites there is enhancement when $n_a < n_b$, but in this geometry the largest envisioned enhancement is much smaller, only approaching 1.5 for an index ratio of 2.0. The experiment is still very important for confirming Bruggeman theory in particular and local-field enhancement theory in general.

Linear Bruggeman Experiment. In this experiment composites were formed with Vycor glass from Corning Incorporated serving as the first component and one of seven fluids serving as the second (*27*). For the linear study fluids were chosen for their range of refractive indices. In the nonlinear study only the fluids with the four largest nonlinearities were tested. To prepare the Vycor, samples were soaked in a warm concentrated hydrogen peroxide solution and then slowly heated to 300° C in

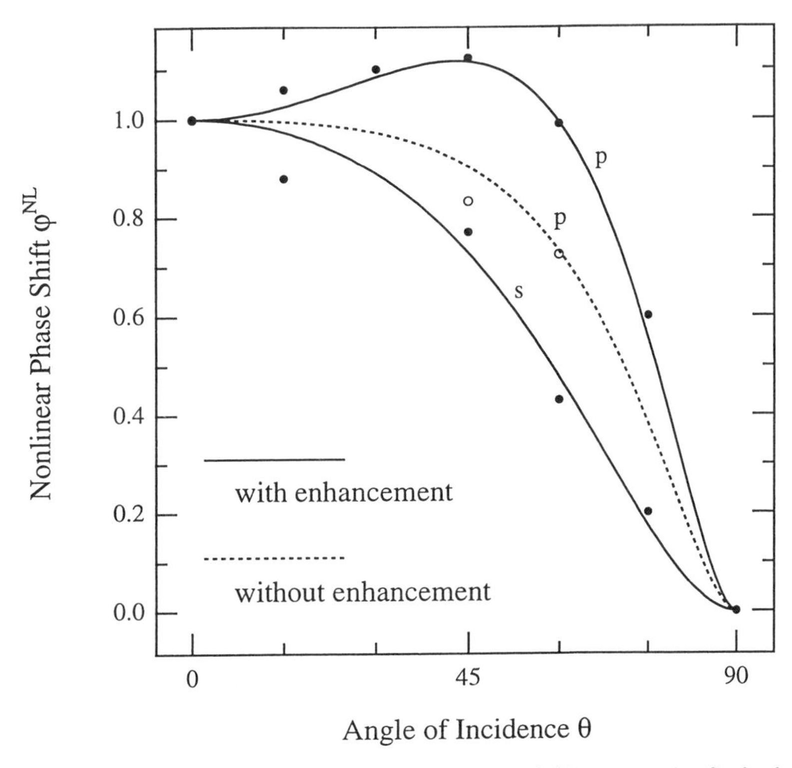

Figure 3. Measured nonlinear response of the PBZT/TiO_2 composite for both *s*- and *p*- polarized light (solid dots) and of a 40 nm film of pure PBZT for *p*-polarized light (open dots). The solid curves show the theoretical predictions and dashed curve shows the expected behavior if there is no local field enhancement of $\chi^{(3)}_{eff}$. (Reproduced with permission from ref. 24. *Phys. Rev. A* **1996**, 53, 2792-2798, Gehr, R. J.; Fischer, G. L.; Boyd, R. W.; Sipe, J. E. Copyright 1996 The American Physical Society. All rights reserved.)

an oxygen environment. After slowly ramping down to room temperature each sample was immediately submerged in its corresponding fluid. The linear refractive index of each composite was measured in a Mach-Zehnder interferometer. The results are shown in Table I. Here the column labeled index (fluid) is the measured index of refraction of the pure fluid. The measured composite index column is the measured index of refraction of the composite. In determining the index of refraction of the Vycor it was fortunate that carbon tetrachloride (CCl_4) was such a good index match to it. Very good agreement was found between theory and experiment, supporting the linear Bruggeman theory.

Table I. Linear Refractive Indices of Bruggeman Constituents and Composites

Fluid	Index (fluid) n_0	Measured Composite Index	Theoretical Composite Index
Air	1.00	1.34	1.327
Methanol	1.32	1.40	1.42
Water	1.33	1.42	1.423
CCl_4	1.46	1.46	1.460
Carbon disulfide (CS_2)	1.63	1.50	1.507
Diiodomethane (CH_2I_2)	1.72	1.51	1.531
Index matching fluid	1.78	1.54	1.547

SOURCE: Reproduced with permission from reference 27. Copyright 1989 Optical Society of America.

Nonlinear Bruggeman Experiment. The Bruggeman model was put to a more difficult test in the nonlinear portion of the Vycor study. Here the low predicted enhancement was partially offset by the ease of sample construction, the simplicity of the experiment, and the straightforward data analysis. Nonlinearities were again measured by the z-scan technique, but there was no need to angle the isotropic composites. Only two measurements were needed to determine the relative enhancement of each composite; a z-scan of the composite, and a z-scan of an equal thickness of the pure liquid. The CCl_4/Vycor composite was also important here; by assuming that there was no local field enhancement (since the indices of refraction of these two components were almost identical), the nonlinearity of Vycor relative to CCl_4 was determined. Having also measured the nonlinearity of CCl_4 relative to each pure liquid, all relative nonlinearities were found. These values were then substituted directly into equation 6. The results of these measurements, along with theoretical predictions based on Bruggeman theory and on simple weighted averaging are shown in Table II.

As can be seen in the last two columns of Table II, the Bruggeman model is more accurate than the weighted average method in predicting the measured nonlinearities of the three composites in question. This section and the one on layered composites have shown that the local field factor can be greater than unity (linear material has a larger dielectric constant) or less than unity (linear material has a smaller dielectric constant). If the local field factor is less than unity the $\chi^{(3)}$ value of the composite is actually lower than that of the starting material, which might be helpful in certain other device applications. Lowering, or even eliminating selected nonlinearities is the subject of the next section.

Table II. Nonlinearities of Bruggeman Composites

Liquid	Experiment $\chi^{(3)}_{glass} / \chi^{(3)}_{liquid}$	Experiment $\chi^{(3)}_{eff} / \chi^{(3)}_{liquid}$	Theory $\chi^{(3)}_{eff} / \chi^{(3)}_{liquid}$	$\sum_i f_i \chi^{(3)}_i$
Methanol	0.62	0.80	0.750	0.726
CCl_4	0.32	0.51	0.51[a]	0.510
CS_2	0.00	0.25	0.231	0.280
CH_2I_2	0.03	0.20	0.222	0.302

[a]Theory and experiment agree by assumption due to data analysis procedure.
SOURCE: Reproduced with permission from reference 27. Copyright 1989 Optical Society of America.

Cancellation of Nonlinear Absorption

Although materials with nonlinearities higher than those of currently existing materials are critical in many planned optical devices, there are applications in the larger field of optics that are hampered by even the low nonlinearites of existing linear-optical materials. In fact one of the beginnings of the field of nonlinear optics was the attempt to lower, or even remove nonlinearities in glass (*28-32*). The destruction of laser rods and optics due to self-focusing motivated this early work. Prevention of catastrophic self-focusing is still an important constraint in the design of high power lasers. This is one of the reasons that the optical components in fusion lasers are so large; the beam must have a large cross section to keep the intensity below that necessary to focus the beam within an optical element. In this and similar situations it would be desirable to remove nonlinearites from materials. A device that could eliminate unwanted nonlinear effects while preserving, or even enhancing the effects needed for its operation could find wide application. One possible solution to this problem is considered in this section, where a composite system was used to lower nonlinear absorption. The large positive nonlinear absorption of a laser dye was canceled by the large negative nonlinear absorption of gold colloid. As the concentration of one of the components is increased, the nonlinear absorption shifts from positive to negative (33). The first component used was itself a composite (in the Maxwell Garnett geometry). It was composed of gold particles with an average diameter of 10 nm in water (*34-37*). The second component used was 1,1',3,3,3',3'-hexamethylindotricarbocyanine iodide (HITC-I) in a methanol solution, purchased from Kodak. This is a laser dye used in the near IR (*38*). Its absorption maximum is at 750 nm.

Nonlinear Absorption. The gold colloid is a saturable absorber. In a saturable absorber low intensity light is absorbed to a greater extent than high intensity light. If the laser pulse length is shorter than or of the order of the excited-state lifetime τ this bleaching or saturation of the absorption can be quite pronounced. In the experiments with concentrated gold colloid suspensions the fractional transmission increased by a factor of five as the incident laser intensity was increased. A common way to describe the process of saturable absorption makes use of the intensity dependent absorption coefficient α. Using this approach α is often represented by

$$\alpha = \frac{\alpha_0}{1 + I/I_{sat}}, \tag{8}$$

where α_0 is the linear or low-intensity absorption coefficient, and I_{sat} is known as the saturation intensity. As the incident intensity is increased the factor I/I_{sat} becomes significant compared to unity and α becomes smaller.

The HITC-I dye that was used as the second component exhibits reverse saturable absorption. Instead of bleaching, some dyes actually became more absorbing at high intensities (*39*). The reverse saturable absorption of the dye used in this experiment begins to saturate at fluences greater than 0.5 J/cm^2 (*40*).

Experimental Cancellation. The experimental arrangement for the experiments in this section was essentially the same as the z-scan set-up used in the experiments above. The only differences were that the aperture was removed and the detector was moved closer to the focal point so that all of the diverging beam impinged on the active area of the detector. The sample was moved through the focal region of the laser beam. The concentration of each of the two components and the intensity of the incident light was adjusted so that far from the focal point there was no detectable nonlinear absorption, but that there was a significant amount near the focal point where the intensity was the greatest. The results of one of the sets of the "open aperture" z-scans is shown in Figure 4. Curve (a) is the z-scan of a sample containing 13 μM of HITC-I dye and no colloid. It shows a large decrease in relative transmitted energy near the focal point. As the intensity was increased, the absorption also increased, this trend corresponds to positive nonlinear absorption. The middle curves are scans of samples that contain the same 13 μM of HITC-I dye, but with concentrations of the gold particles increasing up to a volume fraction of 1.9 x 10^{-6}. The scan of this sample (curve i) has a large transmission peak. In this curve the absorption decreases as the intensity increases, indicative of a negative nonlinear absorption. Curve (f) shows a relatively small nonlinear absorption; the elimination of which was the goal of this experiment. The curves are not centered since the low intensity data from the lens side of the focal point is important for determining the baseline and the detector is located close to the far side of the focal point of the lens to ensure that all the light is collected. The concentrations of water; 73 volume percent, methanol; 27 volume percent, and the HITC-I dye; 13 μM, were held fixed for all of these scans, the variable being the concentration of gold colloid. The lens used in the Z-scans shown in Figure 4 has a focal length of 70 cm. This long focal length allowed the use of more energy for ease in detection, while keeping the intensity below the damage threshold at the focal point. The sample thickness is 1 mm. All of the scans shown in Figure 4 were taken under identical conditions. The experiment was repeated several times with different variation schemes of the concentrations. Each set of data showed the same cancellation of nonlinear absorption. To avoid cluttering the figure, error bars are not shown, but they average about half of one

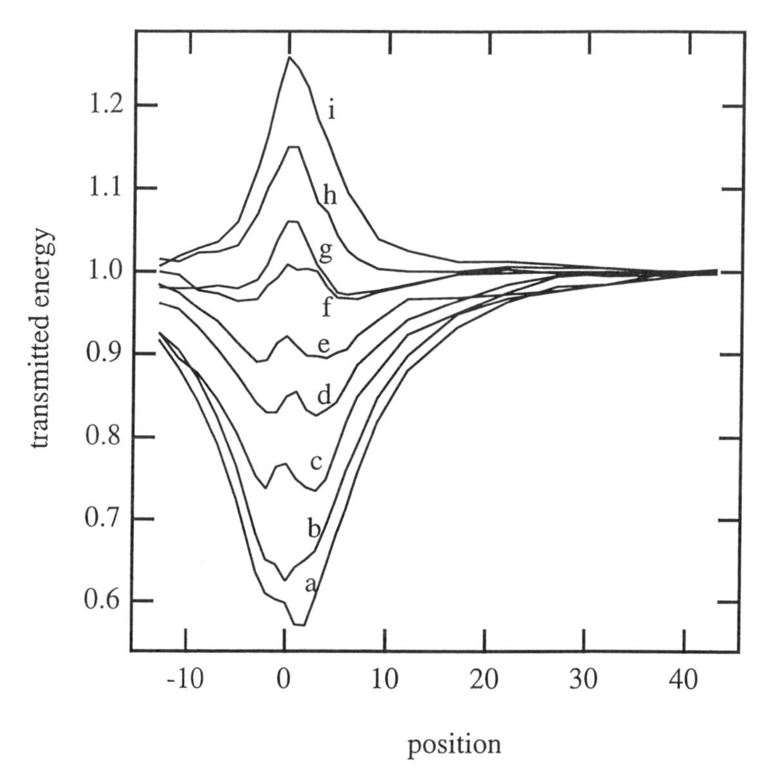

Figure 4. Open aperture Z-scans of gold particles in HITC-I solution showing positive nonlinear absorption (curves a-e) through negative nonlinear absorption (curves g-i). Curve f shows near total cancellation of nonlinear absorption. Position is in cm and relative to the focal point. Transmitted energy is normalized to the low intensity transmission for each sample. (Reproduced with permission from ref. 33. *J. Opt. Soc. Am. B* **1997**, 14, 1625-1631, Smith, D. D.; Fischer, G. L.; Boyd, R. W.; Gregory, D. A. Copyright 1997 The Optical Society of America. All rights reserved.)

percent of the baseline to peak height for the top curve. The small peaks in the lower curves near the focal region of the lens indicate that the intensity dependence of nonlinear absorption of the two components are not the same. Had the dependencies been equal and of opposite sign, perfect cancellation could have been observed.

The Nonlinear Christiansen Filter

The classical Christiansen filter was important as a bandpass filter before the introduction of thin-film dielectric interference filters. This device is different from the composites presented above in that it is an inherently composite device. In the other composites the nonlinearities which were already present in the starting materials are enhanced or reduced by adding a second component. Those composites have improved properties, but the starting materials are also impressive by themselves. On the other hand, one component in the nonlinear Christiansen filter will not filter or limit visible light at all without the presence of the second component. The classical filter is based on light scattering in a heterogeneous medium. Two materials, typically crushed glass and a liquid solution, are combined. Due to dispersions differences, the two materials can be index matched at only one wavelength. At this wavelength light is transmitted. In this experiment the liquid concentrations of acetone and carbon disulfide were adjusted until the solution and the glass were index matched at the laser wavelength of 532 nm. Even at this one wavelength the total indices of refraction are not matched since the nonlinear contribution to the total index of refraction of each component is different. The total index of refraction $n \quad n_0 \quad n_2 I$ is made up of a constant or linear component n_0 and an intensity dependent or nonlinear component n_2. As the intensity was increased the nonlinear index of refraction became larger causing light to scatter at the many component interfaces. Thus the nonlinear Christiansen filter functions as an optical power limiter. Optical limiters with nanosecond response times have been proposed (*41*) and constructed (*42,43*) which utilize thermally induced nonlinear refraction coupled with spatial filtering, but this power limiter has a response time estimated to be 2 ps, the molecular orientational response time of the carbon disulfide.

Filter Theory The attenuation coefficient describing these scattering losses can be calculated in terms of the expression derived by Raman (*44*) to describe a classical Christiansen filter. This expression, modified to allow for arbitrary fill fraction f of the glass particles has the form

$$\alpha = f(1-f)\frac{4\pi^2 d}{\lambda_0^2}\left(n_l - n_g\right)^2 \tag{9}$$

where d is the characteristic size of the glass particles, λ_0 is the vacuum wavelength of the incident radiation, and n_l and n_g denote the refractive indices (including their nonlinear contributions) of the liquid and glass components.

Experimental Studies of the Nonlinear Christiansen Filter. The experimental results (45) are shown in Figure 5. The output energy initially increases linearly with

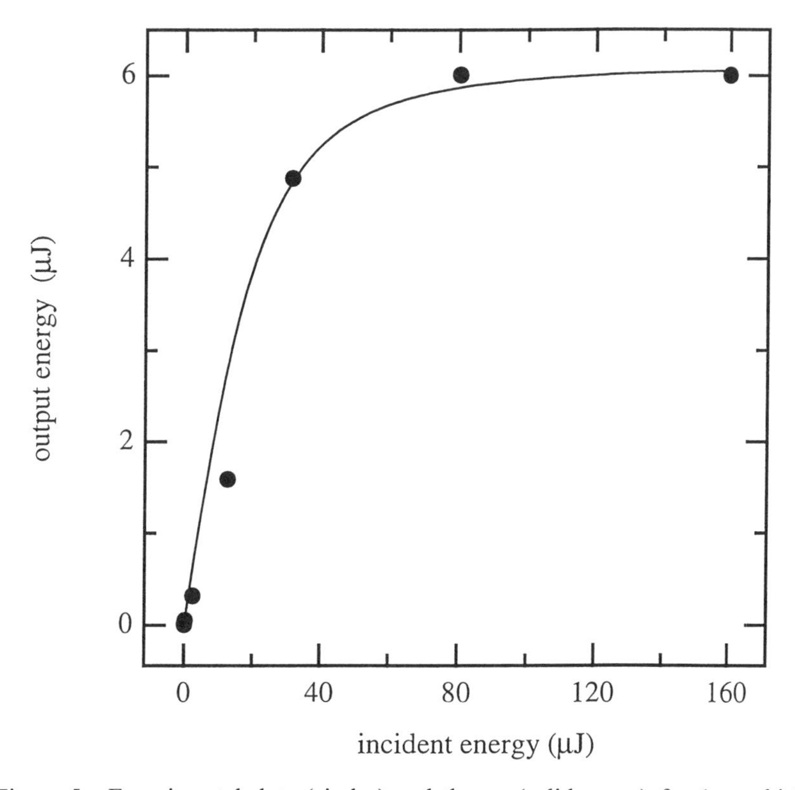

Figure 5. Experimental data (circles) and theory (solid curve) for 1-cm-thick Christiansen filter consisting of 100 μm glass particles in a solution of carbon disulfide and acetone. A frequency-doubled 30 ps Nd:YAG laser was collimated to a diameter of 118 μm at the filter. (Reproduced with permission from ref. 45. *Opt. Lett.* **1996**, 21, 1643-1645, Fischer, G. L.; Boyd, R. W.; Moore, T. R.; Sipe, J. E. Copyright 1996 The Optical Society of America. All rights reserved.)

incident laser energy and eventually saturates at a value of approximately 6 μJ. The solid line is the theoretical fit to the data, calculated through use of equation 9. For simplicity, the transverse variation of the laser intensity and diffractive spreading of the laser beam were ignored so that the theory assumes plane wave propagation. Also, the time evolution of the laser intensity was ignored. Thus, the pulse energy was related to the assumed constant laser intensity through $Q = IAT$ where $A = (\pi/4)d^2$ with $d = 100\mu m$ and $T = 30\ ps$. The calculated transmitted pulse energy was scaled down by a factor of four to account for the measured 25% low-intensity transmission of the cell. The agreement between theory and experiment is seen to be remarkably good.

Acknowledgments
The authors thank collaborators Russell J. Gehr, Thomas R. Moore, J. E. Sipe, David D. Smith, and Samson S. Jenekhe for their many key contributions. Sherry Hultman, Archie Boswell, and Paul S. Danielson, of Corning Inc. provided the Vycor samples as well as much technical assistance.

Literature Cited

(*1*) Service, R. F. *Science* **1995**, *267*, 1918.
(*2*) Smith, P. W. *Phil. Trans. R. Soc. London A* **1984**, *313*, 349.
(*3*) *Nonlinear Photonics*; Gibbs, H.M.; Khitrova, G.; Peyghambarian, N., Eds. Springer Verlag: Berlin, Germany, 1990; Vol. 30.
(*4*) Stegeman, G. I. In *Contemporary Nonlinear Optics*; Agrawal, G. P.; Boyd, R. W., Eds.; Quantum Electronics- Principles and Applications; Academic Press: San Diego, CA, 1992; pp 1-40.
(*5*) Dodenberger, D. C.; Heflin, J. R.; Garito, A. F. *Nature* **1992**, 359, 309.
(*6*) Halvorson, C.; Hagler, T. W.; Moses, D.; Cao, Y.; Heeger, A. J. *Chem. Phys. Lett.* **1992**, 200, 364.
(*7*) Ghoshal, S. K.; Chopra, P.; Singh, B. P.; Swiatiewicz, J.; Prasad, P. N. *J. Chem Phys.* **1989**, 90, 5078.
(*8*) Broussoux, D.; Chastaing, E.; Esselin, S.; Le Barny, P.; Robin, P.; Bourbin, Y.; Pocholle, J. P.; Raffy, J. *Revue Technique Thomson-CSF* **1989**, 20, 151.
(*9*) Garito, A.; Shi, R. F.; Wu, M. *Physics Today* **1994**, May, 51.
(*10*) Lord Rayleigh, *Philos. Mag.* **1892**, 34, 481.
(*11*) Maxwell Garnett, J. C. *Philos. Trans. R. Soc. London* **1904**, 203, 385.
(*12*) Maxwell Garnett, J. C. *Philos. Trans. R. Soc. London* **1906**, 205, 237.
(*13*) Wiener, O. *Physikalische Zeitschrift* **1904**, 5, 332.
(*14*) Bruggeman, D. A. G. *Annalen der Physik* **1935**, 24, 636.
(*15*) Ricard, D.; Roussignol, P.; Flytzanis, C. *Opt. Lett.* **1985**, 10, 511.
(*16*) Agarwal G. S.; Dutta Gupta, S. *Phys. Rev. A* **1988**, 38, 5678.
(*17*) Neeves A. E.; Birnboim M. H., *J. Opt. Soc. Am. B* **1989**, 6, 787.
(*18*) Haus, J. W.; Inguva, R.; Bowden, C. M. *Phys. Rev. A* **1989**, 40, 5729.
(*19*) Sipe, J. E.; Boyd, R. W. *Phys Rev. A* **1992**, 46, 1614.
(*20*) Boyd R. W.; Sipe, J. E. *J. Opt. Soc. Am. B* **1994**, 11, 297.

(*21*) Fischer, G. L.; Boyd, R. W.; Gehr, R. J.; Jenekhe, S. A.; Osaheni, J. A.; Sipe, J. E.; Weller-Brophy, L. A. *Phys. Rev. Lett.* **1995**, 74, 1871.
(*22*) Sheik-Bahae, M.; Said, A. A.; Van Stryland, E. W. *Opt. Lett.* **1989**, 14, 955.
(*23*) Bridges, R. E.; Fischer, G. L.; Boyd, R. W. *Opt. Lett.* **1995**, 20, 1821.
(*24*) Gehr, R. J.; Fischer, G. L.; Boyd, R. W.; Sipe, J. E. *Phys. Rev. A* **1996**, 53, 2792.
(*25*) Zeng, X. C.; Bergman, D. J.; Hui, P. M.; Stroud, D. *Phys. Rev B* **1988**, *38*, 10970.
(*26*) Yu, K. W.; Hui, P. M.; Stroud, D. *Phys. Rev B* **1993**, *47*, 14150.
(*27*) Gehr, R. J.; Fischer, G. L.; Boyd, R. W. *J. Opt. Soc. Am. B* **1997**, 14, in press.
(*28*) Hercher, M. *J. Opt. Soc. Am.* **1964**, 54, 563.
(*29*) Giuliano, C. R. *Appl. Phys. Lett.* **1964**, 5, 137.
(*30*) Chiao, R. Y.; Garmire, E.; Townes, C. H. *Phys. Rev. Lett.* **1964**, 13, 479.
(*31*) Zverev, G. M.; Mikhailova, T. N.; Pashkov, V. A.; Solov'eva, N. M. *JETP Lett.* **1967**, 5, 319.
(*32*) Borrelli, N. F.; Hall, D. W. In *Optical Properties of Glass*; Uhlman, D. R.; Kreidl, N. J., Eds.; American Ceramic Society: Westerville, OH, 1991.
(*33*) Smith, D. D.; Fischer, G. L.; Boyd, R. W.; Gregory, D. A. *J. Opt. Soc. Am. B* **1997**, 14, 1625.
(*34*) Hache, F.; Ricard, D.; Flytzanis, C. *J. Opt. Soc. Am. B* **1986**, 3, 1647.
(*35*) Hache, F.; Ricard, D.; Flytzanis, C.; Kreibig, U. *Appl. Phys. A* **1988**, **47**, 347.
(*36*) Puech, K.; Henari, F. Z.; Blau, W. J.; Duff, D.; Schmid, G. *Chem. Phys. Lett.* **1995**, 247, 13.
(*37*) Miyazoe, Y.; Maeda, M. *Opto-electronics* **1970**, 2, 227.
(*38*) Giuliano, C. R.; Hess, L. D. *IEEE J. Quantum Electron.* **1967**, 3, 358.
(*39*) Giordmaine, J. A.; Howe, J. A. *Phys. Rev. Lett.* **1963**, 11, 207.
(*40*) Swatton, S. N. R.; Welford, K. R.; Till, S. J.; Sambles, J. R. *Appl. Phys. Lett.* **1995**, 66, 1868.
(*41*) Hermann, J. A. *Opt. Acta* **1985**, 32, 541.
(*42*) Leite, R. C. C.; Porto, S. P. S.; Damen, T. C. *Appl. Phys. Lett.* **1967**, 10, 100.
(*43*) Justus, B. L.; Campillo, A. J.; Huston, A. L. *Opt. Lett.* **1994**, 19, 673.
(*44*) Raman, C.V. *Proc. Indian Acad. of Sciences A* **1949**, 29, 381.
(*45*) Fischer, G. L.; Boyd, R. W.; Moore, T. R.; Sipe, J. E. *Opt. Lett.* **1996**, 21, 1643.

Chapter 10

Size-Dependent Electron Dynamics of Gold Nanoparticles

Temer S. Ahmadi, Stephan L. Logunov, and Mostafa A. El-Sayed

School of Chemistry and Biochemistry and Laser Dynamics Laboratory, Georgia Institute of Technology, Atlanta, GA 30332–0400

The electron dynamics of four samples of colloidal gold nanoparticle with average particle diameter of 1.5, 2.5, 3.3, and 30 nm are studied by transient absorption spectroscopy after excitation with 400 fs laser pulses at 600 nm wavelength. Photoexcitation leads to the 'bleach' of the surface plasmon (intraband) and 5d→6sp interband transition. The temporal relaxation of the electron gas of metallic nanoparticles at the peak of plasmon band absorption follows a biexponential kinetics with a 2.5 ps and >50 ps components which are attributed to thermalization of the photoexcited non-fermi 'hot' electron gas through electron-phonon and phonon-phonon interactions, respectively. No direct observation of electron-electron relaxation was possible because this process occurs within the duration of our laser pulse. There is a size-dependent shift in the maximum of steady-state absorption and the transient minimum of the plasmon 'bleach' which is attributed to 'quantum size effect' in the smaller particles.

The existence of free carriers in metallic nanoparticles, in particular gold, silver, and copper, give rise to many interesting linear and non-linear optical properties which have been extensively studied previously (1-7). In metallic nanoparticles, due to their small size and high surface to volume ratio, new optical properties arise which are observed neither in molecules nor bulk metals (8). A specific example is the presence of a strong band in the visible region of the absorption spectrum of colloidal noble metal particles which is attributed to surface plasmon oscillation modes of conduction electrons coupled through their surfaces to the oscillating applied electromagnetic field (9, 10). Colloidal gold particles, with average particle sizes larger than ~ 2nm in diameter, have their plasmon absorption band around 520 nm which is responsible for their red color.

The plasmon-band of colloidal metallic nanoparticles is modeled with Mie theory by treating the particles as 'molecular-like' dipoles (11). All magnetic effects and excitation of electronic multipoles as well as the retardation effects are neglected. Since the sizes of the nanoparticles are much smaller than the wavelength of light and particles occupy a small fraction of the volume, with this approximation, only the dipole modes of the surface plasmon-oscillations are considered.

Investigations involving the interactions of free carriers between themselves and their environment is significant in solid-state physics, e.g. an understanding of the electron-phonon interactions is necessary in electrical and thermal transport in metals as well as the formation of superconducting state (12). Electron-electron interactions play an important role in disordered metallic systems (12). A study of these phenomena in colloidal nanoparticles provides additional information on their size dependence and will prove useful in our knowledge of ferrofluids, catalysis, photocatalysis, and powder metallurgy.

The possibility of generating transient non-equilibrium electron distributions using ultrashort laser pulses have been explored recently in thin metal films employing methods such as thermal assisted multiphoton photoemission (13), single photon photoemission (14), and thermomodulation reflection and transmission (12). The electron gas heat capacity is much smaller than the lattice heat capacity in metals; therefore, one can selectively excite the electron gas and then monitor both spectral and temporal electron thermalization dynamics by various spectroscopic methods (12). The small electronic heat capacity permits temperatures of thousands of degrees to be generated for times up to a few picoseconds following an ultrashort laser pulse excitation (12-14). Because of the high concentration of electrons in metals, it was assumed that thermal relaxation of the electron gas through electron-electron interactions would be instantaneous (12). Electronic temperature changes of ~ 20 K were observed by Sun et al. (12) in gold thin films using transmission and reflection spectroscopy with infrared (pump) and visible (probe) femtosecond pulses. They found evidence for the transient non-fermi electron distribution with an electron-electron scattering lifetime of ~ 500 fs and an electron-phonon relaxation time of 1 ps (12). However, the existence of non-equilibrium (non-Fermi) electron distributions with thermalization times as long as 600 fs have been shown in thin gold films at higher perturbation limits as well, i.e. electronic temperature changes of ~ 400 K (14, 15).

Femtosecond spectroscopy have been applied in the study of the dynamics of metallic nanoparticles recently. Roberti et.al. studied colloidal silver nanoparticle in an aqueous medium by exciting the plasmon band of the colloids at 390 nm (4). They suggested that photoexcited electrons may form solvated electrons in the solid-liquid interface which subsequently relax through channels such as electron-phonon and phonon-solvent interactions. The time constants for the two processes were found to be 2 and 40 ps, respectively (4). Also, a recent study of the thermalization of electrons in copper nanoparticles embedded in glass using femtosecond pump-probe spectroscopy was carried out by Bigot and coworkers (7). Contrary to findings for thin film studies, they found a slower electron-electron relaxation (~ 700 fs) and a slower rate of the electron gas cooling to lattice temperatures due to long lasting energy exchange between plasmon electrons and quasiparticles (7). A more recent study on the fast dynamics of tin nanoparticles considers the role of electron-surface scattering as a cooling channel for electron gas system. The time constant for e-e scattering was measured to be ~100 fs (2).

We present our study of sub-picosecond dynamics of colloidal gold nanoparticles of 1.9, 2.6, 3.2, and 30 nm in diameter. We observe that subpicosecond pulsed laser excitation leads to the 'bleach' of the surface plasmon band (for all particles except 1.9 nm sample) and induces a positive transient absorption at the wings of the 'bleached' plasmon. Both 'bleach' and positive absorptions were found to decay with lifetimes of ~2.5 ps and >50 ps at low laser intensities independent of average particle size. These observations are interpreted in terms of a two temperature model, TTM (16), assuming the formation of a transient non-fermi 'hot' electron distribution within the colloidal particles immediately following the laser excitation pulse. The excitation of electrons are either intraband (electrons promoted to higher energy states within the conduction band) or interband (d-electrons promoted to the conduction band). The equilibration of the 'hot' electrons occurs in two steps: due to electron-phonon scattering (2-3 ps) and due to phonon-phonon relaxation, > 50 ps, (1). In 1.9 nm sample, the cooling processes are the same although there is only interband transition possible since these particles lack a surface plasmon band due to 'quantum size effects'. There is a size dependent shift in the plasmon 'bleach' minimum with respect to the steady state absorption maximum which is also due to 'quantum size effect'. In addition, we found that the lifetime of electron-phonon scattering is laser intensity dependent. This lifetime increases from 2.5 ps at low laser power to 5 ps at high laser intensity. The ratio of the amplitude of slow component to that of the fast component also increased with laser intensity. In analogy to thin metal films, we assert that diffusion of heat from particles to the medium (water or toluene) is a sub-nanosecond process. Therefore, within picosecond time scale we observe an increase of phonon- phonon relaxation amplitude, or in other words a slower loss of heat to the surrounding medium is taking place.

Experimental

Colloidal gold particles with average particle diameter of 30 nm were prepared by the method of Turkevich et. al. which involves the reduction of H_2AuCl_4 by sodium citrate.[17] Smaller gold particles with average diameter of 1.9 and 2.6 nm are passivated by $-SC_{12}H_{25}$ monolayers, and those with average size of 3.2 nm are passivated by $-SC_6H_{13}$ monolayers. Smaller gold particles are dissolved in toluene and have an average size distribution of ± 10 % (18).

The laser system and the transient spectroscopy set-ups were described previously (19). Briefly, the laser system consists of a commercial Coherent Satori dye laser pumped by an Antares mode-locked YAG laser and amplified by regenerative amplifier resulting in 300-400 fs pulses with a repetition rate of 10 Hz. The output energy ranges 700- 1000 μJ at wavelengths between 595 and 615 nm. The optical density of the samples was about 1.0 per 2 mm optical path length at the excitation wavelength. The energy of the pump beam was about 0.1-0.2 mJ at 600 nm. The reference beam had an energy of less than 50 nJ. The reference and probe beams were passed through a monochromator or polychromator and were detected by two photodiodes or CCD detector (Princeton Instruments, EUV-1024, controller ST-130), respectively. Kinetics were analyzed by the least squares method.

Results

The main difference in the steady state UV-visible absorption spectrum of 1.9 nm gold nanoparticles with the spectra of larger size particles is the absence of a plasmon type absorption in the former (Figure 1). The absence of a plasmon band in small metallic particles is attributed to the quantum size effect, i.e. the formation of discrete energy states with substantial separation between the levels in contrast to the band formation in the larger particles and bulk matter. Particles with a bit larger diameter, 2.6 nm, possess a weak plasmon absorption at ~ 517 nm. The 3.2 nm and 30 nm particles have very noticeable plasmon band maxima at 522 nm and 530 nm, respectively.

The gold nanoparticles with average sizes of 1.9, 2.6, and 3.2 were prepared in toluene and possess an onset of absorption at 750 nm (1.7 eV) which is absent in the spectrum of larger particles (30 nm) prepared in water. This absorption continuum has been assigned to 5d→6sp transition at the X point of the Brillouin zone (20). The notion of 'quantum size effect' will be discussed a bit further later.

Transient absorption spectra for 1.9 nm gold nanoparticles are shown in Figure 2. Immediately after the photoexcitation with 600 nm laser pulse a transient absorption band around 470 nm is observed. This absorption decays to the baseline within a few picoseconds without any changes in the spectral profile. The narrow peak at 520 nm results from the Raman gain of the solvent (toluene).

Figures 3 and 4 show the transient absorption spectra of 2.6 nm and 3.2 nm particles. At time zero, a positive absorption at ~ 470 nm, similar to that of 1.9 nm particles is seen. Also, a plasmon band 'bleach' signal with maximum around 550 nm and a positive feature with maximum above 650 nm are observed. The positive absorption at 680 nm is assigned to the absorption of the 'hot' electrons. The lifetime of the plasmon band 'bleach' was found to be 2-2.5 ps (inset for Figure 3) and will be discussed later in the text. There is a considerable shift in the maximum of the plasmon band 'bleach' and maximum of the steady state plasmon absorption of these particles, which will be discussed later (Table 1).

The data for 30 nm gold nanoparticles are given in Figure 5 which are similar to those for smaller size particles (except for the 1.9 nm particles). However, the contribution of the plasmon band 'bleach' relative to the positive absorption at 630 nm and 490 nm is much larger than that for particles with sizes of 2.5 nm and 3.3 nm. The dynamics of the plasmon band 'bleach' recovery has two components, a fast one with lifetime of 2.5 ps and a relatively slow one with lifetime of > 50 ps.

The effect of the laser intensity on the kinetics of the plasmon band "bleach recovery" for 30 nm particles shows that as laser intensity increases, the lifetime of the fast component increases from 2.5 ps at low intensities to 5 ps at high intensities (Figure 6). The ratio of the amplitude of the slow component to the fast component also increases (from ~ 0.1 to ~ 0.4) as the laser intensity increases (Figure 6).

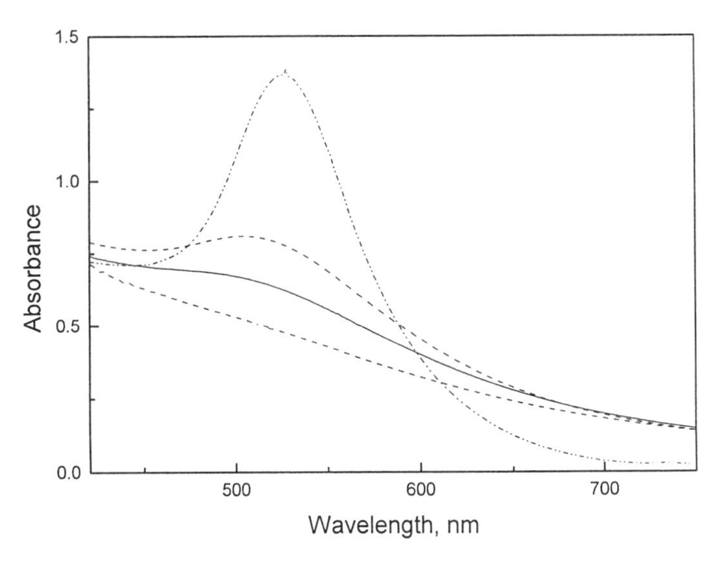

Figure 1. The steady-state absorption spectra of 30 nm particles (dash-dot line), 3.2 nm (dash line), 2.6 nm (solid line), and 1.9 nm (lowest dash line) gold nanoparticles are shown.

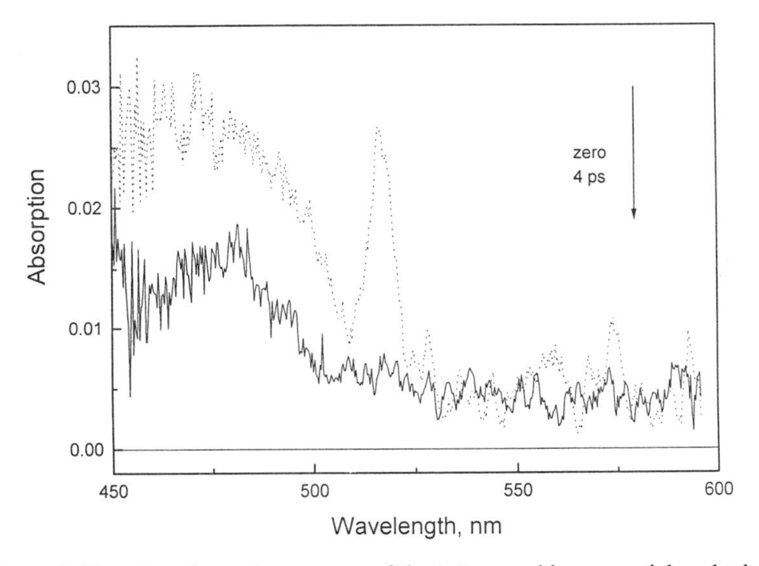

Figure 2. Transient absorption spectra of the 1.9 nm gold nanoparticles obtained immediately after excitation with 600 nm laser pulse (dot line) and at 4 ps delay (solid line).

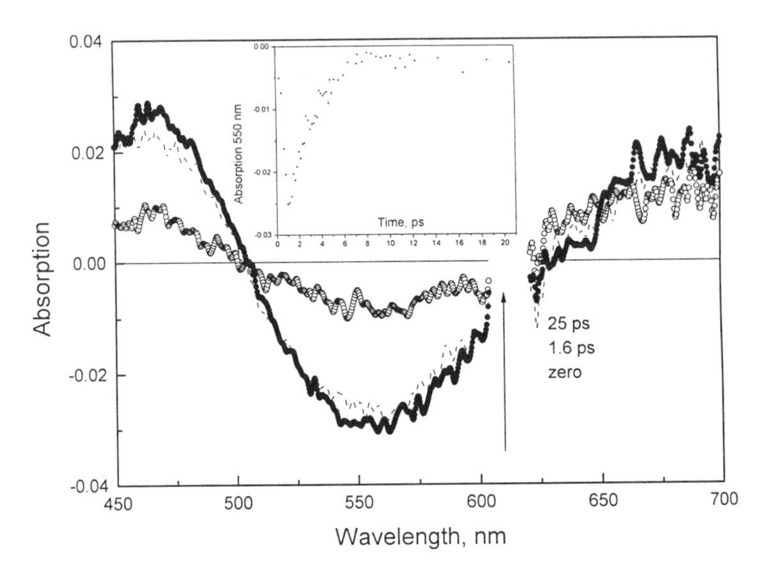

Figure 3. Transient absorption spectra of 2.5 nanoparticles of gold at different delays after excitation by 600 nm laser pulse. Kinetic of absorption changes measured at 550 nm (Inset).

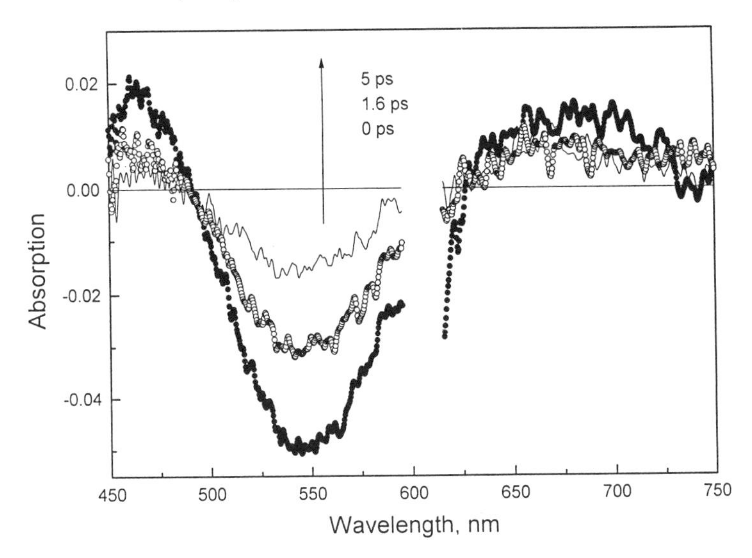

Figure 4. Transient absorption spectra of 3.3 nm gold nanoparticles at delays 0 (solid line), 1.6 ps (dash line), and 5 ps (dot line).

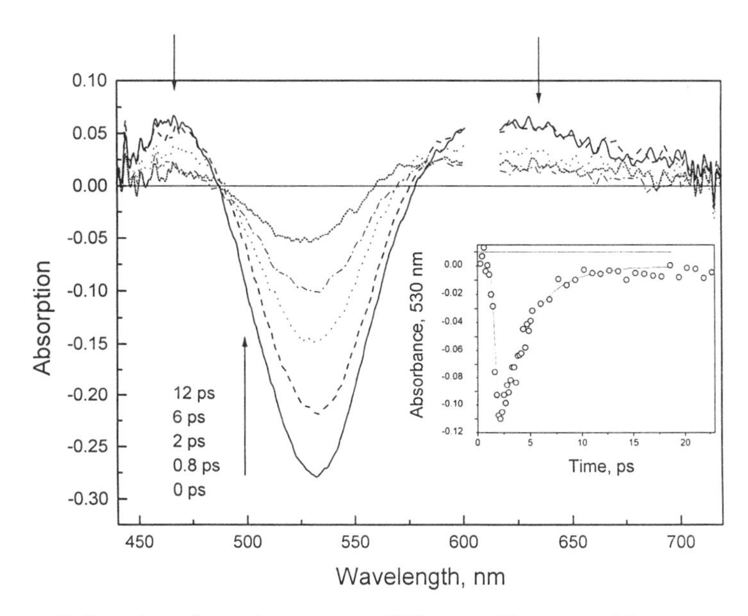

Figure 5. Transient absorption spectra of 30 nm gold nanoparticles measured at 0, 0.8, 2.0, 6.0, and 12 ps after excitation with 600 nm laser pulse. Arrows indicate the change of intensity with time. Kinetic of the absorption changes at 530 nm, solid line is a fit with lifetimes of 2.5 ps and > 50 ps (inset).

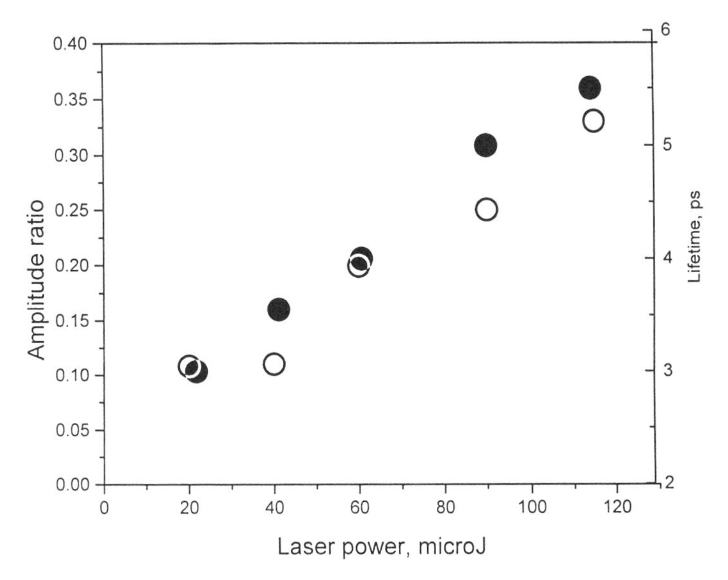

Figure 6. Ratio of the amplitudes of the slow to the fast component and lifetime of the fast component in the kinetics of the absorption changes at 530 nm for 30 nm gold nanoparticles as a function of laser pulse intensity.

Discussion

In metals, the electronic heat capacity (C_e) and lattice (C_l) heat capacities are much different with $C_e << C_l$. Specifically, C_e and C_l values are 1.98 x10^4 $Jm^{-3}K^{-1}$ and 3.5 x 10^6 $Jm^{-3}K^{-1}$ for gold, respectively (14). Therefore, it is possible to selectively excite the electrons by a short laser pulse and to follow the temporal and spectral relaxation dynamics of the heated system by various spectroscopic methods. Let's call the initial temperature of the system T_i, electronic temperature T_e, and the lattice temperature as T_l. Before absorption of the laser photons, $T_i = T_e = T_l$, i.e. the system is in thermal equilibrium and the electrons have a fermi distribution at T_i (16, 21). Immediately following the short laser pulse, the excited electrons will have a whole spectrum of energies, both kinetic and potential, ranging from zero to hν. The potential energies are those of the occupied and unoccupied states of the conduction band. In intraband transitions, electrons are promoted from the lower to higher states of the conduction band, whereas, in the interband transitions, electrons are promoted from the d-band to the formerly occupied states of the conduction band. The excited electrons (at T_e) undergo e-e scattering with the electron gas which subsequently heats up the lattice through e-ph interactions. Heat from the lattice (at T_l) dissipates through ph-ph coupling with the surrounding medium which brings the whole system to a new equilibrium temperature (higher than T_i).

There is a clear correlation between amplitude of the "ground state" plasmon band in different size gold nanoparticles and the laser-induced absorption changes. For particles with strong (weak) plasmon, there is relatively more (less) change in the absorption band immediately following photoexcitation, which suggests that the observed absorption changes are primarily due to the plasmon band 'bleach' for the particles with average sizes of 2.6 - 30 nm (Figures 3, 4, and 5). The interband transition (2.4 eV) in bulk gold metal is close to the energy of the plasmon (2.35 eV) absorption in nanoparticles. Thus in the spectral changes, one should expect contributions from both intraband (plasmon) and interband (5d→6sp) transitions. In 1.9 nm gold nanoparticles (figure 2), the spectral changes are attributed to the transition from the d-band as these particles do not possess any surface plasmon absorption bands. The transient narrow band of 520nm is due to the Raman scattering of the toluene (solvent), as mentioned in the previous section.

In case of small metallic particles embedded in a medium with a dielectric constant of ε_m and assuming spherical particles with radii r, the absorption coefficient for the whole system (particles plus the medium) is given by Mie's formula (9):

$$(1\text{-}T) = (18\pi N V n_0^3/\lambda)\, \varepsilon_2 /((\varepsilon_1 + 2\varepsilon_m)^2 + \varepsilon_2^2) \qquad (1)$$

where N is the particle number density, V is the total volume occupied by the particles, ε_m is the dielectric constant of the medium. The absorption peak occurs when $(\varepsilon_1 + 2\varepsilon_m^2) = 0$. The absorption bandwidth and plasmon damping is directly related to the imaginary dielectric function (ε_2) of gold (9). Our calculations indicate that we excite about 30% of the particles in the samples. Using expression (1) and values of ε_1 and ε_2, the absorption spectrum of the large size gold nanoparticles is calculated (Figure 7). The shape of the calculated spectrum indicates the validity of Mie theory within dipole approximation.

We assigned the spectral changes in the transient spectra of gold nanoparticles mainly to plasmon band 'bleach'. By considering the possibility of both interband and intraband electronic transitions, we calculate the transient spectral changes as follows: Changes in the transmission of gold thin films (16) at low levels of excitation may be written as

$$\Delta T/T = (\delta \ln T/\delta\varepsilon_1)\Delta\varepsilon_1 + (\delta \ln T/\delta\varepsilon_2)\Delta\varepsilon_2 \quad (2)$$

where T indicates transmission. The changes of the real and imaginary part of dielectric function could be calculated through the changes of electron distribution (22). The changes in ε_2 is calculated using the constant matrix element approximation as (16)

$$\varepsilon_2 = 1/(h\nu)^2 \int D(E, h\nu)\Delta\rho(E)\, dE \quad (3)$$

where D(E, hν) is the so-called "joint density of states" with respect to the energy of the final state (E). D gives the density of occupied and unoccupied states separated by the energy of the excitation photon hν (9). D-function is calculated assuming parabolic band structure for the conductive band and d band around L point of the Brillouin zone (16). The change in the imaginary part of dielectric function of gold was calculated by Cardona (23) The change in the real part of dielectric constant is calculated through Kramers-Kronig relationship. Fujimoto et.al. did this calculations for gold thin films (13). Expression (1) combined with (2) reproduces the absorption changes associated with the plasmon due to temperature changes. The shape of the transient signal for gold nanoparticles and thin film (16) are very similar, i.e. negative absorption in the middle with two positive wings at higher and lower energies around the negative peak. However there is a difference in the positions of these two peaks as well as their relative intensities. The signal measured in thin films (center peak is at 510 nm) is only due to the interband transition (16). Whereas in the nanoparticles, the main contribution is from the plasmon band (maximum at 530 nm for the 30 nm particles). Calculation of the absorption changes based on this approach gives good correlation with experiment (Figure 7): the maximum 'bleach' at 530 nm with two positive wings could be reproduced. The experimental observations show wider bands for the 'bleach' and the two transient positive wings, and theoretical calculations give narrower bandwidths for the three transient bands. The difference in the bandwidths is due to the fact that calculations were performed in the limit of the weak perturbation, whereas the energy of the excitation in our experiment (2.04 eV) is relatively high.

As Table I indicates, we observe a red-shift in the position of transient 'bleach' of plasmon band with respect to its absorption maximum in the steady state spectrum. This shift is larger (smaller) for small (large) particles. The origin of this shift is attributed to 'quantum size effect'. Smaller particles have more 'molecule-like' discrete energy levels in contrast to larger particles and bulk matter which possess

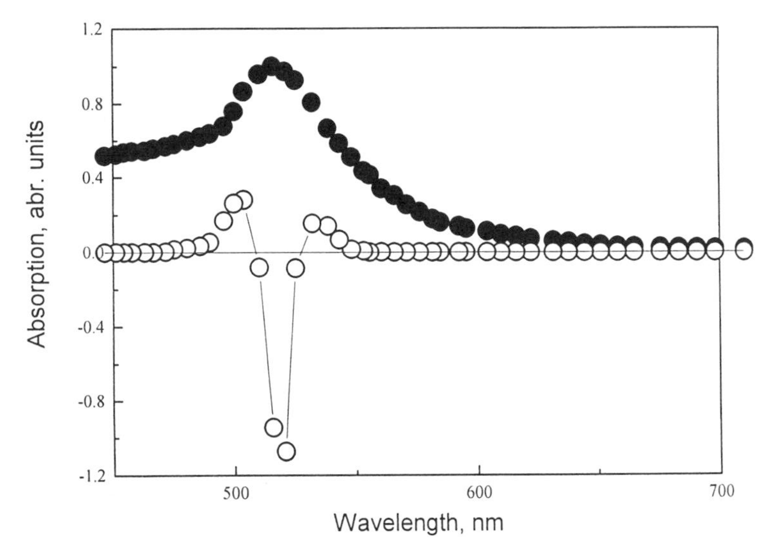

Figure 7. The shape of the plasmon band (filled squares) and transient absorption changes after laser excitation (open circles) calculated on the bases of equations (1-3).

energy bands (24). The mean spacing between electronic levels for a nearly spherical particle is expressed as $\delta \sim E_F/(N^{2/3})$, where E_F is the fermi energy, and N is the number of atoms. In 1.9 nm particles, there are about 200 atoms and δ is equal to ~ 60 meV , which is higher than kT at room temperature. For particles with size 2.6 nm, δ is about 20 meV. The energy spacing decreases as the size of the particles increases. For the 30

Table I. Energies of the steady-state plasmon band maximum, "bleached" plasmon band after laser excitation, and energy splitting between electronic states in gold nanoparticles of different sizes.

Diameter of Gold nano-particle (nm)	Steady-state plasmon band max. (eV)	Laser induced "bleach" of plasmon band (eV)	Energy spacing between electronic states (meV)
1.9	-	-	60
2.6	2.42	2.25	20
3.2	2.39	2.29	10
30	2.36	2.35	0.001

nm particles, δ is about 0.001 meV. A critical size of 5 nm is estimated, from the correlation of δ with size, as the 'transition' size from 'molecule-like' to 'bulk-like' electronic energy states.

The ratio of the amplitude of plasmon 'bleach' to the amplitude of the positive wings in transient spectra (Figures 1-5) increases as size of nanoparticles increases, as shown in Figure 9. One can assume that there is a linear relationship between this ratio and the particle size at diameters less than 5 nm and this ratio is independent of the size at larger than 5 nm. As the size of a particle decreases, the number density of free electrons also decreases which is evident by the presence of a less intense plasmon band in small particles as compared to very intense plasmon absorption in larger particles (see figure 1). Again, 5 nm size may be considered as a transition size from the 'molecule-like' to the 'bulk-like' particles.

We observed an increase in the lifetime of the fast component in the 'bleach' recovery kinetics of the gold nanoparticles as laser intensity was increased. Perhaps, an increase in laser intensity leads to multiphoton absorption for intraband transitions forming very highly excited electrons. In the limit of TTM the heat capacities of the lattice and electrons are well separated, so electron gas and the lattice could be treated independently, interacting through electron-phonon coupling constant (G). G is independent of the temperature beyond the Debye temperature (170K for gold). Our estimates show that there is a temperature rise of 1000K in the electron gas in our experiment. The rate of e-e relaxation increases for highly excited electrons as given by fermi liquid theory, i.e. $\tau_{e-e} \propto (E-E_f)^{-2}$ (25). Since G is constant and τ_{e-e} become shorter for highly excited electrons, the e-ph energy relaxation rate also increases (26). Then, why do we observe an increase in the e-ph lifetime? We also observed that the amplitude of the slow component increases at higher laser intensities, implying a slow loss of heat to the solvent in ps time regime. It is very likely that the rate of heat dissipation to the lattice is far too slow, which leads to accumulation of heat in the other two channels, i.e. e-e and e-ph. Based on TTM, the diffusivity of electron gas is much higher than that for the lattice ($K/C_e >> K/C_l$) (16). This condition leads us to assert that the lattice is not dissipating heat to the medium as fast as it warms up through e-e, and e-ph scattering. Hence, the system is heated much more at higher laser intensity compared to 'normal' intensity, while the cooling processes and cooling channels remain totally the same. The process of lattice cooling happens in nsec time regime in thin films. Perhaps, a similar time is required for the cooling of nanoparticles.

In small gold colloids, the electrons excited by the photons are the free carriers. The intraband transition involves the excitation of electrons from the bottom to higher energy states within the conduction band. Interband transitions involve excitation of electrons from the 5d band to 6sp band around the X or L point of the Brillouin zone (9a). The relaxation of excited electrons occur through electron-electron scattering which is mostly assumed to be instantaneous. We did not observe any direct evidence for the internal relaxation dynamics of 'hot' electron gas through e-e scattering. There are evidence from metal thin film studies that the e-e and e-ph relaxation times are in the order ~500 fs and 1 ps, respectively (16). Some recent work finds the e-e interaction to be as long as 2 ps using the numerical solution of the Boltzman equation (16). Also, as Bigot et al. suggested, in their study of the dynamics of copper nanoparticles in a glass matrix, the contribution of electron-electron interaction and energy relaxation to the lattice occur in picosecond time regime and it must not be ignored (7). In addition, Fann et.al. found e-e thermalization time of 1 ps using photoemission spectroscopy (15). Following the reasoning of Fann et.al. we assert that during the time that electron gas is undergoing thermalization through e-e scattering, some heat is also lost by a parallel channel of e-ph scattering. The Debye temperature of gold is 170K. If one assumes a fermi velocity of ~10^8cm/sec for the free electrons of gold and a mean free path of 30 nm (the diameter of the largest particles in the set), one finds that it takes about 30 fs for each e-ph interaction to take away 85K (half of Debye temperature) of heat from the electron gas to the lattice (15). Therefore, in the first 1 ps after photoexcitaion, there will be 30 phonon collisions occuring parallel to the e-e interactions. The laser pulse duration in our experiment was 400 fs which is much longer than the duration of e-e scattering. In the fast component of the kinetics of 'bleach' recovery, there must be the two processes of e-e and e-ph occurring simultaneously, as our estimates indicate. We are attempting to separate the two parallel channels by using a shorter laser pulse, other metallic particles, and different solvents.

As seen in Figure 8, where absorption changes of the 30 nm gold nanoparticles at different times are normalized, no substantial difference in the band

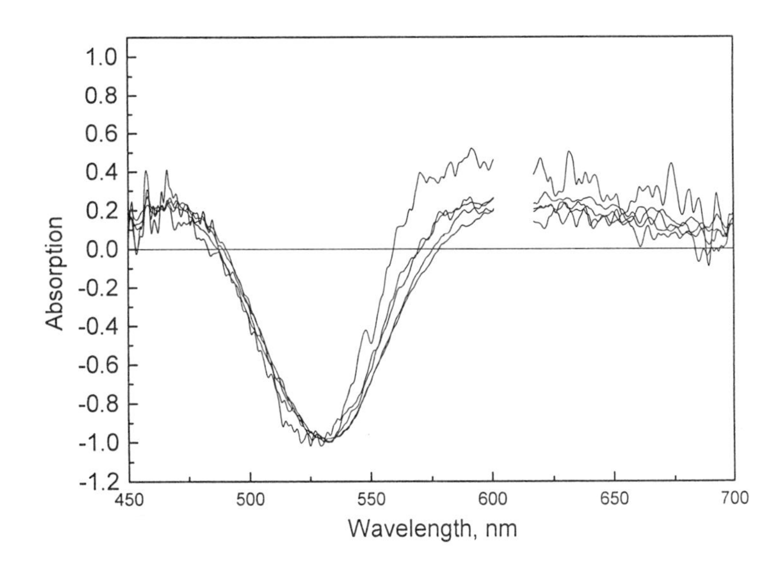

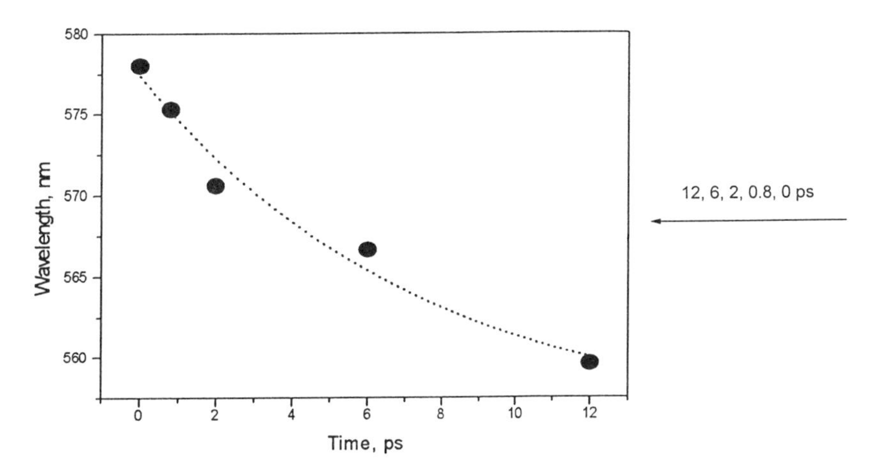

Figure 8. Absorption changes (curves crossing the zero line)of the 30 nm gold nanoparticles at different delay times normalized to their peak intensity. The arrow indicates the direction of the changes with time increase.

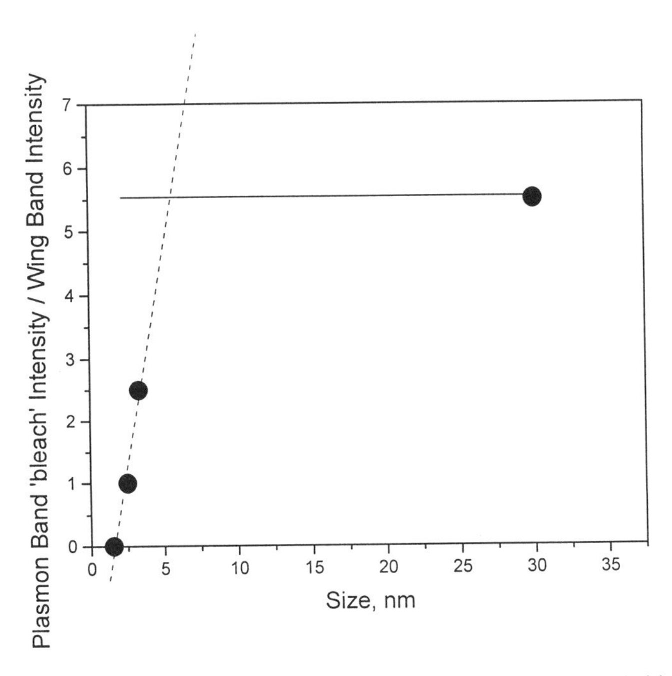

Figure 9. The dependence of the amplitudes ratio of plasmon band 'bleach' to the positive wing as a function of the particle size. The lines indicate the linear relationship and saturation range.

shape within the time-range of the first six picosecond could be observed. There is a small time-dependent shift of the crosspoint of the absorption changes with zero baseline (shown in the inset), which probably reflects the lowering of the electron gas equilibrium temperature which consequently heats the lattice. At 12 ps, the shape of the transient absorption band is no longer symmetic with respect to its center, i.e. the left and right sides of the minimum of the plasmon 'bleach' are not mirror images of each other. There is also a relatively larger positive absorption at ~600 nm. These observations imply a change in the mechanism of the heat loss at longer time (12 ps) as compared to the shorter times (>6 ps). We attribute the asymmetric changes of the 'bleach' band to an onset of the contribution of ph-ph interaction of particles with those of the solvent or, perhaps, it is due to the scattering of excited electrons by surface defects.

Conclusion

It is shown that transient 'bleaching' of plasmon band in small gold particles, with average particle sizes of 2.6, 3.2, and 30 nm, is generated through excitation of plasmon and d-band electrons by ultrashort laser pulses. The excited electrons form a non-fermi, non-equilibrium distribution which undergoes electron-electron and electron-phonon scattering and subsequently heating the lattice. These cooling processes occur in the first 2-3 ps after photoexcitation. The cooling of the lattice takes place in >50 ps and occurs via phonon-phonon scattering. All these relaxation times seem to be size independent. We could not distinguish between the e-e and e-ph relaxation times. Perhaps, the process of e-e occurs within the duration of our laser pulse, i.e. 400fs.

Acknowledgments

We thank Professor Robert Whetten and Dr. Joseph Khoury from the School of Physics, Georgia Institute of Technology, for providing us with Au (in toluene) particles. We thank the Office of Naval Research (Grant no. N00014-95-1-0306) for its financial support of this project.

Literature Cited

1. Ahmadi, T. S.; Logunov, S.; El-Sayed, M. A. *J. Phys. Chem.* **1996**, 100, 8053.
2. Stella, A.; Nisoli, M.; De Silvestri, S.; Svelto, O.; Lanzani, G.; Chyssac, P.; Kofman, R. *Phys. Rev. B* **1996**, 53, 15497.
3. Bloemer, M. J.; Haus, J. W.; Ashley, P. R. *J. Opt. Soc. Am. B* **1990**, 7, 790.
4. Roberti, T. W.; Smith, B. A.; Zhang, J. Z. J. *Chem. Phys.* **1995**, 102, 3860.
5. Heilweil, E. J.; Hochstrasser, R. M. *J. Chem. Phys.* **1985**, 82, 4762.
6. Doremus, R. H. *J. Chem. Phys.* **1964**, 40, 2389.
7. Bogot, J. -Y.; Merle, J. -C.; Cregut, O.; Daunois, A. *Phys. Rev. Lett.* **1995**, 75, 4702.
8. Perenboom, J. A. A. J.; Wyder, P.; Meier, P. *Phys. Rep.* **1981**, 78, 173.

9.(a) Kreibig, U.; Vollmer, M. *Optical Properties of Metal Clusters*; Springer : Berlin, 1995. (b) Papavassiliou, G. C. *Prog. Solid St. Chem.* **1980**, 12, 185.
10. Kerker, M. *The Scattering of Light and Other Electromagnetic Radiation*; Academic: New York, NY, 1969. (b) Bohren, C. F.; Huffman, D. R. *Absorption and Scattering of Light by Small Particles*; Wiley: New York, NY, 1983.
11. Mulvaney, P. *Langmuir* **1996**, 12, 788.
12. Sun, C-K.; Vallee, F.; Acioli, L. H.; Ippen, E. P.; Fujimoto, J. G. *Phys. Rev. B* **1993**, 48, 12365.
13. Fujimoto, J. G.; Liu, J. M.; Ippen, E. P.; Bloembergen, N. *Phys. Rev. Lett.* **1984**, 53, 1837.
14. Fann, W. S.; Storz, R.; Tom, H. W. K.; Boker, J. *Phys. Rev. Lett.* **1992**, 68, 2834.
15. 14. Fann, W. S.; Storz, R.; Tom, H. W. K.; Boker, J. *Phys. Rev. B* **1992**, 46, 13592.
16. Sun, C-K.; Vallee, F.; Acioli, L. H.; Ippen, E. P.; Fujimoto, J. G. *Phys. Rev. B* **1994**, 50, 15337.
17. Enustun, B. V.; Turkevich, J. *J. Am. Chem. Soc.* **1963**, 85, 3317.
18. Whetten, R. L.; Khoury, J.T.; Alvarez, M. M.; Murthy, S.; Vezmar, I.; Wang, Z. L.; Stephans, P. W.; Cleveland, C. L.; Luedtke, W. D.; Landman, U. *Adv. Mater.* **1996**, 5, 428.
19. Logunov, S. L.; El-Sayed, M. A.; Song, L.; Lanyi, J. K. *J. Phys. Chem.* **1996**, 100, 2391.
20. Alvarez, M. M.; Khoury, J.T.; Schaaf, T.G.; Shafigullin, M.N.; Vezmar, I.; Whetten, R. L. *J. Phys. Chem.* **1997** (in press).
21. Elsayed-Ali, H. E.; Juhasz, T.; Smith, G. O.; Bron, W. E. *Phys. Rev. B* **1991**, 43, 19914.
22. Rosei, R.; Antongeli, F.; Grassano, U. M. *Surf. Sci.* **1973**, 37, 689.
23. Cardona, M. In *Modulation Spectroscopy*; Seitz, F.; Turnball, D.; Ehrenreich, H., Eds. Solid State Physics Supplement II; Academic: New York, NY, 1969.
24. Kreibig, U. *J. Phys. F* **1974**, 4, 999.
25. Pine, D.; Nozieres, P. *The Theory of Quantum Liquids*; Benjamine: New York, 1966.
26. Groenveld, R. H. M.; Sprik, R.; Lagendijk, A. *Phys. Rev. B* **1992**, 45, 5079.

Chapter 11

Time-Resolved Carrier Dynamics near the Insulator—Metal Transition

M. J. Feldstein[1], C. D. Keating[2], W. Zheng[3], Y. H. Liau[4], A. G. MacDiarmid[3], Michael J. Natan[2], and N.F. Scherer[4]

[1]Center for Biomolecular Science and Engineering, Naval Research Laboratory, Washington, DC 20375
[2]Department of Chemistry, The Pennsylvania State University, University Park, PA 16802–6300
[3]Department of Chemistry and Laboratory for Research on the Structure of Matter, University of Pennsylvania, Philadelphia, PA 19104–6323
[4]Department of Chemistry, University of Chicago, Chicago, IL 60637

Carrier dynamics in polyaniline and colloidal Au films have been examined using a combined approach of time resolved laser spectroscopy and atomic force microscopy (AFM). These systems exhibit insulator-metal transitions (IMT) in conjunction with synthetic modification of their structure. The relationship between structure and reactivity, in terms of hot carrier lifetimes and transport, has been identified by correlating changes in the dynamics with the directly measured morphology. Physical insight into the processes affecting carrier lifetimes and localization and the nature of the IMT has been derived from an analysis of the experimental data. The results and conclusions presented herein have implications for directing research and development in device applications based on thin film technologies.

I. Introduction

Metallic systems play a prominent role in material science and physical-chemistry research. The principal attribute that defines a material as metallic may be subject to the theoretical perspective,(1) but, charge conduction is clearly a primary characteristic. The specific nature and detailed physics of a material's carrier dynamics and dielectric response has significant implications for the resultant properties of the system. In real systems, structural aspects greatly influence these dynamics as well as their opto-electronic properties, and, consequently, the overall material properties. For example, the optical excitation of collective modes, such as surface plasmon polaritons (SPP), depends strongly on the structural and dielectric properties at the material's interface.(2)

Knowledge of the structure-reactivity relationship of materials and interfaces impacts our understanding of surface phenomenon and dynamics in thin films. This knowledge becomes crucial for the design and development of electronic and opto-electronic components which are shrinking to the nanometer and single electron scale.(3) The traditional, well developed theories have primarily focused on bulk metallic systems and have relied on indirect dynamical measurements.(4) Few have tried to understand, using direct measurements, how nanometer length scale structure effects real time carrier dynamics. Theoretical models which do take into account structural effects can typically only address the relationship between structure and function for idealized or completely random geometries.(5)(6) There are theoretical models capable of calculating the dielectric response and electromagnetic interactions for arbitrary shapes.(7) However, they have not yet been widely applied and, moreover, it may not be trivial to apply these models to amorphous materials, such as conducting polymers.

Direct measurements of both dynamics and structure can facilitate obtaining deeper understanding carrier dynamics and electro-optic properties in thin film metallic systems, such experiments are reported in this paper. Further, the study of insulator-metal transitions (IMTs) can provide insight into the dynamical properties of complex, disordered systems.(8)(9) The subtleties of phenomena like carrier localization and transport in systems such as conducting polymers and colloidal metal films can be probed and elucidated. This chapter reports the results of optically excited carrier dynamics near the IMT in a series of thin films composed of colloidal gold particles 12 nm in diameter and a series of thin protonated polyaniline (PANI) films of varying conductivity. Carrier lifetimes and dynamics have been probed directly by time resolved laser spectroscopy measurements. Additionally, the films' mesoscopic structure has been determined with atomic force microscopy (AFM). Correlations are made between film structure, carrier dynamics and conductivity. The issues of carrier localization, transport, and scattering are addressed.

II. Experimental

Time resolved carrier dynamics were studied by way of transient transmission measurements, performed using a standard pump-probe configuration. The laser system is based on a home-built, cavity-dumped Ti:Sapphire laser(10) which produces sub-20 fs pulses centered at 800 nm. Pulse energy's are in excess of 50 nJ/ pulse (peak powers $>10^6$ W) and a repetition rate of 500 kHz was used. The cavity-dumped pulse is dispersion compensated using a prism-pair sequence and then split 90:10% into a two beams; a strong pump and weak probe. The pump pulse travels along a fixed optical delay line while the probe passes through a variable delay line. The two beams are focused, at near normal incidence, to a common spot on the sample, where each pump pulse has an energy of approximately 15 nJ yielding a fluence of ~0.15 mJ/cm^2, which is in the low fluence regime for metallic systems.(11) The samples were continuously rotated at ~20 Hz about an axis normal to the surface to provide "fresh" sample and avoid excessively heating a single location.

III. Results and Discussion

Films composed of multi-layers of 12 nm Au colloids tethered to the substrate were prepared according to previously published procedures.(12) Figure 1 shows the topography of two such films; a comparison of the two layer and the four layer films shows colloidal aggregation and growth of domain sizes. The Au colloidal films, which vary in thickness, from approximately 10 to 60 nm, and proportionally in domain (or aggregate) size, exhibit structurally dependent hot-electron lifetimes.(13) Specifically, the hot-electron lifetimes vary from 1-3 ps in a predictable manner with the film's growth. These responses were measured by femtosecond optical spectroscopy detecting transient photo-bleach signal arising from hot-electron relaxation via electron-phonon coupling and are shown in Figure 2. In a simplified model, the carrier lifetime, τ, due to electron-phonon scattering is simply related by the electron heat capacity, C_e, as: $\tau = C_e/G$, where G is the electron-phonon coupling constant.

The structural dependence of τ has been accounted for by considering the effect of the domain size on the coupling constant G.(14) Two competing phenomena which determine G have been identified. The first of these, as illustrated in Figure 3, is electron oscillation frequency-phonon resonance detuning (EOPRD) which increases as the domain size, d, decreases due to an increased electron oscillation frequency. The effective oscillation frequency arises from elastic scattering of electrons from the domain boundary. The detuning is significant since without spectral overlap with the phonon spectrum the electron-phonon coupling is ***reduced***. The second phenomenon effecting G is inelastic surface scattering (ISS) which increases as the domain size, d, decreases due to the greater number of carrier-surface collisions. Even though ISS is less effective for energy transfer than electron scatter from bulk phonons, ISS yields ***enhanced*** electron-phonon coupling since it is an additional source of coupling.

The relative contribution of EOPRD and ISS can be determined by first assuming the following functional form: G/G_B = ISS/EOPRD , where G_B is the electron-phonon coupling constant in the bulk limit. ISS can be calculated according to:(15) ISS = $1 + 3(1-P)/8\beta + 7\alpha/5$, where β is a film thickness parameter defined as d/Λ and d is taken as the average film height measured with an AFM, Λ is the effective electron mean free path, P is a specular reflection parameter of the electron from the film surfaces (and is taken to be zero), α is the grain diameter parameter defined as $(R/1\text{-}R)(\Lambda/D)$, where D is the grain size and is fixed at 12 nm for the colloids presently under study, and R is the reflectivity of electrons at grain boundaries and is taken to be ~0.2.(16) The values of ISS for each film, shown graphically in Figure 4, have the expected trend of increasing with decreasing film dimension as collisions with the boundary become more frequent. The values of EOPRD have been solved for and are shown in Figure 4. In these results, the trend anticipated in the work of Tomchuk(14) is found: EOPRD increases with

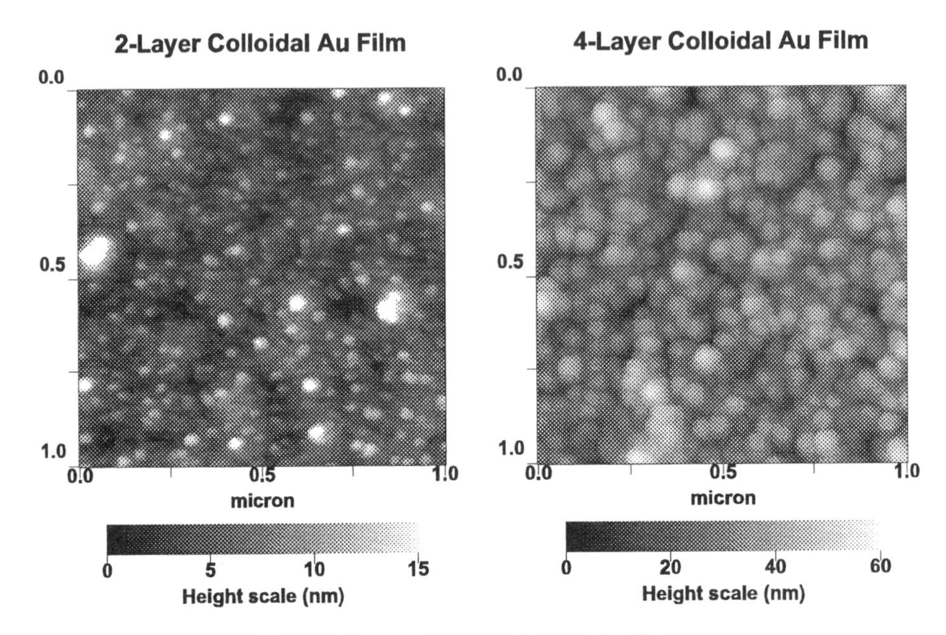

Figure 1 AFM images of Au colloid films.

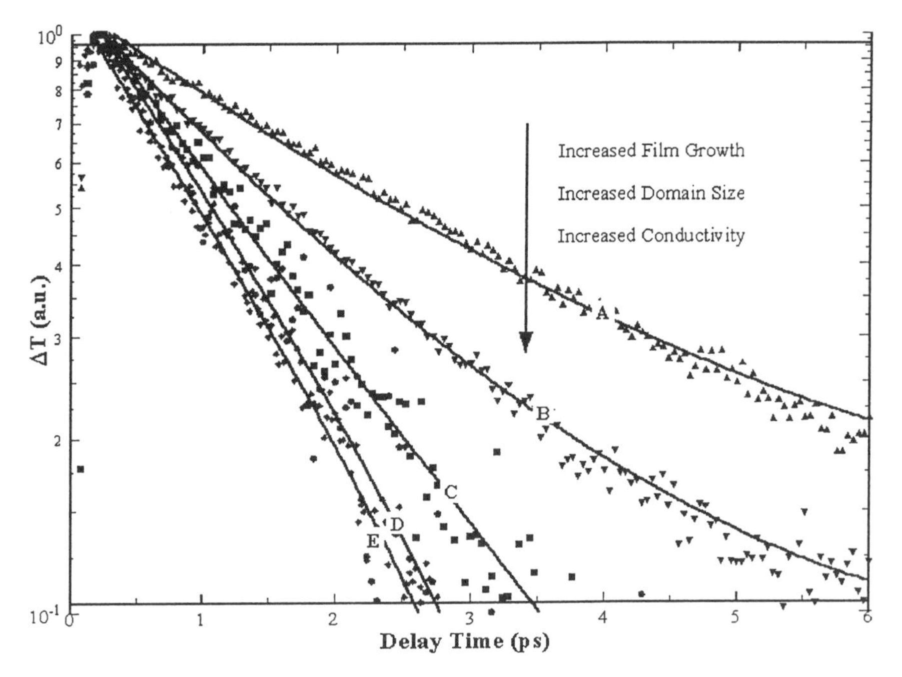

Figure 2 Pump-probe lifetime results for Au colloid films. Conductivity increases with number of layers (2-5) for samples A-E.

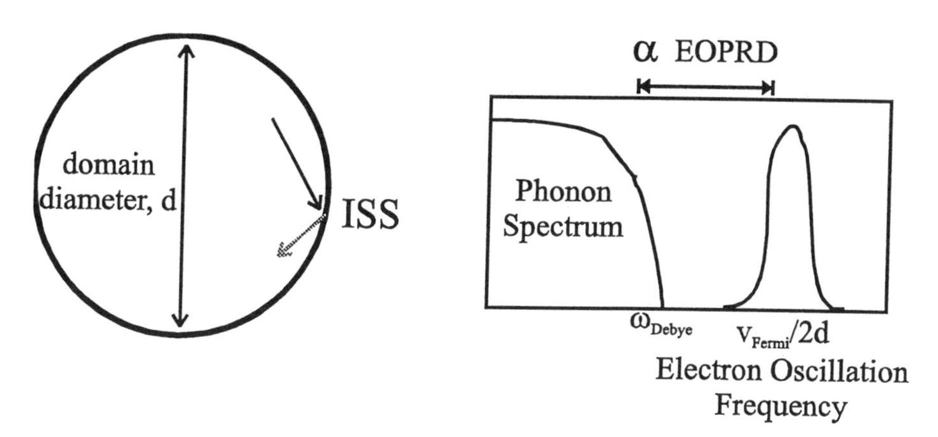

Figure 3 Electron-phonon coupling mechanims.

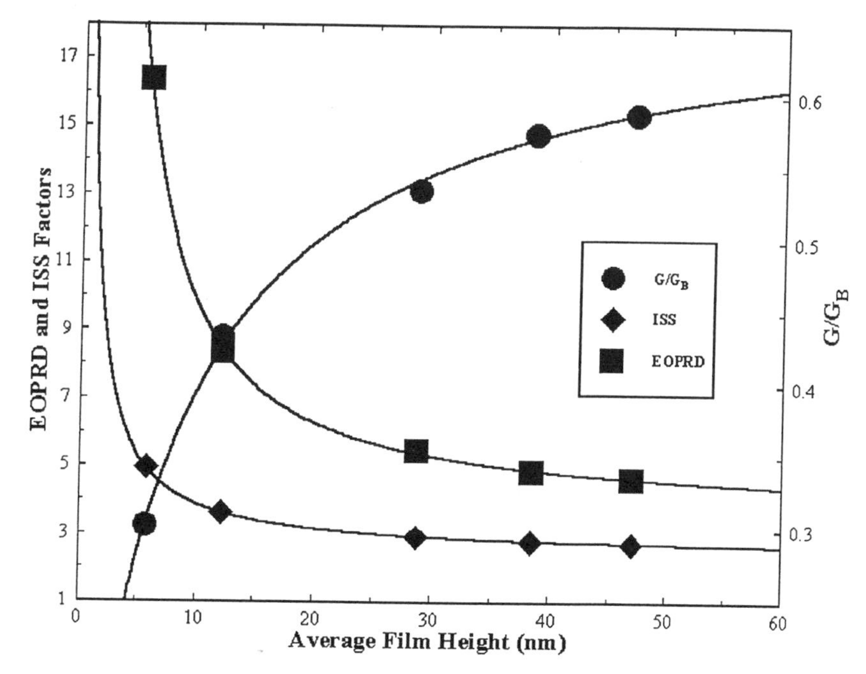

Figure 4 Factors affecting electron-phonon coupling vs. Film height.

decreasing dimensions as the electron oscillation frequency goes farther out of resonance with the upper limit of the phonon spectrum. The calculated EOPRD for the thinnest film begins to approach yet never reaches the 2 orders of magnitude predicted by Tomchuk for nanometer scale metal islands. This is not surprising since the reflectivity factor, R, used to calculate ISS may not appropriately take into account surface defects and traps or the elliptical boundary conditions relevant to the colloids. These factors could yield a reduced value for R and increased coupling leading to a larger theoretically estimated value for EOPRD. Nevertheless, the fact that *both* trends agree with theoretical predictions makes it possible to conclude that the two phenomenon are in competition and that a balance between them yields the effective electron-phonon coupling determined experimentally herein.

It is evident from the present results that the electrons exhibit inter-colloid mobility even though the colloids are not in direct metallic contact. Carrier scattering time and EOPRD, as shown in Figure 4, correlate with film growth by colloidal aggregation. A comparison of EOPRD and the electron oscillation frequency, where the domain size used to determine the electron oscillation frequency is the average film thickness (related to the aggregate size), not the single particle radius, also shows a positive correlation.(17) Thus, if the EOPRD changes with film growth it can only be the case that the electrons are not confined to the volume of a single colloid but are mobile within the aggregated domains. Electron mobility in this system is also consistent with the fact that electrons exhibit enhanced tunneling probability (vs. vacuum) through functionalized alkane chains.(18)(19)(20) However, the electron dynamics in colloidal Au films are different than in bulk metals where electrons are fully mobile. Specifically, a comparison (not shown) of EOPRD with the calculated electron oscillation frequency shows a non-zero intercept which suggests that even for films with the greatest aggregation (*i.e.*, smallest frequency difference) the electrons lack compete mobility and the colloidal size still maintains an influence on the e-ph coupling.(13)

The coupling between aggregated particles has been shown to be of significant importance for the optical and electronic properties of the colloidal films. This coupling takes the form of both the well known dipolar coupling to produce the collective plasmon optical resonance at ~800 nm and the newly reported inter-colloid coupling that increases the hot-electron decay rate due to inelastic electron-phonon collisions. The mesoscopic structure of the films has been measured with scanning probe microscopy techniques and a simple model is reported which accounts for the dependence of the hot-electron lifetime on the films' structure. Finally, the results consistent with the DC electrical conductivity mechanism of percolation based on activated hopping.(13)

Thin (~40 nm) polyaniline emeraldine base films, doped with camphorsulfonic acid and spin cast from chloroform solutions, were prepared in a range of conductivities by "secondary doping" with m-cresol vapor. Figure 5 shows the AFM topography for low (σ = 0.03 S/cm) and high (σ = 0.9 S/cm) conductivity spin cast films. Secondary doping functions through enhanced solvation of the charged (*i.e.*, doped) polymer chains. This process promotes rearrangement into more favorable molecular confirmations and tends to uncoil the polymer chains.(21) The result of secondary doping is better intra-chain

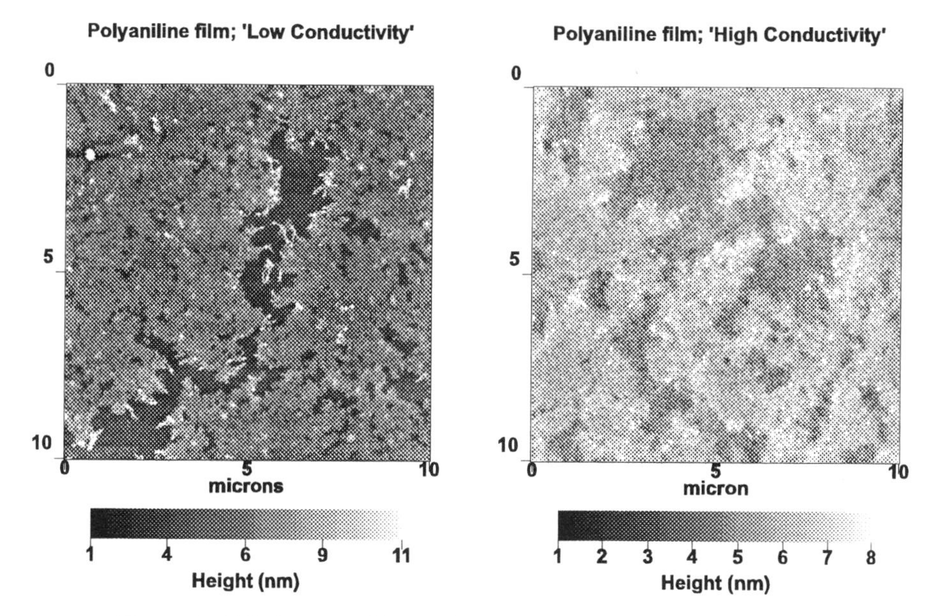

Figure 5 AFM images of low and high conductivity polyaniline films.

homogeneity and inter-chain contact and interaction thereby yielding higher conductivity films. The more highly conductive material might be expected to have a more "smooth" and uniform morphology, as observed in Figure 5, since the degree of crystallinity (or order) increases. By contrast, the disordered low conductivity films would have a more heterogeneous morphology.

Two time scales for optically excited carrier relaxation are observed in optical pump-probe experiments and are shown in Figure 6. However, only the relatively slowly relaxing carriers which have exponenetial decay constants in the picosecond range change as the measured conductivity in polyaniline changes: the results for samples B (σ=0.033 S/cm) and E (σ= 0.940 S/cm) in Figure 6 indicate that the trend in the 'slow' carrier scattering time correlates with the conductivity. The results of Figures 5 and 6 indicate that the lifetime of weakly localized carriers increases with decreasing disorder. These conclusions regarding the primary importance of the slow carriers are rather interesting since it has been shown herein and previously(9) that only a small fraction of the total carrier population is quasi-localized and thereby relax with the longer scattering times. Presumably, these carriers are quasi-1D localized, as identified by Kohlman *et al.*, where the localization is the result of inhomogeneous disorder, such as crystalline defects or impurities.(9) Since the scattering time increases with decreased disorder it seems clear that the localization length also increases and that the quasi-1D carriers contribution to the film's conductivity is based on the percolation mechanism.

Figure 7 depicts the idea that a localization length, L_{loc}, is associated with the carriers in materials governed by the percolation mechanism. When this length exceeds the inter-metallic spacing for a statis-tically significant percent of the film (50% for a 2-D film) an IMT occurs and conductivity rises exponentially. An IMT may occur when either L_{loc} or the density of metallic centers is increased. Associated with increased DC conductivity, σ_{DC}, is an increased carrier scattering time, τ, which according to the Drude model is: $\sigma_{DC}=\omega_p 2\tau/4\pi$, where ω_p is the plasma frequency. The correlation of increased conductivity with increased order is evident in the AFM images of Figure 5. Further, the height correlation function analysis (not shown) of these images shows quantitatively that the low conductivity film is dominated by short range order and large roughness while the high conductivity film exhibits long range order and less roughness indicating a film that is more ordered.

The results of time domain measurements of the carrier dynamics in PANI films near the IMT suggest that scattering times and conductivity are not necessarily directly related as predicted by the Drude model. That is, measuring one property is not sufficient to uniquely determine the other quantity. Rather, the direct measurement of both, as well as the plasma frequency, would provide a better characterization of the dynamical time scales and theoretical predictions of conductivity. Still, predictions from the Drude model do agree within an order of magnitude of the measured values.

IV. Conclusion

The implications of this work may be valuable to direct research and development in device application. For example, from a synthetic perspective, this work confirms the understanding of the need for increased film order to achieve maximal conductivity. Both

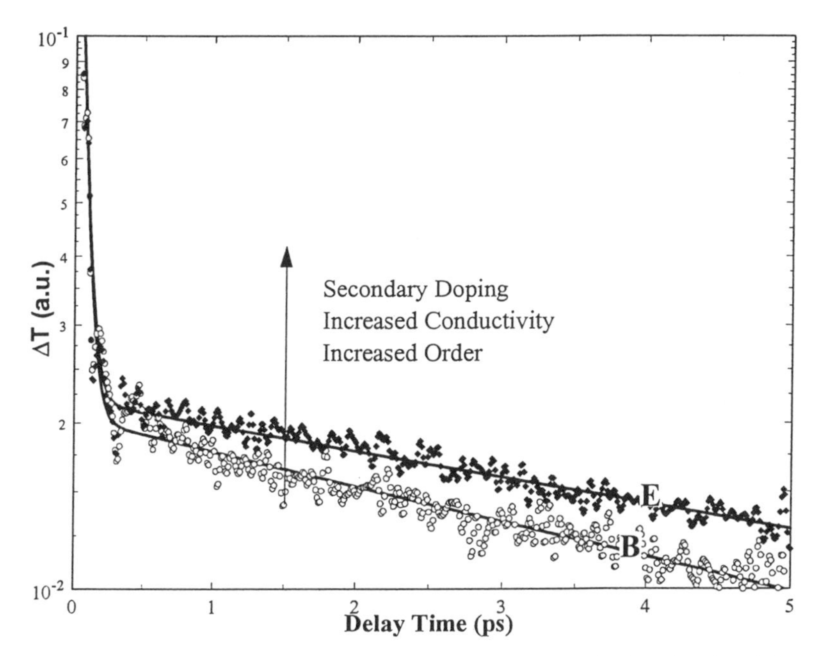

Figure 6 Pump-probe measurement of carrier relaxation in doped polyaniline films.

Figure 7 Cartoon of ordered and disordered regions of polyaniline film.

of the studies of these thin film systems that undergo an IMT have implications for the design and development of devices. For example, the development of opto-electronic devices necessitates a working understanding of the relationship between structural features and performance characteristics. The results reported herein suggest a guide for the design of opto-electronic devices with synthetically tuneable hot-carrier lifetimes. Specifically, since both of these materials exhibit large nonlinear optical responses, the combination of the two could yield the development of devices where the optical response as well as the carrier dynamics are adjustable parameters. Designed control over the system properties (*i.e.*, carrier lifetimes, etc.) could potentially yield better optimized device performance.

Literature Cited

1. C.M. Hurd, *Electrons in Metals*, R.E. Kreiger Publishing Co., Florida, 1981, chapter 2.

2. H. Raether, *Surface Plasmons,* Springer-Verlag: Berlin(1988).

3. J.I. Pascual, J. Méndez, J. Gómez-Herrero, A.M. Baró, N. Garcia, U. Landman, W. Luedtke, E.M Bognachek, H.-P. Cheng, *Science*, **267**, 1753 (195).

4. Sir N.F. Mott, *Conduction in Non-Crystalline Matterials, 2nd Ed.*, Clarendon Press: Oxford (1993).

5. M. Moskovits, *Rev.Mod.Phys.*, **53**(3), 783 (1985).

6. M.I. Stockman, T.F. George, V.M. Shaleav, *Phys.Rev.B,* **44**, 115 (1991).

7. W.H. Yang, G.C. Schatz, R.P Van Duyne, *J.Chem.Phys.*, **103**, 869 (1995).

8. Sir N.F. Mott, *Metal-Insulator Transitions*, Taylor & Francis: London (1990).

9. R.S. Kohlman, J. Joo, Y.G. Min, A.G. MacDiarmid, A.J. Epstein, *Phys.Rev.Lett.*, **77**(13), 2766 (1996).

10. D.C. Arnett, P. Vöhringer, N.F. Scherer, *J.Am.Chem.Soc.*, **117**(49), 12262 (1995).

11. H.E. Elsayed-Ali, T. Juhasz, G.O. Smith, W.E. Bron, *Phys.Rev.B.*, **43**, 4488, (1991).

12. R.G. Freeman, K.C. Grabar, K.J. Allison, R.M. Bright, J.A. Davis, A.P. Guthrie, M.B. Hommer, M.A. Jackson, P.C. Smith, D.W. Walter, M.J. Natan, *Science,* **267**, 1629 (1995).

13. M.J. Feldstein, C.D. Keating, Y.H. Liau, M.J. Natan, N.F. Scherer, *J.Am.Chem.Soc.*, in press 1997.

14. S.A. Gorban, S.A. Nepijko, P.M. Tomchuck, *Int.J.Electronics*, **70**(3), 485 (1991).

15. T.Q. Qui, C.L. Tien, *Int.J.Heat.MassTransfer*, **167**, 25 (1988).

16. J.W.C. de Vries, *Thin.Sol.Films.*, **1988** 167, 25.

17. The frequency difference is calculated as: $v_F/\langle H\rangle - \omega_D$ where v_F is the electron (Fermi) velocity, $\langle H\rangle$ is the average film height, as determined by AFM, and ω_D, the Debye frequency, is 2.2×10^{13} s^{-1}.[14]

18. Finklea, H. O.; Hanshew, D. D.; *J.Am.Chem.Soc.*; **1992**, 114, 3173.

19. Smalley, J. F.; Feldberg, S. W.; Chidsey, C. E. D.; Linford, M. R.; Newton, M. D.; Liu, Y-P; *J.Phys.Chem,* **1995**, 99, 13141.

20. The through-bond electron tunneling rate constant decay parameter has been found to be on the order of 1 per methyl group.[18,19] Given the limiting rate constant, $k_{n=0} = 6 \times 10^8$ sec^{-1}, the number of methyl groups in the crosslinker, the number of carriers per colloid as 5×10^4, and assuming a first order rate equation, one would expect an electron transfer rate on the order of 100 femtoseconds. If the number of crosslinkers between adjacent colloids is taken into account an even high rate would be predicted. Thus, the rate of electron mobility is significantly faster than the e-ph coupling times reported herein and it is reasonable to predict that electron dynamics within a given colloid are sensitive to the presence of adjacent colloids.

21. A.J. Epstein, A.G. MacDiarmid, *Synth.Met.*, **69**, 179 (1995).

Chapter 12

Changes in Thin-Metal-Film Nanostructure at Near-Ambient Temperatures

Shane E. Rorak, Alan Lo, Rex T. Skodje, and Kathy L. Rowlen[1]

Department of Chemistry and Biochemistry, University of Colorado, Boulder, CO 80309–0215

Post-deposition changes in the nanostructure of a thin Ag film (5 nm) on a mica substrate were studied using a combination of experimental and theoretical techniques. Atomic force microscopy was used to quantitatively monitor nanostructure as a function of time at temperatures in the range of 30 - 56 °C. The island size distribution function at these near-ambient temperatures was found to be of a scaling form which increased with time as a power law. Experimental results for this low temperature annealing process are most consistent with an island-island coalescence coarsening mechanism. Island coarsening was found to be temperature dependent up to a crossover time, t_x. Based on an Arrhenius analysis of island count prior to t_x, the coarsening process has an activation energy of 13 ± 2 kcal/mol. After t_x, a temperature-independent asymptotic size distribution and island count was observed.

The structure and properties of metal films on nonmetal surfaces have been the object of considerable recent interest (1). It has become increasingly clear that the morphology of a metal film can drastically affect its optical and chemical characteristics. A salient example is the surface enhanced Raman scattering (SERS) effect. It is well appreciated that the magnitude of surface enhancement is critically dependent on the shape and size of the metal islands on the surface (2). As another example, the performance of a supported metal catalyst is strongly influenced by the distribution of metal islands (3,4). The morphology of a metal film grown or deposited on a solid surface is determined by many factors, including the growth conditions as well as the chemical and physical characteristics of the substrate (1,5,6). However, the factors governing the growth of the film may be insufficient to determine its actual morphology since the film may continue to evolve in time after the deposition phase is complete. Post-deposition reorganization of the film may result from surface processes like atomic diffusion. This reorganization phenomenon

[1]Corresponding author

is interesting as a fundamental problem in surface kinetics. Furthermore, it is of great practical importance to understand the conditions under which the film is stable in order create films with well-defined and reproducible properties.

Reorganization of the metal film can be induced by post-deposition treatment, such as high temperature thermal annealing (7). In the case of supported metal catalysts, thermal annealing is very often used to alter catalyst performance. Thermal annealing of SERS-active thin metal films has also been utilized as a means to improve performance (2). Under most circumstances it is assumed that the annealing rate for thin metal films at room temperature is negligibly slow. However, it has been noted that vapor-deposited thin Ag films must be allowed to "stabilize" after deposition at room temperature (8). Previous work in our laboratory indicated that when heated only slightly above room temperature, thin films of Ag reorganize on a time scale of minutes to hours following deposition (9).

This work investigates Ag films which have been vapor-deposited with an average thickness of 5 nm on a mica substrate. The growth mode involved appears to be Volmer-Weber, i.e. growth through the formation of three-dimensional islands (1). This mode of growth is consistent with the assumption that the Ag-Ag interaction is much stronger than the Ag-substrate interaction. In order to study the stability of thin Ag films, atomic force microscopy (AFM) is used to probe film morphology as a function of time. For each image obtained at a single time and temperature, a three-dimensional image analysis routine was used to quantitatively extract the number density of metal islands and to obtain the distribution functions for island height and radius (10). Thus, the distribution function can be explicitly followed versus time. This provides a time-dependent characterization of the film morphology which is essential for an analysis of the kinetics of the coarsening process. Although scanning probe microscopy has been used to measure surface diffusion on a variety of surfaces (11-14), this is the first *in-situ*, quantitative determination of metal film evolution at low temperatures. In this study, Ag-island size distributions were found to exhibit dynamical scaling behavior upon annealing, consistent with island coalescence. Interestingly, it was observed that the scaling phenomena terminates, leaving a temperature-independent (over the range 30-100 °C) distribution of islands. Thus, it is concluded that low-temperature annealing is sufficient to stabilize a freshly deposited metal film at room temperature.

Overview of Coarsening Kinetics

The theoretical description of thin film growth and the subsequent kinetics of adlayer coarsening (or clustering) has been the subject of considerable recent interest (1,4,5,6,15). A quantitative model for the kinetics of growth and coarsening is difficult to formulate since, typically, the adatoms can lie in many energetically distinct environments leading to a wide range of rates for the fundamental processes. A promising approach to the analysis of film growth, and to the study of aggregation phenomena in general, has been to employ the concepts of self-similarity and scaling (15). Of particular significance for the present analysis is the successful application of scaling theory to island growth for metal films by Chaiken and co-workers (16). The essence of the scaling hypothesis is that a single length scale, say $\bar{h}(t)$ is sufficient to describe the condition of the film (15). Thus, if another scale is defined, e.g., the mean separation between islands, l, then $\bar{h}(t)$ and l must be proportional as time elapses (17).

Several physical mechanisms have been proposed to describe cluster growth on surfaces and in other environments. Of these, Ostwald ripening is probably the most well known (18-20). In Ostwald ripening, islands are located at fixed spatial positions but are allowed to exchange mass with a dilute gas phase through an evaporation-condensation process. For the present problem, the "gas phase" would be Ag atoms confined to the two-dimensional space of the mica substrate. Islands smaller than a critical radius shrink and disappear while bigger islands grow larger. The critical radius grows in time by a power law. The scaling exponent and time-independent distribution function depend on whether the evaporation event or surface diffusion of gas atoms constitute the rate determining step. If the rate of mass transport is limited by adatom surface diffusion, then the average island size varies as $t^{1/4}$. If the mass transport mechanism is limited by a surface barrier at the substrate-island interface then the average island size varies as $t^{1/3}$. The theory of Ostwald ripening also predicts the island size distribution function, $f\left[h/\bar{h}(t)\right]$, for the various cases (18,19).

A second class of growth scenarios are provided by coalescence models (16,21-23). In this case, the islands are mobile on the surface and growth occurs as the islands collide and coalesce. It can be assumed that the reverse process of spontaneous cluster break-up is negligible and the growth is thus irreversible. The scaling properties in coalescence models are determined by the dependence of the cluster diffusion coefficient on cluster size (larger clusters should diffuse more slowly) and on any effective interactions that might occur between the clusters. If the diffusion coefficient, D_k, for an island of mass k, defined as the number of constituent atoms, scales as (23)

$$D_k \sim k^{-\alpha}, \tag{2.1}$$

then the coarsening exponent is $\beta = 1/(3\alpha + 3)$. For an activated diffusion process, there is of course also a factor $\exp(-E_a/k_bT)$ multiplying D_k. The value of α, in turn, depends on the mechanism through which the island diffuses (for example, evaporation-condensation or periphery diffusion). From numerical simulations, the typical range for β_α has been found to be between 1/6 and 1/2. Coalescence models have been successful in describing the annealing of Al clusters on Al alloy films. (16,24)

The island size distribution function may be obtained for coalescence driven coarsening if we assume the kinetics are modeled by the Smoluchowski (16) equation,

$$\frac{dn_k}{dt} = \frac{1}{2}\sum_{j=1}^{k-1} K_{j,k-j} n_j n_{k-j} - n_k \sum_{j=1}^{\infty} K_{j,k} n_j \tag{2.2}$$

where n_k is the number of islands of mass k, and $K_{i,j}$ is the collisional rate constant (or kernel) for the reaction of an island of mass i with an island of mass j. The total number of clusters is

$$N(t) = \sum_{k=1}^{\infty} n_k(t) \tag{2.3}$$

In the present problem, the islands are sufficiently large so that the index can be regarded as nearly a continuous variable, i.e. $n_k \approx n(k)$. The first term in the Smoluchowski equation contains all the contributions due to collisions that form an island of mass k, while the second term describes the irreversible loss of islands of mass k. The use of the Smoluchowski equation involves certain assumptions. First, island formation is due to binary collisions of islands of mass i with islands of mass j to irreversibly form a new island of mass $i + j = k$. Thus, islands cannot split into smaller islands. Second, the reaction is essentially second order with the collision rates of the islands proportional to n_i and n_j. Lastly, the spatial distribution of islands is assumed to be random, with the restriction of no overlapping clusters. Then for Brownian diffusion, the Smoluchowski kernel on a 2-dimensional substrate is

$$K_{i,j} \sim \left(i^{-\alpha} + j^{-\alpha}\right)D_o \tag{2.4}$$

where D_o is a temperature dependent, mass independent constant. It has been shown (16) that the asymptotic cluster size distribution obtained from the Smoluchowski equation with a kernel of the form of Eq. (2.4) can be expressed in the scale invariant form

$$f(\xi) = \frac{3w}{\Gamma(\alpha+1)}\xi^{3\alpha+2}e^{-\xi^3} \tag{2.5}$$

where

$$w = \frac{\Gamma(\alpha+4/3)}{\Gamma(\alpha+1)} \tag{2.6}$$

and

$$\xi = \frac{wh}{\bar{h}(t)} \tag{2.7}$$

Thus, the solutions form a family of distribution functions characterized by a single dimensionless parameter, α, which can be optimized to the data.

Experimental

Thin Ag films were deposited onto freshly-cleaved Ruby mica (S & J Trading) or quartz from Ag powder (5-8 mm, 99.9+%, Aldrich) using a Denton Vacuum DV-502 vapor-phase depositor. Quartz substrates were cleaned by sonicating successively in hexane, chloroform, acetone and methanol (Mallinkrodt). A pressure of ~ 5 x 10^{-6} Torr was maintained during depositions. Film thickness and deposition rate were monitored using a quartz crystal microbalance (Syncon Instruments). The Ag films for this study were deposited at a rate of 0.02 ± 0.005 nm/sec. Optical absorption spectra were obtained on a Hewlett-Packard 8452A diode array spectrophotometer using the appropriate bare substrate as the blank. *In-situ* optical absorbance measurements of Ag annealing were achieved by adhering the sample to a flat, custom-built heating element (Heater Designs, Inc.) with a 0.45 cm hole in the center. The optical beam was passed through the center of the sample. A thermocouple was used to measure the temperature at the center of the sample and the experiment was performed under nitrogen.

A Digital Instruments Nanoscope E with etched single-crystal-silicon tips was used for contact mode atomic force microscopy. The etched single-crystal-silicon tips have a radius of curvature between 5 and 20 nm and a spring constant between 0.01 and 0.3 N/m. All AFM images were acquired at a constant force less than 10 nN and a scan rate of 4.1 Hz. Tip induced perturbation of the Ag surface morphology was not apparent at forces less than ~ 19 nN (25). The AFM system was heated by placing an 80-175 W (E-Z Heat Corp.) band heater around the support base of the tube scanner. AFM *in-situ* Ag annealing studies were achieved by first equilibrating the AFM system to the desired temperature and then quickly positioning the sample on the scanner for imaging. The film temperature was continuously monitored with a fine wire K-type thermocouple (Omega) that was in direct contact with the surface. The AFM was enclosed in a custom-built Lucite chamber which was continuously purged with nitrogen (prepurified, >99.998%, Praxair, Inc). The relative humidity inside the sample chamber typically decreased from ~ 23 % at the start of an experiment to only a few percent at 4 hours. It was previously determined that the measured island characteristics were not dependent on humidity over this range (25). High temperature annealing was performed in a separate enclosure under nitrogen at 100 ^{o}C for 45 min. Long-term annealing studies were performed by periodically imaging a 5 nm Ag film on mica maintained at ~ 43 ^{o}C over a period of 4 days. The homogeneity of a 5 nm Ag film was determined by quantifying 1 μm x 1 μm images obtained over an area of ~ 20 mm^2. From 25 measurements, the relative standard deviation in mean island count, height and radius was 7 %, 12 % and 6 %, respectively.

Calibration of the AFM at elevated temperatures has been previously described (17). At the temperatures used in this study, the instrument sensitivity decreased, resulting in an artificially low measured height; however, the height correction factors used were typically less than 1 nm.

The morphology of the Ag films was quantified using an algorithm described in Ref. 10. The images consisted of a grid of 512 x 512 height values. Each raw AFM image was corrected for any tilt in the x-y plane through automatic leveling, and lowpass filtered to reduce noise. Image boundary effects were eliminated from the analysis by excluding 51 data points from each edge of the image. The numerical algorithm employed a 5 x 5 grid of data points to identify local maxima for use in the island count. Once an island maximum is located, the height is calculated as the difference between the maximum and the average of the ten lowest points on the image. Island radius is obtained by fitting the islands to hemispheroids (10). For each 1 μm x 1 μm image, between 600 and 1500 Ag islands were measured for height, radius and the distribution in these values.

Results and Discussion

Figure 1 shows top-view AFM images of a 5 nm Ag film on mica at three stages during the annealing process at 42 °C. The coarsening of the film results in a decrease in the number of islands, *N*, and an increase in the island height and radius. For evaluation of coarsening kinetics it is important to utilize the most accurate measurement of the film condition. It was previously demonstrated that for these films *N* and island height, *h*, are the most reliable measurements (25). Therefore, the following quantitative discussion focuses on the results obtained for *N* and *h*.

Kinetics. Figure 2 shows the time-dependence of the island count, $N(t)$, for three temperatures T = 35, 42 and 48 °C. The coarsening process terminated at all temperatures considered here; i.e., at long times the island count approached a fixed value, $N(\infty)$. The annealing time was extended to four days in order to verify that the coarsening had indeed ceased. In addition, a significantly higher temperature anneal at 100 °C for 45 min revealed an asymptotic island density of approximately 750 islands per μm^2, the same asymptotic value obtained for a 35 °C anneal for 240 min.

The island count exhibited power law decay up to a crossover time, t_x, beyond which the limiting value was approached, i.e.,

$$N(t) \rightarrow \begin{cases} C(T)t^{-\beta_N} & t < t_x \\ N(\infty) & t > t_x \end{cases} \tag{4.1}$$

where β_N is the coarsening exponent for island density, $C(T)$ is a temperature dependent constant, and $N(\infty)$ is the asymptotic density. $N(\infty)$ was experimentally determined by averaging the value of *N* for the final 40 min of each annealing experiment, which is well beyond the crossover time, t_x. The quality of the power law fit to $N(t)$ is demonstrated in Figure 2b, in which the results are plotted as $\ln[N(t) - N(\infty)]$ vs $\ln(t)$. The function $N(t) - N(\infty)$ is plotted in order to correct for the relatively large asymptotic density of islands (~ 750 μm^2). It is apparent that the island count does show power law decay at early times. At later times, deviation from the power law decay is observed. The transition time, t_x, is defined by the intercept of two linear regressions to the data, as shown in Figure 2b. The values for t_x at temperatures of 30, 35, 42, and 48 °C are shown in Table I. Consistent with the picture that the coarsening proceeds more rapidly as the temperature is increased, t_x decreases as the temperature increases. Also shown in Table I are the coarsening exponents β_N for each temperature, determined by applying a linear regression to $\ln[N(t) - N(\infty)]$ vs $\ln(t)$

From dynamic scaling arguments (15), an increase in temperature is the same as a contraction of the relevant time scale. Thus, the activation energy can be extracted by rescaling the time variable for $N(t)$ as

$$\tau = \exp\left(\frac{-E_a}{k_B T}\right)t. \tag{4.2}$$

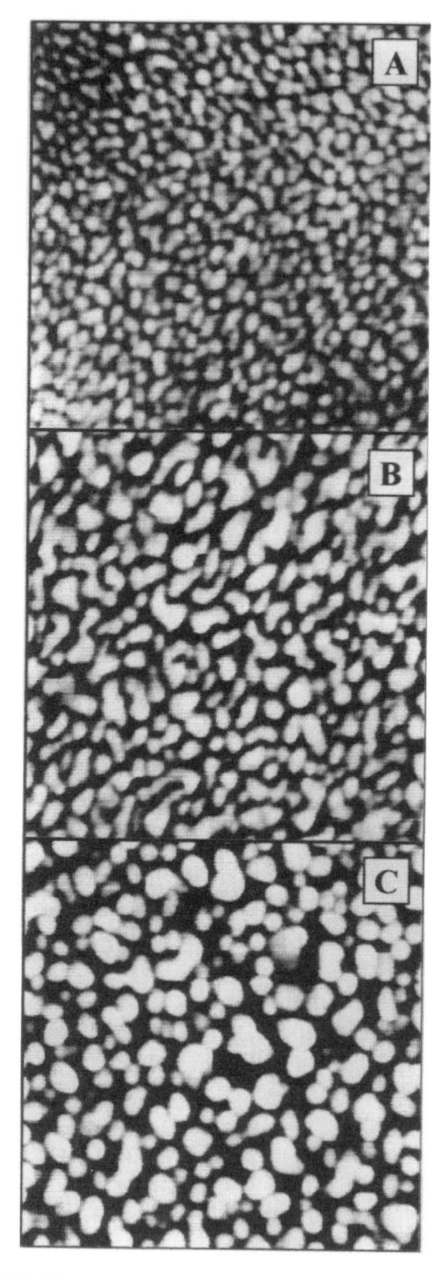

Figure 1. *In-situ* AFM images (500 nm x 500 nm) of a 5-nm thick Ag film on mica at various stages of annealing at 42 °C. (A) is the film prior to annealing, (B) is after 13 minutes of annealing and (C) is after 240 minutes of annealing. The height scale from black (background) to white is 20 nm. Adapted from Ref. 17.

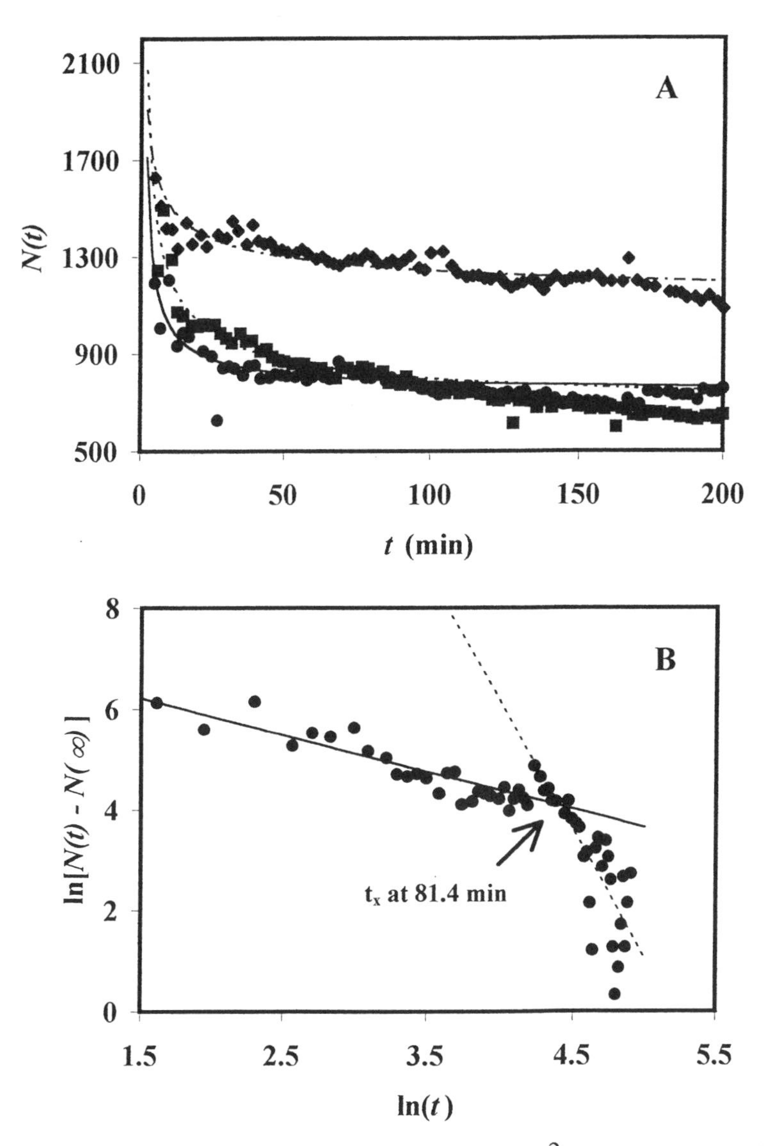

Figure 2. (A) The number of islands, $N(t)$, in a 0.81 μm^2 region, plotted against annealing time at 30 °C (diamonds), 35 °C (squares), and 42 °C (circles). In (B) the data for the 42 °C anneal are plotted as $\ln[N(t) - N(\infty)]$ vs $\ln(t)$. The lines correspond to least squares fits to the data. The crossover time t_x is determined from the intersection of the two lines. Adapted from Ref. 17.

This interpretation of the scaling behavior of the coarsening process is consistent with the Smoluchowski equation (2.5), combined with the Brownian diffusion kernel given in Eq. (2.4). The activation energy, E_a, is extracted from the experimental annealing data through a numerical determination of the time scaling factor required to make the decay profiles, $[N(t) - N(\infty)]/N(0)$, equal at all temperatures (see Ref. 17). Using a least squares fitting procedure, the data at all temperatures was rescaled to the 35 °C data set. The Arrhenius plot of the rescaling factor is shown in Figure 3. The activation energy obtained from the slope of the Arrhenius plot is 13 ± 2 kcal/mol. at times less than t_x.

Table I. Coarsening Exponents

T (°C)	t_x	β_N	β_h
30	165 ± 27	0.13 ± 0.04	0.12 ± 0.02
35	115 ± 22	0.18 ± 0.01	0.30 ± 0.03
42	81.4 ± 34	0.26 ± 0.02	0.15 ± 0.01
48	22.5 ± 10	0.38 ± 0.09	0.15 ± 0.03
Avg.		0.24 ± 0.03	0.18 ± 0.01

Coarsening Mechanism. The height distribution for the three images shown in Figure 1 are given in Figure 4a in the form of histograms with a bin size of 0.15 nm. Qualitatively, there was a clear trend toward larger heights at longer times and the RMS spread of the distribution grew proportionately wider at longer times. A similar trend was also observed in the distribution functions for the island radii. Over a period of 4 hours at 42 °C, the average Ag island height on mica increased by ~300 %. The average island radius increased by ~ 35 %. Qualitatively, the rate of change in height was slower than the rate of change in radius, indicating a change in the island shape during the annealing process.

The physical significance of the activation energy depends, of course, on the mechanism of coarsening. As mentioned in the Introduction, the coarsening exponent, β, can also provide insight into the mechanism. As in the case of $N(t)$, the height and radius distribution functions approached a fixed asymptotic form. That is, the height distribution function can be expressed by the time-independent form, $f[h/\bar{h}(t)]$, where $\bar{h}(t)$ is the mean height. In Figure 4b the three height distribution functions shown in Figure 4a are replotted as a function of the rescaled variable. The resulting functions are identical within statistical error. Hence, even though $\bar{h}(t)$ approached a limiting value, the shape of the distribution was scale invariant. Furthermore, as with $N(t)$, the quantity $\bar{h}(t)$ was observed to exhibit power law behavior up to a cross-over time, t_x (see Ref. 17). Thus, the coarsening exponent for the process can also be

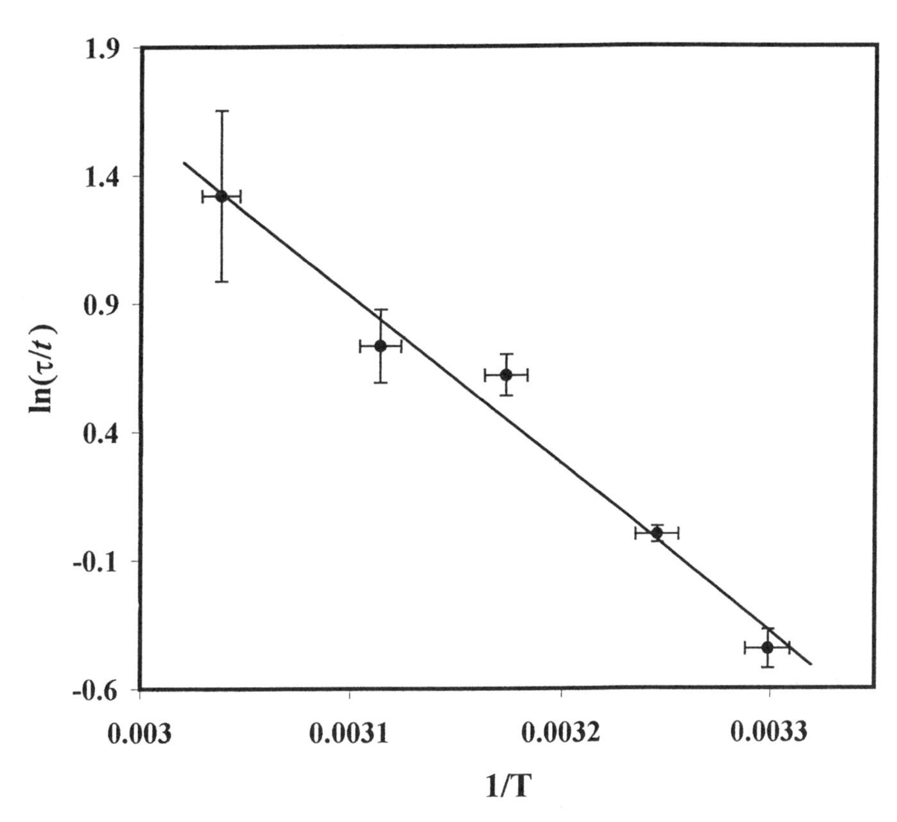

Figure 3. An Arrhenius plot using the time rescaling variable τ, and the annealing data for 30, 35, 42, 48, and 56 °C. From a least squares fit, the slope corresponds to an activation energy of 13 ± 2 kcal/mol. Adapted from Ref. 17.

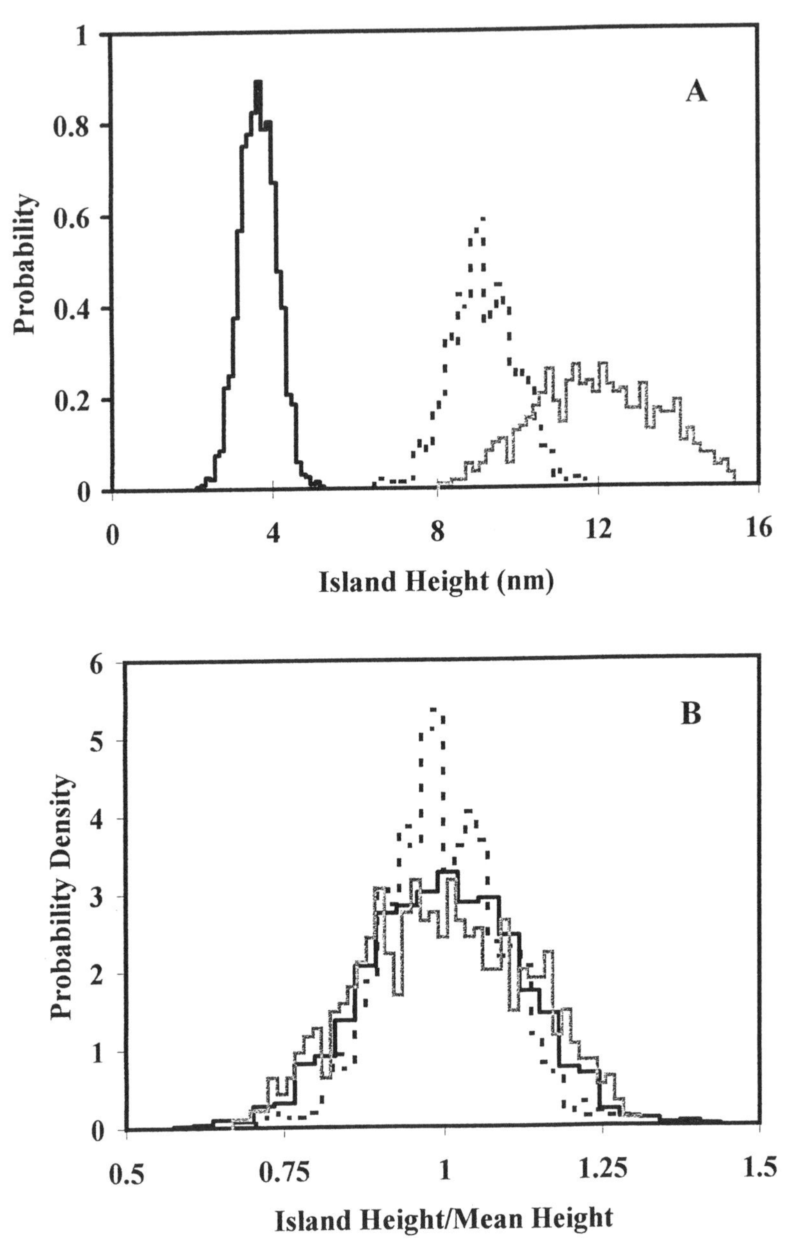

Figure 4. The height histograms for the AFM images shown in Figure 1 are plotted in (A). From left to right the histograms correspond to before annealing, 13 min of annealing, and 240 min of annealing at 42 °C. Based on AFM calibration at elevated temperatures, a correction factor of + 0.9 nm was added to the 13 min and 240 min data sets. In (B), the same three histograms are rescaled to the time-invariant form $f(h,t) \sim f\left[h/\bar{h}(t)\right]$. Adapted from Ref. 17.

extracted from a plot of $\ln\left[\bar{h}(t)-\bar{h}(\infty)\right]$ vs. ln*(t)*. The values of β obtained in this manner, β_h, are given in Table I for comparison with β_N values. The smallest β value predicted for Ostwald ripening is 0.25, whereas the predicted range for coalescence is ~ 0.16 - 0.5. The average value for β_N is 0.24 ± 0.03 and the average value for β_h is 0.18 ± 0.01. A t-test indicates that these two mean values are significantly different. However, it should be noted that β_h is less than the minimum predicted by Ostwald ripening, while neither value is inconsistent with the range specified for coalescence. The difference in the two values of the coarsening exponent probably reflects a breakdown in the proportionality between island number and island height during the annealing process (recall that island shape changes during the process) (17).

Since both Ostwald ripening and coalescence models predict an island size distribution function, a comparison of the two models with the experimentally observed height distributions is made. In Figure 5, the height distribution for a Ag film after 15 min of annealing at 42 °C is shown. The two dashed curves are the normalized theoretical distributions for an Ostwald ripening model (18). The narrower distribution results if surface diffusion is the rate limiting step, while the broader distribution results if atomic evaporation from the island is the rate limiting step. Significant discrepancies are observed between the predicted and experimental results. In contrast, the solid line is a one parameter fit to the solution of the Smoluchowski equation, Eq. (2.5). It is clear from Figure 5 that the coalescence model provides the best agreement with experiment. Further evidence for the coalescence mechanism may be found by examination of the AFM images directly (see ref. 17). Larger islands appear relatively immobile while smaller islands appear to coalesce into adjacent islands.

Interpretation of the Activation Energy. If the coalescence model is an accurate description of the coarsening mechanism, then the activation energy reflects the temperature dependence of the diffusion coefficient for island diffusion. If the Ostwald ripening picture holds, then the activation energy is likely related to the temperature dependence of the atomic evaporation rate from the islands. In the Ostwald model, the fundamental atomic process involves atoms breaking away from the Ag-islands. For the coalescence models, the island diffusion mechanism may either be evaporation-condensation (EC) or periphery diffusion (PD) (23). In the EC mechanism the island diffusion is due to the change in the center of mass occurring when atoms break away from (or attach to) the Ag-islands. In this case, the fundamental activated process is likely to be the same as with Ostwald ripening, i.e., Ag-atoms breaking away from the islands. For PD the center of mass migrates as Ag-atoms migrate around the boundary of the Ag-islands. It is interesting to note that activation energies close to the value determined in this work were reported for the Ag-island diffusion and Ag-atom evaporation on Ag substrates (26,27).

Summary and Remaining Questions

The central results of this study can be summarized as follows: First, the *in situ* time-dependent AFM analysis demonstrated clearly that Ag films on mica exhibit irreversible coarsening at low temperatures, 30 - 56 °C. Second, the island size distribution function assumes a time-independent scaling form parameterized by a

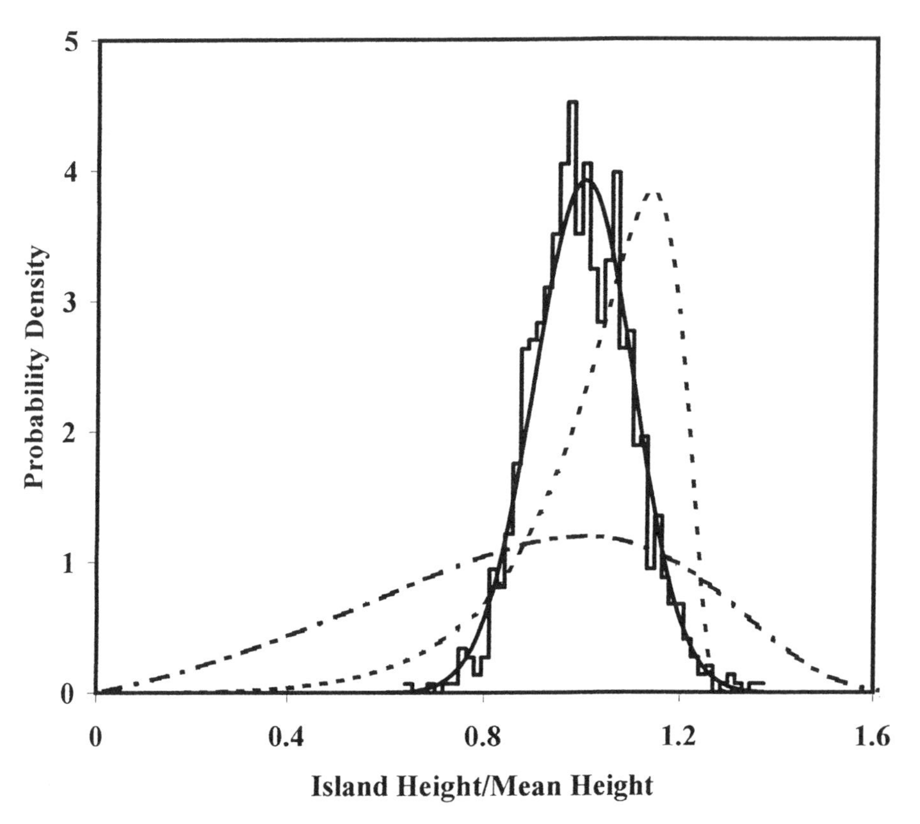

Figure 5. The rescaled island height histogram for a 5 nm Ag film on mica, 15 min after annealing at 42 °C, is shown along with predicted curves for Ostwald ripening and coalescence. The solid line corresponds to a one parameter fit to a solution of the Smoluchowski equation, Eq. (2.5). The dotted line is the distribution for the Ostwald ripening model with mass transport limited by surface diffusion. The dotted-dashed line is the distribution for the Ostwald ripening model where the rate limiting step is atomic evaporation from the Ag island. Adapted from Ref. 17.

mean island size which grows as a power law in time up to a cross-over time, t_x. Third, the scaling behavior terminates beyond t_x and a stable asymptotic island distribution is obtained. Fourth, both the form of the observed island size distribution function and the AFM images provide evidence that the mechanism of coarsening is island-island coalescence. Finally, the coarsening of the Ag film is an activated process with an activation energy of 13 ± 2 kcal/mol.

A potentially useful practical conclusion of this work concerns the preparation of stable thin films. The results obtained here suggest that a newly deposited film at room temperature will undergo coarsening over a period of hours to days. Thus, any physical or optical properties of the film would not be stable during this time. Our results suggest that if the film were to be subject to a slightly elevated temperature, the coarsening process could be greatly accelerated yielding a stable film much more quickly. Furthermore, the final island size distribution is independent of the temperature at which the annealing takes place in the range 30 - 56 °C.

Even though there is evidence to support the diffusion-coalescence model for the Ag-film coarsening, further work is needed to fully establish this interpretation. In particular, the coarsening kinetics exhibit several characteristics that are atypical of standard scaling models. First, the coarsening terminates at a time t_x instead of continuing indefinitely. We are exploring two possible explanations for this phenomenon. It is conceivable that there is a low density of "pinning sites" on the mica substrate which inhibit the diffusion of islands. Using a Ag coated AFM tip and a clean mica surface, the binding forces are being systematically mapped in search of particularly strong binding sites. On the other hand, the cessation of coarsening may be the result of chemical equilibrium between the forward coalescence mechanism and a backward island breakup mechanism. Using Monte Carlo simulations based on empirical potential functions, the viability of the breakup mechanism is being assessed. An additional atypical behavior is that a pronounced shape change in the islands occurs with the average island becoming more prolate with time. This implies that there may be two time scales for coarsening, one to establish a steady state island shape and another for the diffusion-coalescence. Finally, at the 5 nm thickness level there is a very high island density on the substrate which leads to some connectivity of the islands, a small diffusion path for a given island, and presumably a fairly strong correlation in the size distribution of neighboring islands. To simplify the interpretation of the coarsening phenomenon, a set of kinetic measurements and simulations are underway on a 0.1 nm (0.3 monolayer) Ag film subject to low temperature annealing. At this level of coverage, the islands are well separated and diffusion paths can be monitored in real time.

Preliminary studies have been conducted to address the role of substrate in post-deposition annealing at near-ambient temperatures. The first experiment to indicate that the coarsening process in not equivalent for Ag on other substrates is presented in Figure 6. In this experiment, the optical absorbance of 5 nm Ag films simultaneously deposited onto both mica and quartz was monitored as a function of annealing time at 50 °C. It is obvious from this study that temperature does affect Ag on quartz but to a lesser extent than Ag on mica. AFM studies of Ag on quartz do indicate a change in the nanostructure of the films. The height histograms for this particular case are shown in Figure 7. For a 5 nm Ag film simultaneously deposited on mica and quartz, and annealed at 48 °C for 240 min the island density decreases by

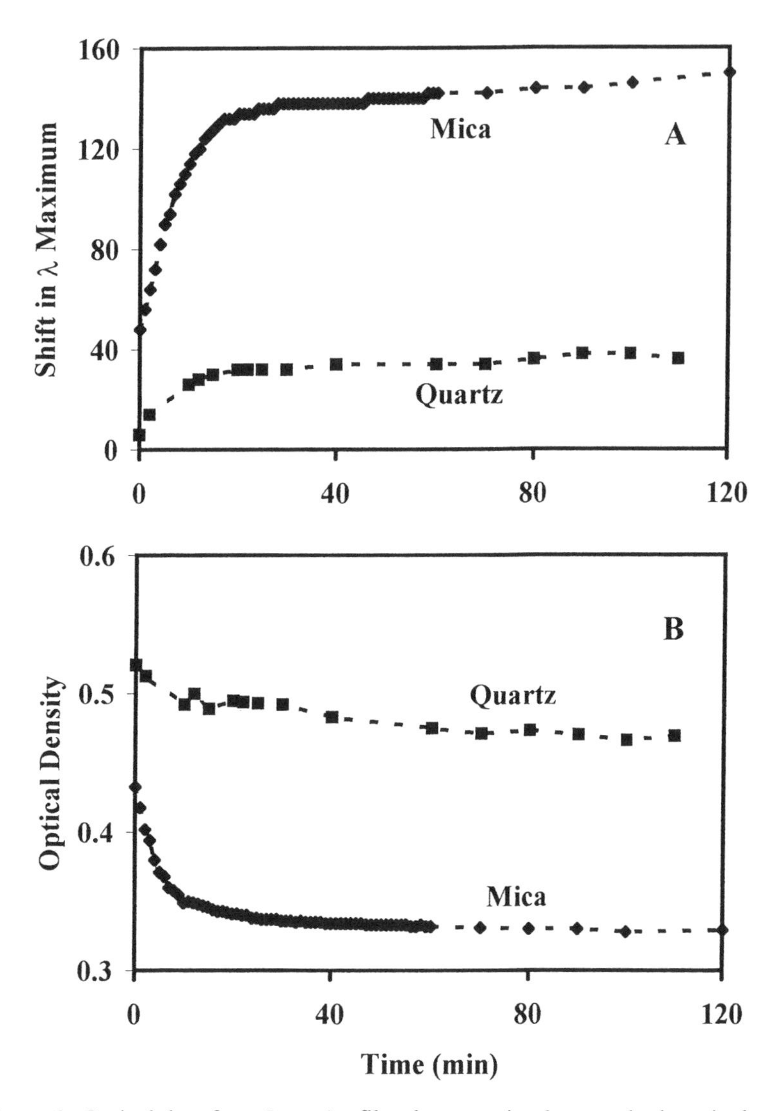

Figure 6. Optical data for a 5-nm Ag film that was simultaneously deposited onto quartz and mica and annealed at 50 °C. The shift in wavelength at the maximum absorbance is shown in (A), and the change in optical density is shown in (B). Adapted from Ref. 17.

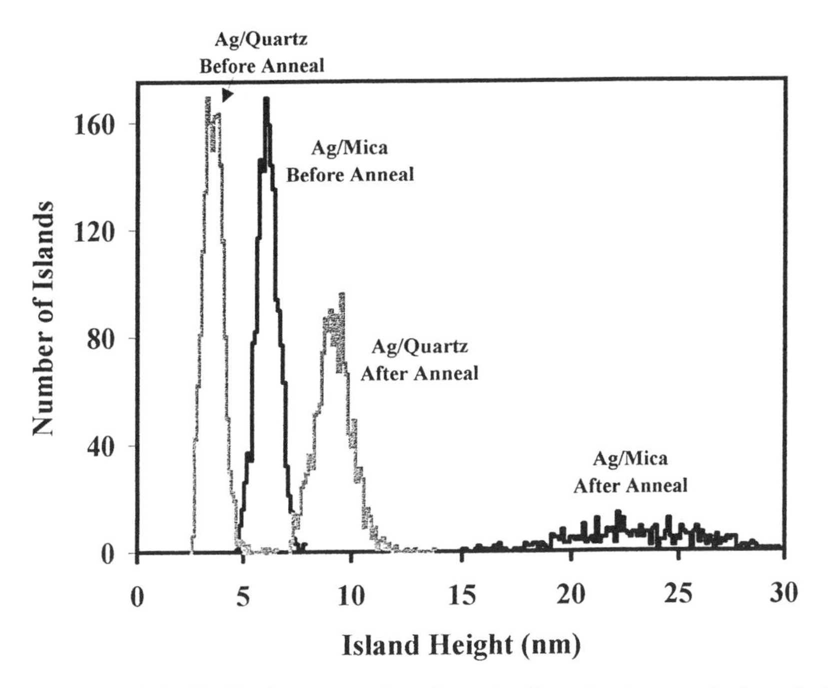

Figure 7. Height distribution curves for a 5-nm Ag film, simultaneously deposited onto quartz and mica, before and 240 min after annealing at 48 °C.

~70% and 20%, respectively. Clearly, substrate plays an important role in establishing the equilibrium that dictates the asymptotic value for both $N(t)$ and $\bar{h}(t)$.

Acknowledgments

This work was supported by grants from the National Science Foundation, CHE-9311638 and CHE-93215434. We also acknowledge support from the Petroleum Research Fund through grant number 27176-AC6.

References

1) D. L. Smith, *Thin-Film Deposition*; McGraw-Hill, New York, 1995.
2) Van Duyne, R. P; Hulteen, J. C. and Treichel, D. A. *J. Chem. Phys.* **1993**, *99*, 2101.
3) *Metal-Support Interactions in Catalysis, Sintering, and Redispersion*; Stevenson, S. A.; Dumesic, J. A.; Baker, R. T. K.; Ruckenstein, E., Eds.; Van Nostrand Reinhold, NY, 1987.
4) Wynblatt, P., In *Growth and Properties of Metal Clusters*; Bourdon, J., Ed.; Elsevier, Amsterdam, 1980, p. 15.
5) Venables, J. A.; Spiller, G. D. T.; Hanbucken, M. *Rep. Prog. Phys.* 1984, *47*, 399.
6) Meakin, P. *Phys. Rep.* **1993,** *235*, 189.
7) Levine, J. R.; Cohen, J. B.; Chung, Y. W. *Surf. Sci.* **1991,** *248*, 215.
8) Tzeng, C.; Lue, J. T. *Surf.Sci.* **1989,** *216*, 579.
9) Semin, D. J. and Rowlen, K. L. *Analy. Chem.* **1994,** *66*, 4324.
10) Roark, S. E.; Semin, D. J.; Lo, A.; Skodje, R. T.; Rowlen, K. L. *Analy. Chim. Acta* **1995,** *307*, 341.
11) Lin, T. and Chung, Y. *Surf. Sci.* **1989,** *207*, 539.
12) Tong, W. M.; Snyder, E. J.; Williams, R. S.; Yanase, A.; Segawa, Y. and Anderson, M. S. *Surf. Sci. Lett.* **1992,** *227* L63.
13) Porath, D.; Goldstein, Y.; Grayevsky, O. M. *Surf. Sci.* **1994,** *321*, 81; Porath, D.; Bar-Sadeh, E.; Wolovelsky, M.; Grayevsky, A.; Goldstein, Y.; Millo, O. *J. Vac. Sci. Technol. A* **1995,** *13*, 1165.
14) Schimmel, T.; Bingler, H. G.; Franzke, D.; Wokaun, A. *Adv. Mater.* **1994,** *6*, 303.
15) Barabasi, A. L. and Stanley, H. E. *Fractal Concepts in Surface Growth*, Cambridge University Press, Cambridge, 1995.
16) Chaiken, J. and Goodisman, J. *Thin Solid Films* **1995,** *260,* 243.
17) Semin, D. J.; Lo, A.; Roark, S. E.; Skodje, R. T.; Rowlen, K. L. *J. Chem. Phys.* **1996,** *105*, 5542.
18) Chakraverty, B. K. *J. Phys. Chem. Solids* **1967,** *28*, 2401.
19) Jain, S. C. and Hughes, A. E. *J. Material Sci.* **1978,** *13*, 1611.
20) Zinke-Allmang, M., In *Kinetics of Ordering and Growth at Surfaces*; Lagally, M. G., Ed.; Plenum Press, New York, 1990, p. 455.
21) Botet, R. and Jullien, R. *J. Phys. A* **1984,** *17*, 2517; Villarica, M.; Casey, M. J.; Goodsman, J.; Chaiken, J. *J. Chem. Phys.* **1993,** *98*, 4610.
22) Meakin, P. *Physica* **1990,** *A 165*, 1.
23) Sholl, D. S. and Skodje, R. T. *Phys. Rev. Lett.* **1995,** *75*, 3158; Sholl, D. S. and Skodje, R. T. *Physica A.* **1996,** *231*, 631; Sholl, D. S.; Fichthorn, K. A.;and Skodje, R. T. accepted, *J. Vac. Sci. Tech.*
24) Aceto, S.; Chang, C. Y.; Vook, R. W. *Thin Solid Films* **1992,** *21,* 80.
25) Roark, S. E.; Semin, D. J.; Rowlen, K. L. *Analy. Chem.* **1996,** *68*, 473.
26) Morgenstern, K.; Rosenfeld, G.; Poelsema, B.; Comsa, G. *Phys. Rev. Lett.* **1995,** *74*, 2058.
27) Nelson, R. C.; Einstein, T. L.; Khare, S. V.; Rous, P. J. *Surf. Sci.* **1993,** *295*.

Chapter 13

Synthesis of Doped Fullerene Clusters and Boron–Nitrogen Tubules Using Laser Ablation

Z. Charles Ying[1], Jane G. Zhu[1], R. N. Compton[2,3], L. F. Allard, Jr.[3], R. L. Hettich[3], and R. E. Haufler[4]

[1]Department of Physics, New Mexico State University, Las Cruces, NM 88003
[2]Departments of Chemistry and Physics, University of Tennessee, Knoxville, TN 37996
[3]Oak Ridge National Laboratory, P.O. Box 2008, Oak Ridge, TN 3781
[4]Comstock Inc., 1005 Alvin Weinberg Drive, Oak Ridge, TN 37830

A variety of nanostructured materials, including clusters, nanoparticles, nanotubules, and thin films, can be produced by using the laser-ablation technique. This paper reports our synthesis of (i) metal endohedral fullerene clusters, (ii) nitrogen doped fullerene clusters, and (iii) boron-nitrogen nanotubules and nanoparticles employing the laser-ablation technique. These novel materials have been characterized by electron microscopy, mass spectrometry, and x-ray photoemission spectroscopy. The boron-nitrogen nanoparticles and nanotubules have been separated from other laser-ablation products, including boron particles and boron-nitride flakes, using a hydrogen-peroxide treatment.

Fullerene molecules are a class of carbon clusters with close-shell geometric structures. The fullerene molecule with the highest symmetry (icosahedral symmetry I_h) is C_{60}, where all sixty carbon atoms are located at equivalent sites of a "sphere" of 0.7 nm diameter.

Fullerene molecules were discovered in 1985 in experiments of molecular beams produced by supersonic expansion of carbon vapor created by laser ablation of a graphite target (*1*). Only microscopic quantities of fullerenes were available at that time. In 1990, production of macroscopic quantities of fullerenes was achieved by arc discharge using two graphite rods as electrodes (*2*). It was found in the following year that fullerene synthesis in macroscopic quantities can also be achieved using laser ablation of solid graphite at a high temperature (*3*).

A fullerene molecule can be modified in many ways. One may add an atom either inside (figure 1a) or outside of the fullerene cage (figure 1b), forming

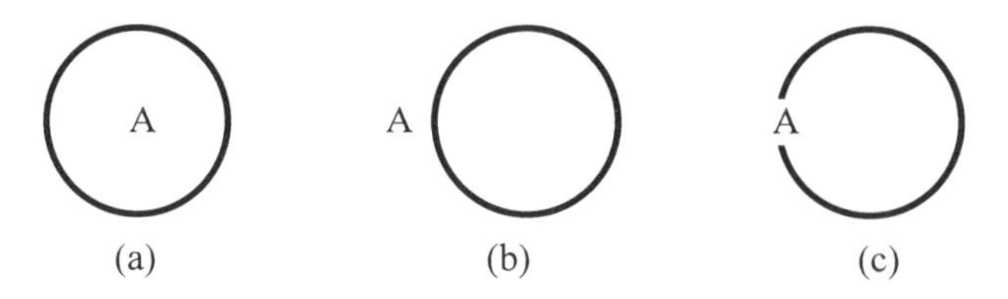

Figure 1. Modification of a fullerene molecule by placing an atom (a) inside, (b) outside, or (c) at the cage.

endohedrally or exohedrally doped species, respectively. One may also replace a carbon atom at the fullerene cage by a foreign atom, creating an intrahedrally-doped molecule. Examples of endohedral fullerenes that have been synthesized and purified are Ar@C_{60} (*4, 5*), La@C_{82} (*6, 7*), Y@C_{82} (*8*), and Sc_3@C_{82} (*9*). Hydrogen (*10*) and fluorine atoms (*11*) have been attached to the outside of a fullerene cage structure. Intrahedrally doped fullerene species, including NC_{59} (*12*), BC_{59} (*13*), BC_{69} (*13*), have been synthesized and purified very recently.

Another route of fullerene modification is to replace carbon atoms entirely. One may view a fullerene molecule as a graphite sheet rolled into a closed "sphere" by adding 12 pentagons. Solid graphite consists of layers of carbon atoms in a hexagonal arrangement (figure 2a). There are many other solids that have similar structures. Hexagonal boron nitride, for example, is a solid with a layered crystal structure (figure 2b). In each layer, boron and nitrogen atoms are alternatively located in a hexagonal pattern. The geometric and physical properties of hexagonal boron nitride are very similar to those of graphite. The lattice constants of boron nitride are $a = 0.251$ nm and $c = 0.668$ nm. The values for graphite are $a = 0.246$ nm and $c = 0.671$ nm. It might therefore be the case that boron and nitrogen form fullerene-like molecule and/or nanotubules (*14, 15*).

In this article, we describe our recent experimental effort in production and characterization of fullerene-based materials. The systems include endohedral and intrahedral fullerene species and boron-nitrogen nanotubules and nanoparticles. These

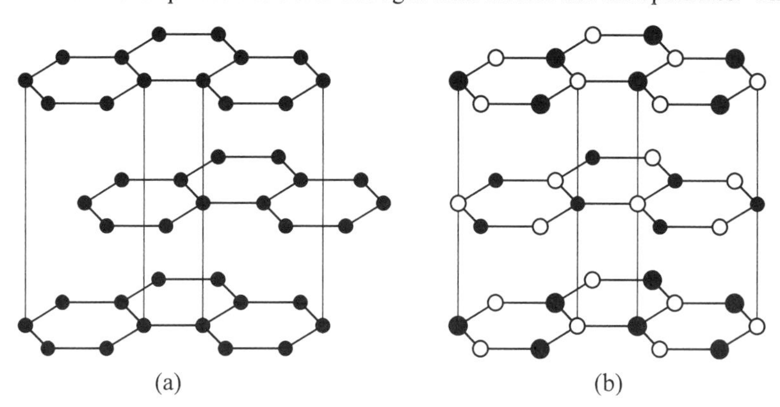

Figure 2. Crystal structure of (a) graphite and (b) hexagonal boron nitride.

materials were produced using the laser-ablation technique, which provided a well-controlled production environment.

The major experimental techniques for materials characterization in this study include transmission electron microscopy (TEM) for structure characterization, x-ray photoemission spectroscopy (XPS) for chemical composition determination, and mass spectrometry for molecular weight measurements. The mass spectrometry experiments employed a Fourier-transform ion-cyclotron resonance mass spectrometer (FTMS) and a time-of-flight mass spectrometer (TOFMS).

Experimental Procedures

The laser-ablation apparatus used in our experiments is shown in figure 3. The laser source was the second-harmonic output (532 nm) of a Nd:YAG laser operated in the nanosecond mode at a repetition rate of 10 Hz. The laser beam was directed to a solid target surface, at an incident energy of 0.2 J per pulse. The laser-beam size at the target surface varied from 1 to 3 mm in diameter.

The target was placed inside a quartz tube, evacuated to ~10 mTorr and then back filled with a gas. During laser ablation, the gas flowed continuously and the target was heated to 1100°C using a tube furnace. A fraction of the ablated material was deposited on the quartz-tube wall in front of the target (position A in figure 3). Another fraction of materials was carried out of the heated zone by the gas flow and condensed near the furnace edge (position B in figure 3), where it was much cooler. These materials were collected for analysis after laser ablation (*16*, *17*).

A high-resolution TEM with a field-emission gun (Hitachi 2000FX) was used. The atomic images and diffraction patterns were recorded using a charge-coupled device.

The photoemission measurements were performed using a Perkin-Elmer instrument. The x-ray source was provided by the aluminum Kα lines. The photoelectrons were analyzed by a hemispherical electrostatic analyzer.

The mass analysis employed a Fourier-transform mass spectrometer (Finnigan FTMS-2000) and a linear time-of-flight mass spectrometer (Comstock Inc. L-TOF-160-5KV). In both cases, the fullerene molecules were desorbed from a sample disk and simultaneously ionized by a pulsed laser beam of UV wavelength (355 nm for FTMS and 337 nm for TOFMS). Positive ions were detected.

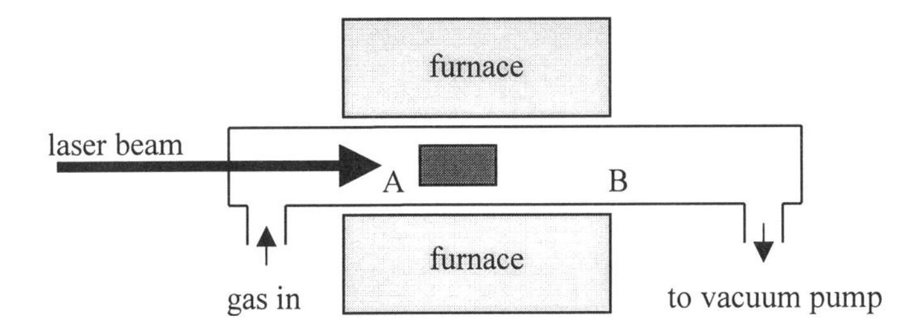

Figure 3. Schematics of the laser-ablation apparatus.

Endohedral fullerene: La@C_n

Endohedral fullerenes with one or more metal atoms inside the carbon cage were synthesized by laser ablation of graphite-metal mixture. Synthesis of lanthanum doped fullerenes, for example, was carried out using targets made of graphite and lanthanum oxide or lanthanum carbide. The optimal molar ratio between carbon and lanthanum atoms is about 100:1 (*16*). A vacuum sublimation procedure was used to transfer the synthesized fullerene molecules onto a stainless-steel disk for further analysis. This procedure minimized the exposure of the samples to air and eliminated any solvent, often used in fullerene material collection.

Figure 4a exhibits a mass spectrum for the molecules synthesized by laser ablation of graphite-lanthanum mixture. The most abundant species in the spectrum are two empty fullerene molecules, C_{60} at 720 amu and C_{70} at 840 amu, and a number of lanthanum-doped molecules La@C_n (n = 60–84). Of these endohedral fullerenes, La@C_{60}, La@C_{70}, La@C_{74}, and La@C_{82} (at 859, 979, 1027, and 1123 amu, respectively) stand out, and their abundance decreases as the molecular weight increases.

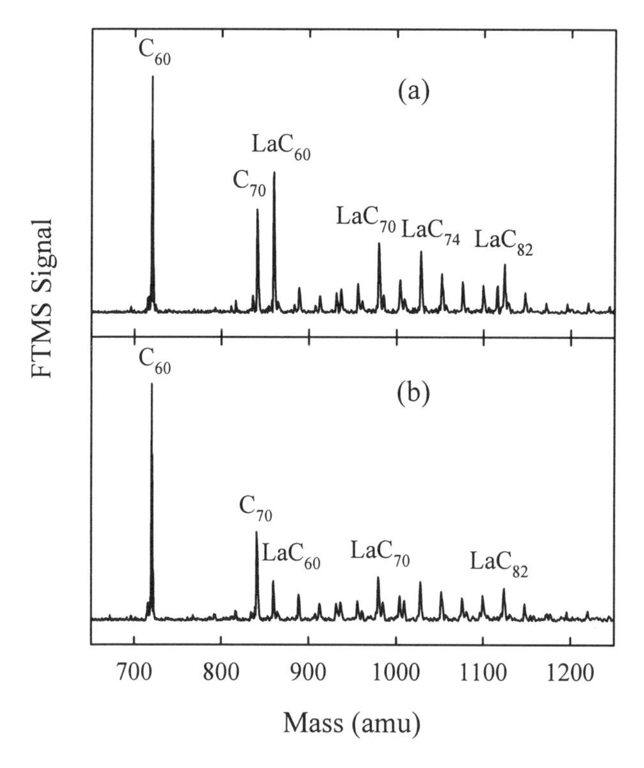

Figure 4. Fourier-transform mass spectra of fullerene molecules synthesized by laser ablation of graphite-lanthanum mixture. The mass analysis was performed (a) 3 days and (b) 66 days after the material synthesis.

The observation of La@C_{60} and La@C_{70} in high abundance in our experiments is in contrast to the cases where the endohedral fullerenes were extracted using toluene. The latter method yields La@C_{82} and La@C_{74}, but no La@C_{60} (*6, 7*). A possible explanation of this discrepancy is that La@C_{60} is more reactive than the heavier lanthanum-doped fullerenes. Figure 4b shows a mass spectrum for the same sample used in figure 4a, but recorded 2 months later. Comparing these two spectra, the signal of La@C_n (n = 70, 74, 82, etc.) decreases by a factor of 2 after two months. In contrast, the La@C_{60} signal diminishes by nearly a factor 4, presumably due to reaction with water or oxygen in the air. The data indeed suggest that La@C_{60} is more reactive than other lanthanum-doped fullerenes.

In addition to La@C_n, Ho@C_n (*16*), Ca@C_n (*18*), and Sc_3@C_n have been synthesized using the laser ablation method in our laboratory.

Intrahedral fullerene: NC_{n-1}

Since the discovery of the arc discharge of graphite in helium as an efficient method of fullerene production (*2*), attempts have been made for synthesis of fullerene molecules doped with nitrogen atoms using arc discharge of graphite in a nitrogen (*19, 20*) or pyrrole (C_4H_5N) atmosphere (*21*). The results of these studies, however, are inconclusive due to the presence of a large number of species in the samples and/or limited resolution of the mass spectrometers. Recently, Hummelen *et al.* have isolated and characterized a dimer of the azafullerenyl radical, biazafullerenyl $(C_{59}N)_2$, using a condensed-phase organic synthesis method (*12*). They were also successful in producing the $C_{59}N^+$ ions in the gas phase using fast-atom-bombardment mass spectroscopy of a cluster-opened N-methoxyethoxy methyl ketolactam. Electrospray mass spectroscopy of the solution sample showed a strong mass peak at 722 amu ($C_{59}N^+$) and a weaker peak at 1445 amu ($C_{118}N_2^+$). They argued that the latter peak is not due to molecular clustering in the gas phase but rather to the existence of the dimer in solution.

By performing laser ablation of pure graphite under nitrogen environment, we have found that nitrogen-doped fullerene material was readily synthesized. The presence of nitrogen was confirmed by XPS measurements and the molecular identity of the doped fullerenes was revealed by mass analysis.

A x-ray photoemission spectrum is shown in figure 5. The spectrum exhibits a strong peak at 284.6 eV, due to the carbon 1*s* level, and a weak peak at 400.7 eV, assigned to the nitrogen 1*s* level. The observation of this nitrogen peak indicates the presence of nitrogen atoms in the sample. There are three possible sources of nitrogen: (a) air contamination, (b) nitrogen molecules trapped in the space between the fullerene molecules, and (c) nitrogen atoms incorporated into the fullerene molecules. The first two possibilities can be ruled out by performing test experiments and by examination of the experimental procedure (*17*). That leaves the last possibility, that the nitrogen atoms are incorporated into fullerene molecules.

If the sample is 100% pure of $C_{59}N$, the molar ratio between the nitrogen atoms and carbon atoms is 1:59. The XPS peak ratio between nitrogen and carbon should be 1:35 after taking XPS sensitivity factors into consideration. The observed XPS peak ratio is 1:300. The data suggest that undoped fullerene molecules are also present in the sample and ~10% of the fullerene molecules are doped with nitrogen.

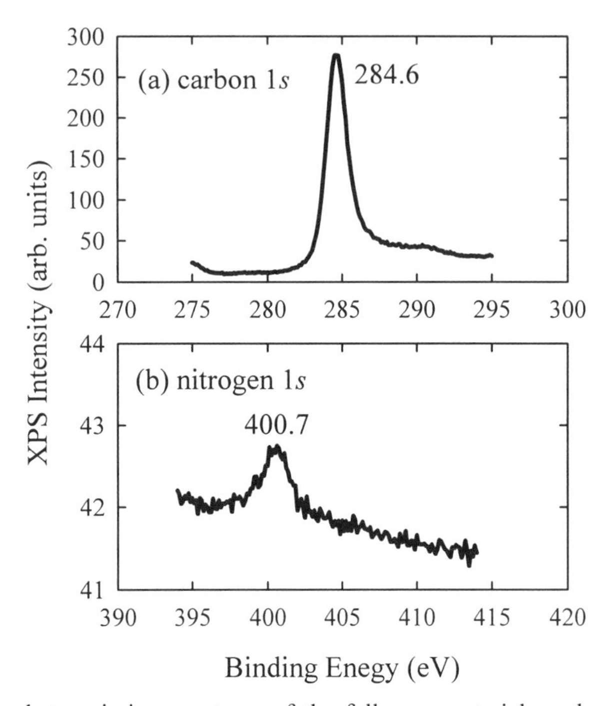

Figure 5. X-ray photoemission spectrum of the fullerene material produced by laser ablation of graphite under a nitrogen atmosphere.

Figure 6 shows a time-of-flight mass spectrum of the material produced by laser ablation of graphite under a nitrogen environment. The spectrum, exhibiting a long progression of fullerene peaks separated by 24 amu (a C_2 unit), is similar to a spectrum of all-carbon fullerenes synthesized by laser ablation of graphite. The high-fullerene signal is somewhat stronger in the present case. A close examination of each peak, which actually consists of several subpeaks separated by 1 amu, reveals a crucial difference.

An expanded plot around the C_{60} region of the same spectrum is shown in the upper left corner of figure 6. The peak at 720 amu is due to $^{12}C_{60}$. The peak at 721 amu can be attributed to $^{13}C^{12}C_{59}$. Its intensity agrees with the expectation (67% of the 720-amu-peak intensity) based on the natural abundance of ^{12}C (98.9%) and ^{13}C (1.1%). The peak at 722 amu has contribution from $^{13}C_2{}^{12}C_{58}$, but its intensity should be only 22% of that of the 720-amu peak. In contrast, the observed intensity of the 722-amu peak is higher than that of the 720-amu peak, suggesting the presence of another species. A logical candidate for this species is NC_{59}. In order to confirm this assignment as well as to rule out other possibilities such as hydrogen attachment, laser ablation experiments were also performed using isotopically labeled nitrogen $^{15}N_2$. Its TOF mass spectrum exhibits a low intensity peak at 722 amu but an intense peak at 723 amu. This is exactly what would be expected from an increase of nitrogen-atom

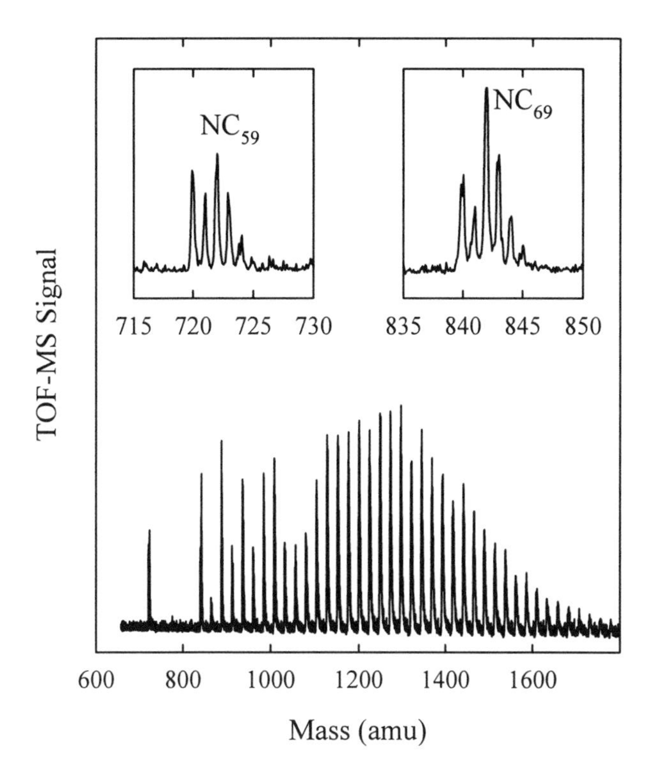

Figure 6. Time-of-flight mass spectrum of fullerene molecules produced by laser ablation using N_2 as the carrier gas.

mass by 1 amu. There can be little doubt that what is observed in the mass spectra are $C_{59}N^+$ ions.

Nitrogen substitution also occurs for other fullerenes (*17*). The upper right corner of figure 6, as an example, presents the mass spectrum in the C_{70} region. The data shows the presence of $C_{69}N$ as well as C_{70}: The peaks at 840 and 841 amu are due to C_{70}. The major contribution to the peaks at ≥842 amu is $C_{69}N$. The presence of multiple peaks is due to the natural abundance of ^{13}C.

It should be noted that the observed mass spectrometry peaks represent positive ions produced by laser desorption and ionization of the sample material. There might not be any neutral $C_{n-1}N$ existing in the sample since they would be free radicals. The observed $C_{n-1}N^+$ ions could be due entirely to fragmentation of neutral $(C_{n-1}N)_2$ during the ionization process, as suggested by Hummelen *et al.* in the case of n = 60 (*12*). Indeed, we have observed $C_{118}N_2^+$ in mass spectrometry experiments (*17*). It is likely that the $C_{59}N^+$ ions observed in our study come from the breaking of $C_{59}N$ dimers. We have also searched for dimer ions of larger nitrogen-doped fullerenes, $(C_{m-1}N)_2^+$ (m>60), but found no evidence for their existence. The state of larger nitrogen-doped fullerenes in the sample requires further investigation.

Boron-nitrogen nanotubules and nanoparticles

The third class of nanoscaled materials produced using the laser-ablation technique is not made of carbon, but boron and nitrogen. The experiments were done by ablating a solid target of hexagonal boron nitride. Because of white color of the target material, the laser beam was focused more tightly in this case (1 mm beam diameter at the target surface) in order to realize efficient ablation. The material produced during laser ablation was brown in color. The sublimation method, which was used in our fullerene experiments, yielded no observable material on a sample disk in this case. The vapor pressure of the produced boron-nitrogen material must be very low. The boron-nitrogen material was collected by brushing the quartz tube.

A x-ray photoemission spectrum of the as-produced material is shown in figure 7a. The spectrum exhibits a nitrogen 1*s* peak at 397.8 eV, but two boron 1*s* peaks of approximately equal intensity at 186.4 and 190.2 eV. The positions and relative intensities of the nitrogen peak and one of the boron peak at 190.2 eV match exactly those of hexagonal boron nitride. The good match indicates that a fraction of the product consists of flakes of hexagonal boron nitride. The second boron peak at 186.4 eV can be assigned to elemental boron.

High-resolution TEM study revealed the presence of several species in the product. There are a large number of crystalline flakes of hexagonal boron nitride, in consistent with the XPS observation. There are many straight "lines" and round particles, as shown in figure 8. High-resolution imaging of one of these "lines" shows

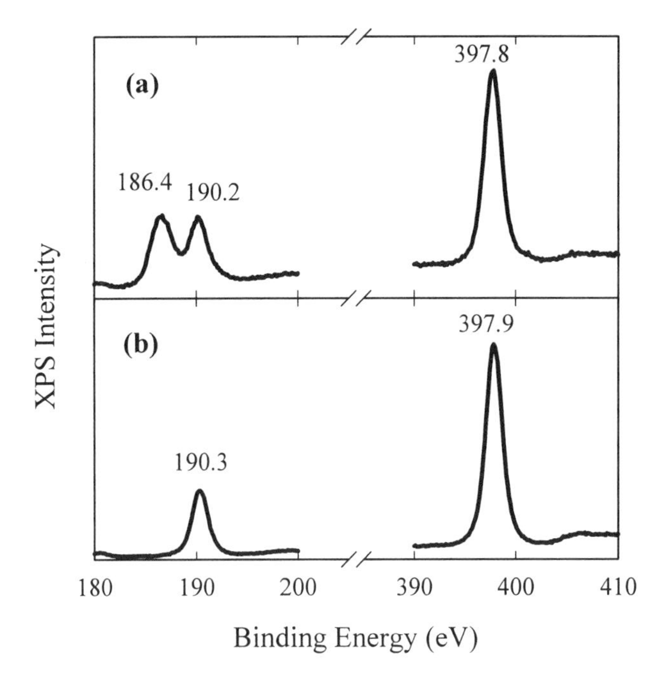

Figure 7. X-ray photoemission spectrum of the boron-nitrogen material produced by laser ablation of boron nitride. (a) As produced and (b) after H_2O_2 treatment.

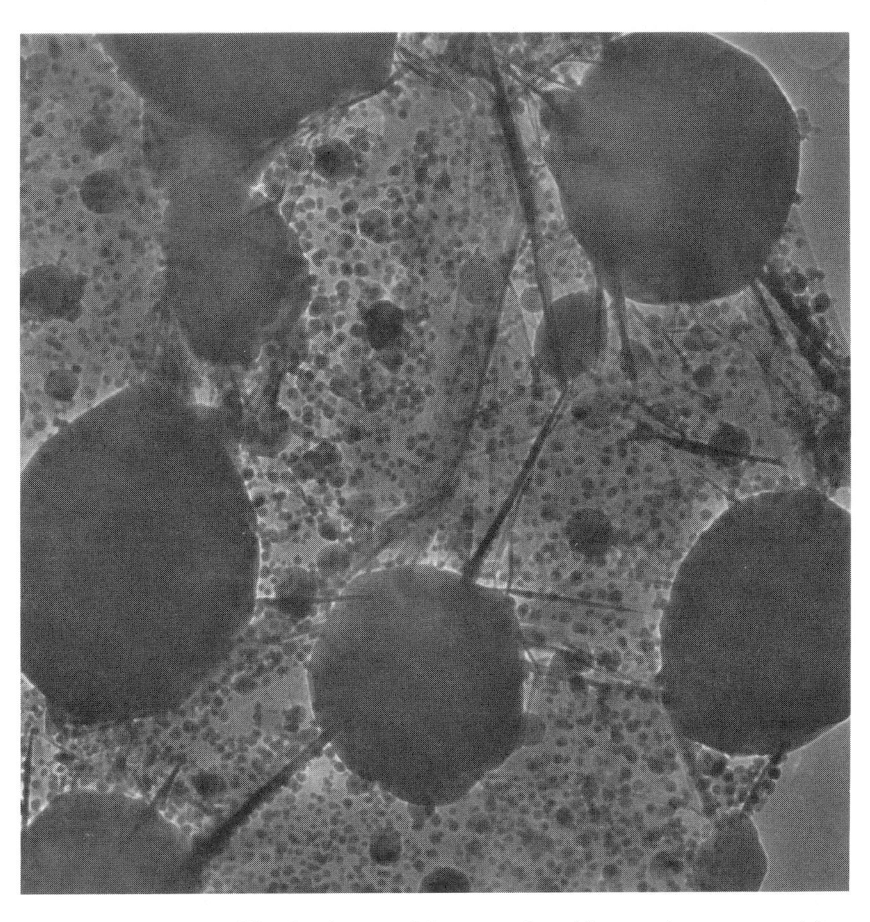

Figure 8. A low-magnification image of the as-produced boron-nitrogen material. The image area is 1.77 μm × 1.77 μm.

lattice fringes separated by 0.33 nm (figure 9). This value matches the distance of two adjacent planes of hexagonal boron nitride. Tilting the TEM specimen in two orthogonal directions by 10 degrees did not substantially change the TEM contrast of these "lines" in the low-magnification image or the lattice fringes in high resolution. This observation strongly suggests that these "lines" are actually nanotubules made of boron and nitrogen. It is noted that these nanotubules are filled to the center; hollow nanotubules have not been found in our experiments. The number of BN layers ranges from 4 to ~20. The ends are round in shape, which is similar to carbon nanotubules but different from one recent report of boron-nitrogen nanotubules (*15*).

Figure 10 is a TEM image of a large particle, which has a multilayer shell. The spacing between the two adjacent shell layers is 0.33 nm, identical to that of bulk boron nitride. Therefore, the particle imaged in figure 10 is surrounded by a multilayer shell of hexagonal boron-nitride structure.

High-resolution imaging of small particles often shows well-defined lattice fringes, with spacing as large as 0.9 nm. These crystals cannot be explained by boron nitride because of the large lattice spacing. The candidate for these crystallites is elemental boron, as suggested by the XPS data. There are three known phases of elemental boron. One is tetragonal ($a = 1.01$ nm and $c = 1.42$ nm). The other two are rhombohedral: rhombohedral-α ($a = 0.491$ nm and $c = 1.257$ nm) and rhombohedral-β ($a = 1.093$ nm and $c = 2.38$ nm). An analysis of the lattice imaging suggests that the small particles are likely to be in the rhombohedral-β phase of elemental boron.

The above TEM results show at least four types of materials present in the product of laser ablation of boron nitride: nanotubules, large shelled particles, small boron crystallites, and boron-nitride flasks. These materials can be separated in a simple procedure. Adding water solution of hydrogen peroxide into the laser-ablation product, the color of the solution changed to light yellow and the color of the sediment changed from brown to white over a period of several days. An XPS spectrum of the sediment (figure 7b) exhibits only two peaks, which are assigned to hexagonal boron nitride. TEM imaging of the materials suspended in the solution shows nanotubules, large shelled nanoparticles, and boron-nitride flakes (to a lesser extent), but no small boron particles. Apparently, hydrogen peroxide dissolved particles of elemental boron, forming boric acid:

$$2B + 3H_2O_2 \rightarrow 2BH_3O_3.$$

The large boron-nitride flakes settled at the bottom of a vial. The nanotubules, shelled nanoparticles, and small boron-nitride flakes were suspended in the solution. The chemical treatment thus largely separated nanotubules and shelled nanoparticles from other laser-ablation products.

The exciting results above have prompted us to search for the molecular forms of boron-nitrogen clusters (e.g., $B_{36}N_{24}$) that are similar to fullerenes. Neither positive nor negative ions of any boron-nitrogen clusters were observed by using mass spectrometry. The filtered solution of the laser-ablated material is transparent and lacks any absorption at optical wavelengths as short as 200 nm. These results suggest that boron-nitrogen clusters were either not made in our experiments or unstable in air. Another possibility is that the electron affinity of the clusters is negative and the ionization potential is high, making their observation by mass spectroscopy impossible or very difficult.

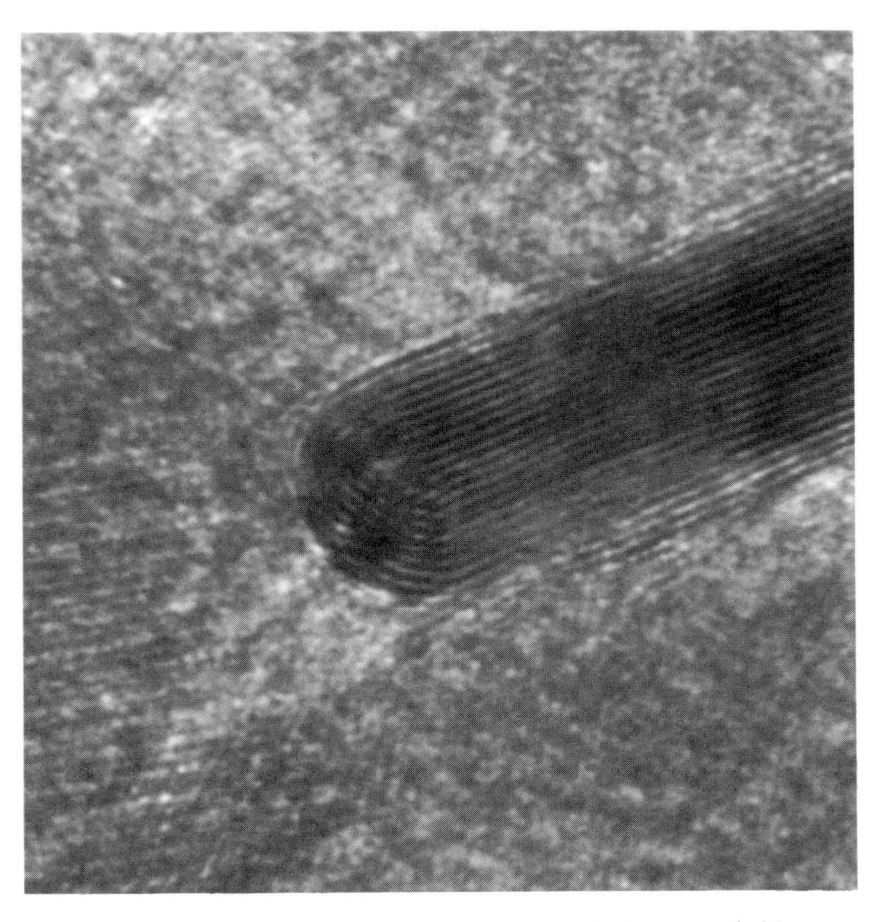

Figure 9. An image of a part of a boron-nitrogen tube. The image area is 25 nm × 25 nm.

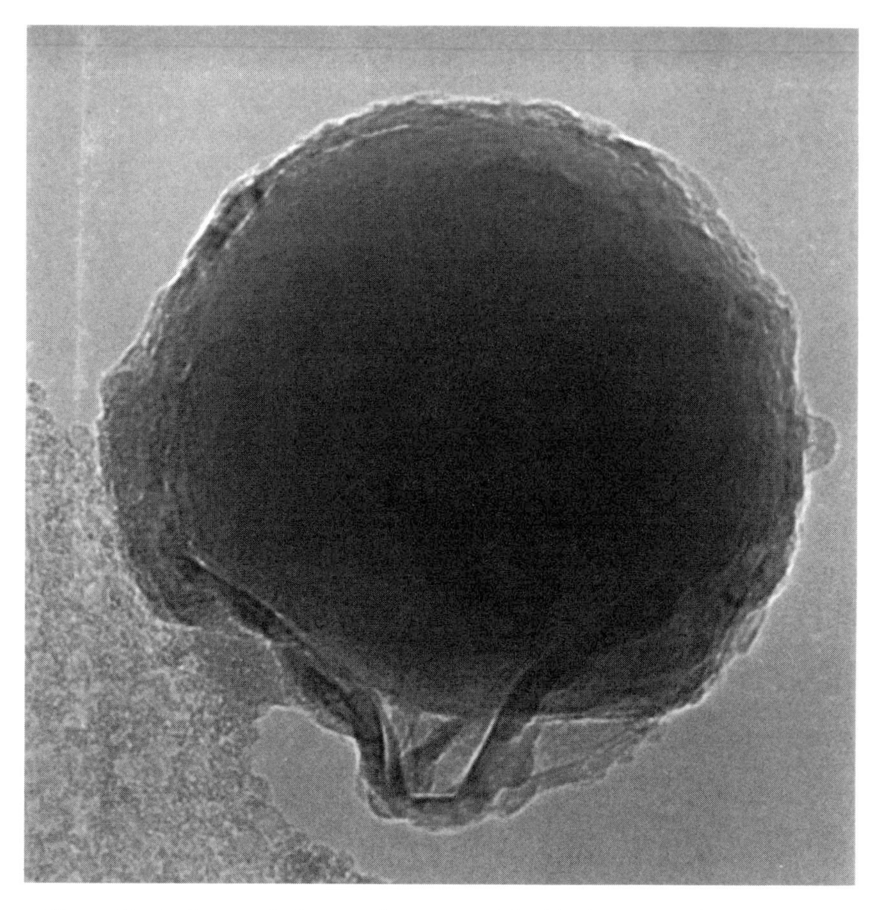

Figure 10. An image of a boron-nitrogen particle. The image area is 147 nm × 147 nm.

Conclusions

Laser ablation has been used to synthesis several fullerene-related materials. These include endohedral fullerenes doped with lanthanum atoms, nitrogen-doped intrahedral fullerenes, and boron-nitrogen nanoparticles and nanotubules. The nanostructured materials have been identified and characterized using electron microscopy, mass spectrometry, and photoemission spectroscopy. This and other studies (*22*, *23*) have shown that the laser ablation is a viable technique for synthesis of exotic-fullerene and nanotubules.

Acknowledgments

This research was sponsored in part by the U.S. Department of Energy under contract DE-AC05-96OR22464 with Lockheed Martin Energy Research Corporation.

References

1. Heath, J. R.; O'Brian, S. C.; Zhang, Q.; Liu, Y.; Curl, R. F.; Kroto, H. W.; Tittel F. K.; Smalley, R. E. *J. Am. Chem. Soc.* **1985**, *107*, 7779.

2. Krätschmer, W.; Lamb, L. D.; Fostiropoulos, K.; Huffman, D. R. *Nature* **1990**, *347*, 354.

3. R. E. Haufler, Chai, Y.; Chibante, L. P. F.; Conceicao, J.; Jin, C.; Wang, L.-S.; Maruyama, S.; Smalley, R. E. *Mater. Res. Soc. Symp. Proc.* **1991**, *206*, 627.

4. Saunders, M.; Jiménez-Vázquez, H. A.; Cross, R. J.; Mroczkowski, S.; Gross, M. L.; Giblin, D. E.; Poreda, R. J. *J. Am. Chem. Soc.* **1994**, *116*, 2193.

5. DiCamillo, B. A.; Hettich, R. L.; Guiochon, G.; Compton, R. N.; Saunders. M.; Jiménez-Vázquez, H. A.; Khong, A.; Cross, R. J.; *J. Phys. Chem.* **1996**, *100*, 9197.

6. Kikuchi, K; Suzuki, S.; Nakao, Y.; Nakahara, N.; Wakabayashi, T.; Shiromaru, H.; Saito, K.; Ikemoto, I.; Achiba, Y. *Chem. Phys. Lett.* **1993**, *216*, 67.

7. Yamamoto, K.; Funasaka, H.; Takahashi, T.; Akasaka, T. *J. Phys. Chem.* **1994**, *98*, 2008.

8. Kikuchi, K; Nakao, Y.; Suzuki, S.; Achiba, Y.; Suzuki, T.; Maruyama, Y. *J. Am. Chem. Soc.* **1994**, *116*, 9367.

9. Shinohara, H.; Inakuma, M.; Hayashi, N.; Sato, H.; Saito, Y.; Kato, T.; Bandow, S. *J. Phys. Chem.* **1994**, *98*, 8597.

10. Jin, C.; Hettich, R.; Compton, R.; Joyce, D.; Blencoe, J.; Burch, T. *J. Phys. Chem.* **1994**, *98*, 4215.

11. Jin, C.; Hettich, R. L.; Compton, R. N.; Tuinman, A.; Derecskei-Kovacs, A.; Marynick, D. S.; Dunlap, B. I. *Phys. Rev. Lett.* **1994**, *73*, 2821.

12. Hummelen, J. C.; Knight, B.; Pavlovich, J.; González, R.; Wudl, F. *Science* **1995**, *269*, 1554.

13. Muhr, H.-J.; Nesper, R.; Schnyder, B.; Kötz, R. *Chem. Phys. Lett.* **1996**, *249*, 399.

14. Boulanger, L.; Andriot, B.; Cauchetier, M.; Willaime, F. *Chem. Phys. Lett.* **1995**, *234*, 227.

15. Loiseau, A.; Willaime, F.; Demoncy, N.; Hug, G.; Pascard, H. *Phys. Rev. Lett.* **1996**, *76*, 4737.
16. Ying, Z. C.; Jin, C.; Hettich, R. L.; Puretzky, A. A.; Haufler, R. E.; Compton, R. N. in *Fullerenes, Recent Advances in the Chemistry and Physics of Fullerenes and Related Materials*, Kadish, K. M.; Ruoff, R. S. Ed.; Electrochem. Soc. Proc.; The Electrochemical Society: Pennington, 1994, Vol. 94-24; pp. 1402-1412.
17. Ying, Z. C.; Hettich, R. L.; Compton, R. N.; Haufler, R. E. *J. Phys. B*. **1996**, *29*, 4935.
18. Lahamer, A. S.; Compton, R. N. (to be published).
19. Pradeep, T.; Vijayakrishnan, V.; Santra, A. K.; Rao, C. N. R. *J. Phys. Chem.* **1991**, *95*, 10564.
20. Yu, R.; Zhan, M.; Cheng, D.; Yang, S.; Liu, Z.; Zheng, L. *J. Phys. Chem.* **1995**, *99*, 1818.
21. Glenis, S.; Cooke, S.; Chen, X.; Labes, M. M. *Chem. Mater*. **1994**, *6*, 1850.
22. Guo, T.; Nikolaev, P.; Thess, A.; Colbert, D. T.; Smalley, R. E. *Chem. Phys. Lett*. **1995**, *243*, 49.
23. Guo, T.; Nikolaev, P.; Rinzler, A. G.; Tománek, D.; Colbert, D. T.; Smalley, R. E. *J. Phys. Chem.* **1995**, *99*, 10694.

Chapter 14

Thin-Film Formation by Laser-Assisted Molecular Beam Deposition

Robert L. DeLeon, Paras N. Prasad, and James F. Garvey

Department of Chemistry, University of New York at Buffalo, Buffalo, NY 14260–3000

Laser Assisted Molecular Beam Deposition (LAMBD) is a pulsed laser deposition technique which utilizes a train of gas pulses to precisely control the chemistry and transport of species to then be deposited on a substrate. The LAMBD technique has been used to grow films on a variety of substrates and to ablate a variety of target materials including: metals, metal oxides, metal halides, Si, C and SiC. In addition, a variety of more complex composite films such as Cu embedded polymers, $YBa_2Cu_3O_{7-x}$ and organic doped TiO_2 and Al_2O_3 have also been produced. These films have been analyzed by Raman scattering, SEM, EDX, FTIR, ESCA, NMR, mass spectrometry and surface profile measurements to determine film structure and composition. These results demonstrate the potential of using LAMBD sources as a means of depositing a variety of high quality single and multiple component films.

In the LAMBD technique, a target is ablated and gas pulses are passed over the target ablation plume to sweep the species towards a substrate. The pulsed gas cools and mixes with the ablation species which allows this technique to deposit flat, high quality films. This technique provides the high species fluxes and precise control of the process chemistry difficult to achieve by other deposition methods. The LAMBD process has advantages over other deposition techniques for a number of different types of materials. The present paper will concentrate on Cu, Si, C, SiC, TiO_2, Al_2O_3 and $YBa_2Cu_3O_{7-x}$.

Initial work focused on the generation of copper metal (*1*), copper oxide (*1*) and copper iodide (*2,3*) thin films. Copper ablation is efficient and the chemical formation of copper oxide and copper iodide make a convincing demonstration of the rapid reaction of the carrier gas with the ablated Cu metal. Eventually this work led to the production of copper-embedded polymers (*4*).

Silicon and other Group IV elemental films have been prepared by a variety of techniques, including molecular beam epitaxy (MBE) and chemical vapor deposition (*5*). Many of the standard growth techniques have been successful (*6*) for high temperature substrates heated to 350-1000 C. However, laser ablation methods typically suffer from poor film quality. Deposition of Si and C by MBE is extremely difficult since the low vapor pressure of these materials requires very high temperatures and makes precise flux control difficult. The high flux rate and precise control over the deposition process make the LAMBD process much better for the Group IV elements.

High temperature superconductors (HTS) have been prepared successfully by conventional pulsed laser deposition (*7-9*). However, typically for materials such as $YBa_2Cu_3O_{7-x}$ only the heavy elements Y, Ba and Cu are deposited by laser ablation. The oxygen is driven into the film by a harsh, high temperature annealing process at 800-1000 C. Such high temperature treatments tend to cause cracking and other physical defects in the film. In the LAMBD process, the proper oxygen content of the film can be achieved in a one-step process without annealing.

Thin films containing organic materials are useful in a number of applications in the areas of electronics and photonics (*10-14*). One important class of materials which can be produced by LAMBD is non-linear optical (NLO) materials for photonics applications. The key is to generate mechanically stable films in which the organic molecules are embedded in an inorganic matrix using a low temperature process which does not decompose the delicate organic component. Approaches such as sol-gel formation have disadvantages such as cracking, aging and insufficient densification. The LAMBD technique has been used to deposit a variety of probe organic dye molecules in inorganic matrixes such as TiO_2 and Al_2O_3 (*15*).

Experimental

A sketch of the LAMBD source is shown in Figure 1. The beam from a Lambda Physik EMG 150 pulsed excimer laser operated at 248 nm is focused into the source. Laser pulse energies are varied depending upon the target and application. Materials easy to ablate, such as C and Si, can be removed by pulse energies of only 5 mJ or less. Cu and Al are intermediate and require pulse energies on the order of 100 mJ. Ti and $YBa_2Cu_3O_{7-x}$ are difficult to ablate and require pulse energies approaching 200 mJ. The beam is focused to an estimated spot size less than 0.1 mm by a two lens optical system onto a cylindrical target rod. A variety of targets have been employed in the past including metals (*1-4,17*) , non-metals (*18*), composites (*1*) and even pressed powders (*19*). The present paper will review our work on Cu, Si, C, SiC, Al_2O_3, TiO_2 and $YBa_2Cu_3O_{7-x}$ films in detail. The target rod is rotated and translated to expose a fresh target surface during each individual laser pulse. However, aside from resurfacing the rods between runs, the targets are not pre-treated to remove oxides. A train of ~ 2 ms gas pulses is directed over the target which serves to entrain the ablation species in the flow. When an inert gas, such as He, is used the gas pulse

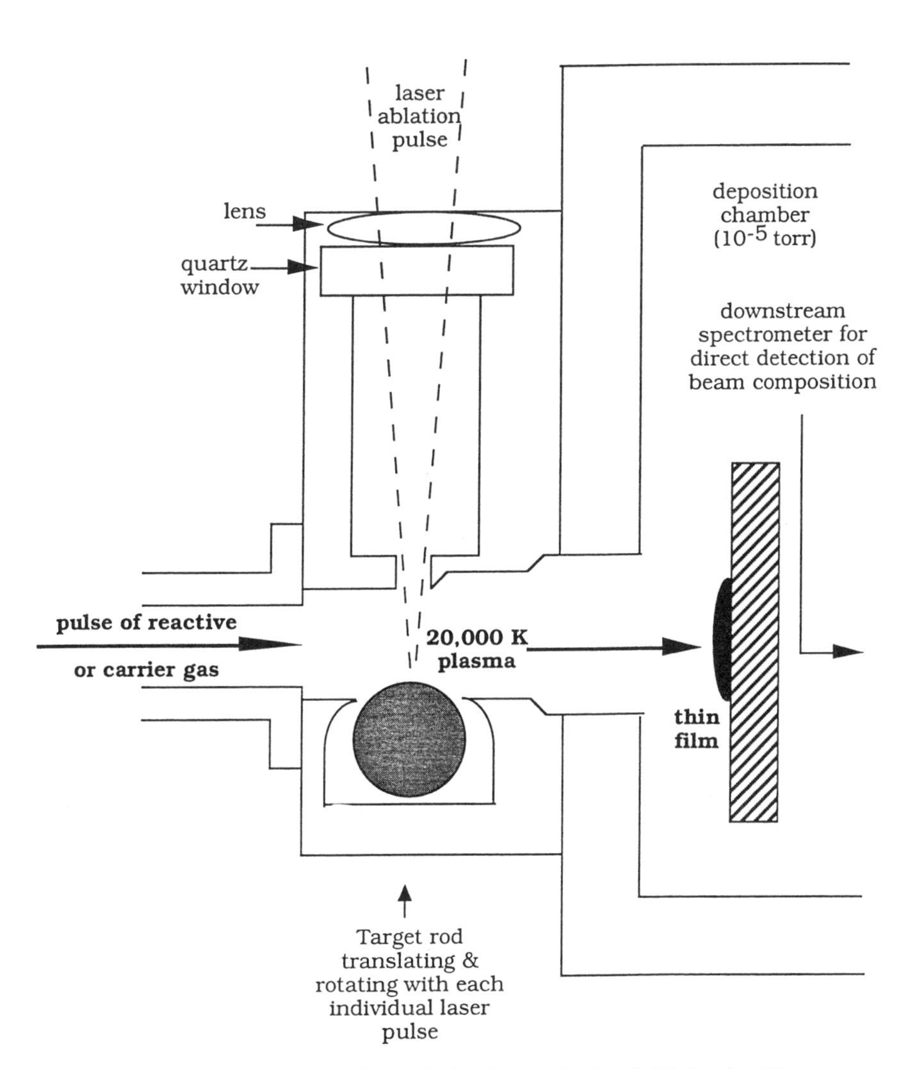

Figure 1. Schematic drawing of the Laser Assisted Molecular Beam Deposition source apparatus.

simply serves to transport the ablation species to the substrate. Previous programs have also successfully used reactive gases (*2,3*) or gases seeded with a variety of organic molecules (*15,16*). When a reactive gas, such as oxygen is used, gas phase chemistry in the ablation plume can be induced to produce desirable products. The ablation species and the gas mix and are cooled by expansion. The expanded gases are directed downstream for deposition onto a room temperature substrate. The substrate is located in a vacuum chamber pumped down to a base pressure of 10^{-5} T by a diffusion pump. Optical emission spectra of the ablation plasma were obtained by dispersing the plasma radiation with a 0.25 m monochromator, detecting with a photomultiplier and processing the signal with a box car averager. The observation of optical emission from the plasma is used to monitor the ablation process. Emission intensity has been correlated with the ablation flux (*18*) and has been used to measure the ablation plasma temperature which is on the order of 20,000 K (*19*).

Results

Cu, CuO and CuI Films

Early deposition of demonstration films by the LAMBD method concentrated on copper target ablation because of its relative efficiency. Initially studies were made on the effects of varying parameters such as carrier gas identity and laser pulse energy (*1,2*). Figure 2A shows an SEM image of a copper metal film grown on a glass substrate at a high pulse energy of 280 mJ/pulse using He as an inert carrier gas. Note the formation of many circular droplet-like structures. These structures are caused by the liquefaction of macroscopic portions of the target rod followed by transport of the droplets to the deposition surface (*20*). The laser pulse energy was reduced by 25% to 210 mJ to obtain the film characterized by the SEM shown in Figure 2B. Note the complete elimination of droplet formation. We infer that this difference in gross morphology is due to a change in the average particle size leaving the source. Lower laser pulse energy translates to lower plasma temperatures causing micro-droplets to become nano-droplets.

Copper metal films grown using He as the carrier gas show the bright, highly reflective surface characteristic of pure copper surfaces. However, if O_2 gas is used, the appearance of the films changes drastically. The films take on the non-reflective rusty brown appearance of CuO films. ESCA and EDX measurements confirmed that the chemical nature of the films grown in the O_2 gas are different from those grown in He gas. To study reactive chemistry within the LAMBD process we chose CuI as the test molecule. I_2 vapor was seeded into the carrier gas by flowing He through a heated U-tube containing iodine crystals. Copper target rods were ablated similar to the copper metal deposition experiments. The I_2 reacts with the ablation plasma to form CuI which is then deposited as a thin film. The CuI films were extensively analyzed by EDX, SEM, Raman, UV absorption, and ESCA (*2,3*). Figure 3 shows a comparison of the ESCA spectra of a vacuum-evaporated high purity CuI film with a film grown by LAMBD. The two spectra are essentially identical. Both show Cu^+ as the majority Cu species, I^+ as the I species and trace amounts of oxygen and a second Cu oxidation

A

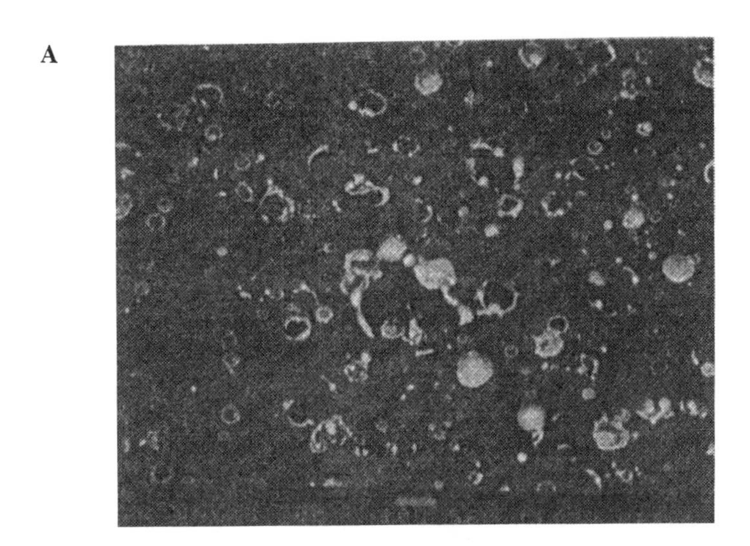

B

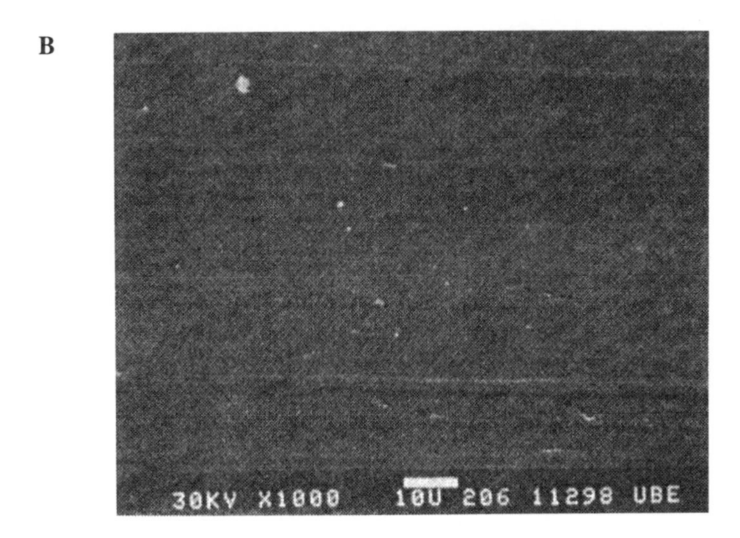

Figure 2. SEM images of Cu metal films grown by the LAMBD technique. A) On the top is a film grown at a laser pulse energy of 280 mJ/pulse. B) On the bottom is a film grown at a laser pulse energy of 210 mJ/pulse. (Reproduced with permission from reference 1. Copyright 1991 Materials Research Society.)

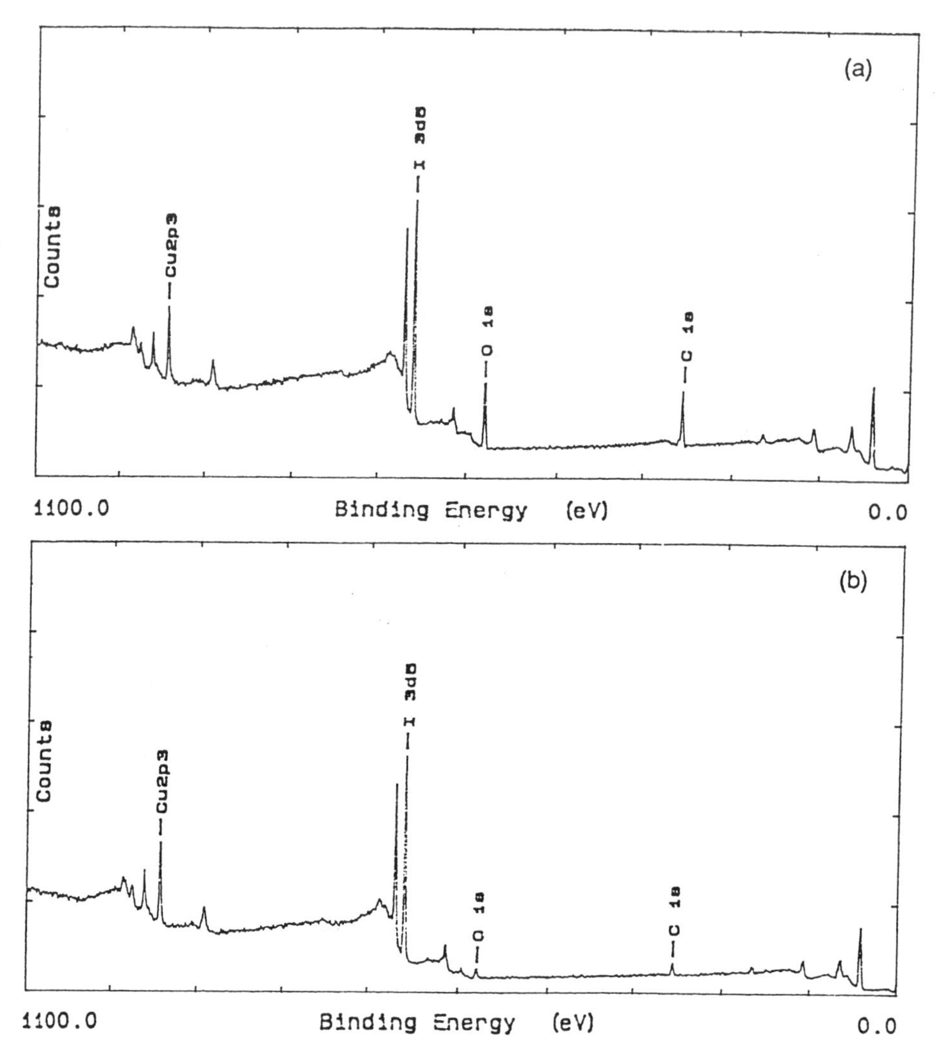

Figure 3. Electron Spectroscopy for Chemical Analysis (ESCA) survey spectrum of A) LAMBD film and B) CuI powder. (Adapted from reference 3.)

state. EDX spectra show predominantly Cu and I with small amounts of Si presumably from the silica substrate. UV-visible absorption spectra and Raman spectra of the two types of film are identical. SEM images comparing the vacuum-evaporated CuI films with low laser pulse energy LAMBD films show fewer surface inhomogeneities in the LAMBD films. The CuI work is a convincing demonstration that the LAMBD process is capable of formation of multielement films by *in situ* chemical reaction.

$YBa_2Cu_3O_{7-x}$ High Temperature Superconductors

A brief, early attempt was made to grow $YBa_2Cu_3O_{7-x}$ superconducting films by the LAMBD technique (*1*). A commercial $YBa_2Cu_3O_{7-x}$ rod was used as the target and initially He was used for the carrier gas. However, it was determined that this approach did not produce the correct stoichiometry. Using He/O_2 mixtures or pure O_2 produced the correct elemental composition. Yet, the films did not show high temperature superconductivity. Further analysis of the films by ESCA and EDX indicated the presence of a small but persistent aluminum impurity. The presence of this impurity alone is sufficient to destroy the superconductivity of the films. The aluminum impurity was traced to the use of an aluminum source block for the ablation process. The source block has been redesigned and rebuilt to widen the ablation channel and to minimize deposition on the entrance window. Current plans are also to construct a stainless steel source with a ceramic liner on the ablation channel which should be much more resistant to unwanted source ablation. The newly designed source will guarantee the elimination of the aluminum impurity. Further attempts will be made to produce high temperature superconducting thin films by the LAMBD technique.

Group IV Films

Si Films. The majority of the effort on Group IV films was made to grow and characterize Si films. Si films were grown on either pure Si wafers etched with HF or silica glass substrates. An SEM image of a thin Si film grown on a Si wafer is shown in Figure 4. The film was cleaved to show the edge in Figure 4. The SEM images and supporting surface profile measurements show smooth, flat Si films which can be grown to the thickness of several microns. The image shows that embedded within the film are a number of round drop-like features which are due to the formation of liquid droplets during the high pulse energy ablation process. The origin of these droplets is well known (*20*) and solutions to the elimination of their formation are also well known. For the case of copper films shown in Figures 2a and 2b a 25% reduction of the laser pulse energy was sufficient to essentially eliminate all droplet formation. Therefore, a series of films were grown at lower laser pulse energies. These lower pulse energy films showed reduced droplet formation.

Chemical characterization of the Si films was also performed. EDX spectra showed only one peak which is assignable to Si, but a high resolution ESCA spectra showed Si peaks somewhat shifted from that of pure elemental Si but not as far as the expected SiO_2 position. The presence of some oxidation on the Si surface is expected in the light of the 10^{-5} baseline pressure. A Raman spectrum of a Si film deposited on a silica substrate shows a peak centered at 522 cm^{-1} which is the zone center optic phonon mode associated with single crystal Si. This signal may be due to the drop-like structures seen in Figure 4. Si films were grown on salt plates and analyzed by FTIR. The main absorption feature is at 1020 cm^{-1}. We attribute this to SiO_x where $x<0.2$.

Figure 5 shows the emission spectrum obtained from the Si plasma. The presence of both atomic Si and Si^+ are detected. Most importantly, the emission spectrum is a powerful tool for monitoring the ablation process.

C Films. C films have been grown on Si and polystyrene substrates. Films grown on silica tend to flake off when the film layer gets thicker than 1 micron. Surface profiles and SEM images show smooth, flat films with much less droplet formation than for the Si films. Measurement of C film resistance was > 100 M-ohm, indicating that the films are amorphous and not graphite-like. The ESCA spectra of C on a Si wafer shows mainly C with some Si and O. The Si and O may be from the substrate or contamination. Emission spectra on the C ablation plasma showed only a single peak at 193 nm due to atomic carbon.

SiC Films. SiC films were deposited on Si and polystyrene substrates. SEM images of SiC on polystyrene show smooth, flat films with less droplet formation than the Si films. The ESCA spectrum shows a peak at 102.3 eV. Which is shifted from the Si-Si bond. We attribute this peak to Si-C bonds. Thus the LAMBD process has produced Si-C films, even using a room temperature substrate. The SiC plasma emission spectrum was taken from 300-600 nm. It is similar to the Si spectrum shown in Figure 5. This indicates the presence of atomic Si and Si^+ in the plasma. No attempt has been made yet to observe the C vacuum UV line at 193 nm during the SiC ablation.

Copper-embedded Polymeric Films

Laser ablated metal can react with organic vapor to generate polymeric materials which can be deposited as thin films. The LAMBD apparatus was used to ablate copper target rods using a mixture of acetylene and acetone as the carrier gas which was expanded into the copper ablation plasma (*4*). The vapor mixture contained 99.6% acetylene dissolved in acetone. Average laser pulse energies of 160 mJ/pulse were used. The films grown were black in color and SEM's indicated that the film surface was inhomogeneous with scattered island-like structures. Films were dissolved in solvents and analyzed by mass spectrometry and NMR. The mass spectrum indicates a pattern which is characteristic of a (poly)hydrocarbon. The NMR indicates the presence of only methylene and methyl hydrogens. The acetone is likely to be the source of the extra hydrogens since no polymer is generated when acetone was not present. ESCA analysis of a film indicates the presence of copper, carbon and oxygen. The binding energy of the copper $2p_{3/2}$ (932.9 eV) and the oxygen 1s (531.9 eV) photoelectron peaks indicate that neither oxygen nor carbon is chemically bonded

Figure 4. SEM image of the edge of an Si film grown on an Si wafer. (Reproduced with permission from reference 18. Copyright 1997 Materials Research Society.)

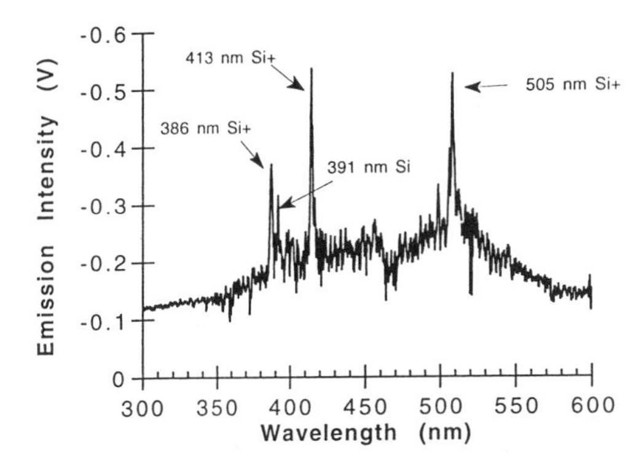

Figure 5. Emission spectrum observed during the growth of an Si film. (Reproduced with permission from reference 18. Copyright 1997 Materials Research Society.)

to the copper. This suggests that nanoparticles of copper metal are physically embedded within the film as opposed to being chemically bonded species. No polymer was generated when the acetylene-acetone mixture was expanded into laser ablated carbon generated when carbon was used as the target. This suggests that the formation of the polymeric film is not simply due to thermal effects of the plasma but is catalyzed by the presence of the copper metal. These polymeric films with embedded metal nanoparticles may have interesting electronic and optical properties. We plan to extend the LAMBD studies to other metals and other organic starting materials.

Al_2O_3, TiO_2 and SiO_2 Inorganic:Organic Composites

Thin films of inorganic:organic composites are of considerable interest in applications for photonic and electrooptic devices. Inorganic glasses such as Al_2O_3, TiO_2 and SiO_2 possess excellent optical properties which make them suitable as host materials. Many organic compounds are excellent nonlinear optical (NLO) materials but exhibit considerable optical losses and pure films of such materials are not mechanically stable. Unfortunately, inorganic:organic composites cannot be prepared directly because the high temperatures required to process the inorganic glasses would decompose the organic component.

LAMBD was used to fabricate inorganic:organic composites of a variety of compounds to determine whether the organic compounds can be embedded in an inorganic matrix without degradation and what properties such composites would have (*15,16*). In reference 15, three main composites were fabricated. TiO_2:copperphthalocyanine (TiO_2:CuPCN), Al_2O_3:6-propionyl-2-(dimethylamino)naphthalene (Al_2O_3:Prodan) and Al_2O_3:Rodamine 6G (Al_2O_3:R6G). The CuPCN was chosen because of its NLO properties, Prodan was chosen because its fluorescence spectrum is sensitive to its local microenvironment (*21,22*) and R6G was chosen because its fluorescence spectrum is sensitive to its state of aggregation (*23*). The organic vapor was introduced by flowing O_2 gas through a heated U-tube containing the organic solid. Ti or Al target rods were used and the films were grown on silica substrates. The LAMBD films were analyzed by UV-visible, FTIR, EDX and fluorescence spectroscopy. The fluorescence measurements were performed using an SLM 48000 MHF spectrofluorometer using a xenon arc lamp as the excitation source.

UV-visible spectra of LAMBD TiO_2:CuPCN films show the characteristic Q-band of CuPCN. The UV-visible spectrum obtained by scraping the LAMBD film off of the substrate and dissolving it in pyridine is identical to a pyridine solution of CuPCN indicating that there is no detectable degradation of the CuPCN during the LAMBD deposition process. FTIR spectra of scraped TiO_2:CuPCN LAMBD films dissolved in pyridine are also identical to those of CuPCN in pyridine, again indicating that the CuPCN is embedded in the TiO_2 matrix without decomposition.

The peak of the fluorescence emission in R6G sol-gels shifts from 546 nm to 568 nm depending on the state of aggregation (*23*). Very dilute sol-gels, which possess only R6G monomers, exhibit a fluorescence spectrum peaked at 546 nm. More concentrated sol-gels where aggregation occurs are measurably red shifted. As

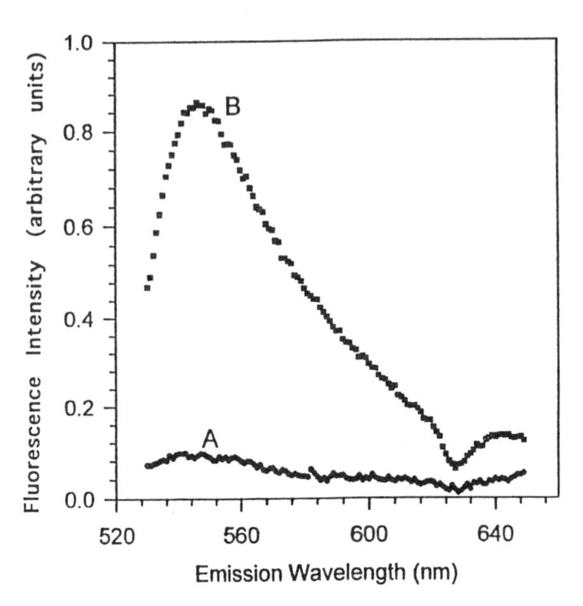

Figure 6. Fluorescence spectra of A) an Al_2O_3 film and B) an Al_2O_3:R6G film. (Reproduced with permission from reference 15. Copyright 1996 American Chemical Society.)

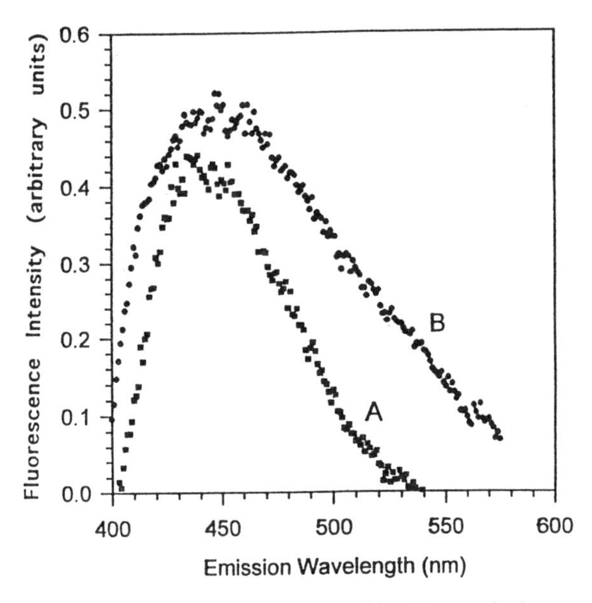

Figure 7. Fluorescence spectra of A) a thin film made by spreading a chloroform solution of pure Prodan on a silica substrate and of B) a LAMBD film of Al_2O_3:Prodan on a silica substrate. (Reproduced with permission from reference 15. Copyright 1996 American Chemical Society.)

shown in Figure 6, a LAMBD Al_2O_3:R6G film shows a fluorescence peak at 546 nm. This indicates that the R6G dye molecules are mononmeric and are not aggregated.

The fluorescence spectrum of Prodan is highly sensitive to the physiochemical properties of its local environment (*21,22*). Its emission maximum changes from 401 nm in cyclohexane to 530 nm in H_2O. As shown in Figure 7, the fluorescence spectrum of a LAMBD film of Al_2O_3:Prodan is somewhat red shifted and substantially broadened compared to a film made by spreading a solution of Prodan on a silica substrate. A film made by vacuum deposition of Prodan on a LAMBD Al_2O_3 film, not shown in Figure 7, has an intermediate width. The FWHM of the Prodan embedded Al_2O_3 LAMBD film, the Prodan coated Al_2O_3 film and the solution of Prodan are 5720 cm^{-1}, 4800 cm^{-1} and 3670 cm^{-1}. This indicates that the Prodan senses a more polar environment within the Al_2O_3 embedded film due to interaction with the host Al_2O_3 matrix. Thus the Prodan molecules are truly embedded within the matrix as opposed to being adsorbed on the surface.

Initial efforts to demonstrate NLO activity in the TiO_2:CuPCN films by four wave mixing were negative (*19*). Highly concentrated and reasonably thick CuPCN films are required to show optical activity. However, the LAMBD source block has recently been redesigned and rebuilt greatly improving the window deposition problems which hampered previous TiO_2 film growth. Initial TiO_2 films grown with the new source block are much improved in quality and in thickness. The TiO_2:CuPCN films grown using the new source block should be greatly improved and hopefully will have demonstrable optical activity.

Summary

Laser Assisted Molecular Beam Deposition (LAMBD) has been used to grow a variety of novel and potentially useful thin films. The present paper has discussed a few materials drawn from a variety of applications in order to demonstrate the flexibility of the technique. For refractory materials, such as Group IV compounds, LAMBD has advantages over other conventional deposition techniques and is capable of generating *in situ* chemistry which is critical for one-step synthesis of complex materials. Furthermore, LAMBD can be used as a low-temperature synthesis of high temperature materials, such as inorganic glasses, which makes it possible to embed temperature sensitive organic compounds in rugged transparent matrices. The flexibility of the LAMBD chemistry, and the ease with which new compounds can be quickly synthesized for analysis, insure that the LAMBD technique should be useful for a broad variety of applications. The LAMBD technique is currently being developed commercially in conjunction with AMBP Tech Corp., a subsidiary of Structured Materials Industries , Incorporated.

Acknowledgments

We gratefully acknowledge financial support from the National Science Foundation Solid State Chemistry program contract number #DMR9213907, the Air Force administered BMDO contract number F33615-96-C-5467, and the Office of Naval Research through an AASERT grant.

Literature Cited

1. Herron, W. J. and Garvey, J. F., In *Clusters and Cluster-Assembled Materials*, Averback, R. S., Bernholc, J. and Nelson, D. L., Eds.; Mater. Res. Soc. Proc. 206 Boston Ma, **1990**, p 391.
2. Wijekoon, W. M. K. P.; Lyktey, M. Y. M.; Prasad, P. N. and Garvey, J. F., *J. Appl. Phys.*, **1993**, *74*, 5767-5771.
3. Wijekoon, W. M. K. P.; Lyktey, M. Y. M.; Prasad, P. N. and Garvey, J. F., *J. Phys. D: Appl. Phys.*, **1994**, *27*, 1548-1555.
4. Wijekoon, W. M. K. P.; Stry, J. J.; Prasad, P. N. and Garvey, J. F., *Langmuir*, **1995**, *11*, 27.
5. See for example: Journal Vacuum Science and Technology, Molecular Beam Epitaxy Workshop Proceedings.
6. Grove, G. A. *Physics and Technology of Semiconductor Devices*, John Wiley & Sons, NY, NY, 1967.
7. Wu, X. D.; Inam, A.; Venkatesan, T. V.; Chang, C. C.; Chase, E. W.; Barboux, P.; Tarascon, J. M. and Wilkens, B., *Appl.Phys. Lett.* , **1988**, *52*, 754.
8. Otsubo, S.; Maeda, T.; Minamikawa, T.; Yonezawa, Y.; Morimoto, A. and T. Shimizu, *Japan. J. Appl. Phys.,* **1990**, *29*, L133.
9. Koren, G.; Baseman, R. J.; Gupta, A.; Lutwuche, M. I. and Laibowitz, R. B., *Appl. Phys. Lett.*, **1989**, *55*, 2450.
10. *Chemical Processing with Lasers*, Bauerle, D., Ed.; Springer Series in Material Science, Springer, New York, 1986; Vol. 1.
11. *Laserl Processing of Thin Films and Microstructures*, Boyd, I. W., Ed.; Springer Series in Material Science, Springer, New York, 1987; Vol. 3.
12. *Beam Epitaxy-Crystallization of Semiconductor Film From Atomic and Molecular Beams*, Benedek, G.; Martin, T. P. and Pacchioni, G.; Eds.; Springer Series in Material Science, Springer, New York, 1987; Vol. 7.
13. Ulman, A., *An Introduction to Ultra Thin Organic Films from Langmuir-Blodgett to Self-Assemblies*, Academic, San Diego, 1991.
14. *Langmuir-Blodgett Films,* Roberts, G.; Ed.; Plenum, New York, 1990.
15 DeLeon, R. L.; Wijekoon, W. M. K. P.; Narang, U.; Hall, M. L.; Prasad, P. N. and Garvey, J. F., *J. Phys. Chem.*, **1996**, *100*, 10707.
16. Wijekoon, W. M. K. P.; Lyktey, M. Y. M.; Prasad, P. N. and Garvey, J. F., *Appl. Phys. Lett.*, **1995**, *67*, 1698.
17. Wijekoon, W. M. K. P.; Prasad, P. N. and Garvey, J. F., *SPIE Proceedings Laser Induced Thin Film Processing*, **1995**, *2403*, 143.
18. DeLeon, R. L.; Sun, L.; Charlebois, J.; Garvey, J. F.; Forsythe, E. W. and Tompa, G. S., *Materials Research Society Symposium Proceedings*, Symposium held in Boston MA, 1996.
19. Unpublished data.

20. Christy, D. R.; and Huber, G., *Pulsed Laser Deposition of Thin Films*, John Wiley & Sons, NY, NY, 1994.
21. Weber, G. and Farris, F. J. *Biochemistry*, **1979**, *18*, 3705.
22. Narang, U.; Jordan, J. D. and Prasad, P. N., *J. Phys. Chem.*, **1994**, *98*, 8101.
23. Narang, U.; Bright, F. V. and Prasad, P. N., *Appl. Spectrosc.*, **1993**, *47*, 229.

Chapter 15

Synthesis and Physical Properties of Semiconductor Nanocrystals Formed by Ion Implantation

Jane G. Zhu[1], C. W. White[2], S. P. Withrow[2], J. D. Budai[2], R. Mu[3], and D. O. Henderson[3]

[1]Department of Physics, New Mexico State University, Las Cruces, NM 88003
[2]Oak Ridge National Laboratory, Solid State Division, Oak Ridge, TN 37831
[3]Physics Department, Fisk University, Nashville, TN 37208

Nanocrystals of elemental and compound semiconductor materials have been fabricated using the ion implantation technique. The nanocrystal formation is controlled by the implantation and annealing conditions, as exemplified by the nanocrystals of Si, Ge and GaAs formed inside SiO_2 matrices. Strong red photoluminescence (PL) peaked around 750 nm has been observed in samples containing Si nanocrystals in SiO_2. Difference in the absorption bandgap energies and the PL peak energies is discussed. The dielectric properties of SiO_2 with Ge nanocrystal inclusions have been studied using infrared reflectance spectrometry. Cavity mode predicted by the Maxwell-Garnett theory has been observed in the samples implanted with a high dose of Ge. Significant influence of implantation sequence on the formation of compound semiconductor nanocrystals are demonstrated with the GaAs in the SiO_2 system. Optical absorption measurements show that Ga particles have already formed in the as-implanted stage if Ga is implanted first. A single surface phonon mode has been observed in the infrared reflectance measurement from samples containing GaAs nanocrystals.

Ion implantation is a very useful technique for altering the near-surface properties of a wide range of materials. Desired elements can be injected into a solid in a controlled and reproducible manner by ion implantation [*1*]. Supersaturated impurity concentrations can be produced by high-dose implantation. Subsequent annealing leads to precipitation and the formation of nanocrystals that are encapsulated in the host material. Semiconductor nanocrystals and quantum dots have attracted considerable interest due to potential applications in novel optoelectronic devices [*2,3*]. Quantum confinement effects are expected when the sizes of the nanocrystals are smaller than the exciton diameters. For optical applications, for example, visible luminescence can be obtained from group IV nanocrystals due to quantum confinement effects [*4*]. Nonlinear optical properties can also be significantly enhanced for the nanocrystals compared to the bulk material [*5*].

We have studied the formation and properties of elemental (group IV) and compound semiconductor (III-V and II-VI) nanocrystals in different host materials [*6,7,8,9*]. The nanocrystal sizes can be controlled by the ion implantation dose and annealing temperatures. The physical properties have been studied by a variety of microstructure and optical property characterization techniques. Examples reported in this paper include Si, Ge and GaAs nanocrystals formed in amorphous SiO_2 matrices.

Experimental details

The semiconductor nanocrystals were formed by ion implantation of appropriate semiconductor species into a SiO_2 layer on (100) silicon substrate, followed by subsequent thermal annealing. The SiO_2 layer was ~ 750 nm thick, formed by thermally oxidizing a (100) Si wafer. Samples were also prepared by ion implantation into fused silica substrates (Corning 7940). Ion implantation was done at room temperature with doses in the range of (3–30) × 10^{16} ions/cm^2. The implant energies were chosen to put the peak concentration at the middle of the oxide layer. Samples were annealed isochronally for 1 h under Ar + 4%H_2 ambient at atmospheric pressure. The annealing temperatures were within the range of 600°C to 1100°C.

The nanocrystalline structures were investigated by transmission electron microscopy (TEM) and X-ray diffraction. All the TEM specimens were prepared in cross sections, since the concentration distribution from ion implantation is a function of depth. Depth profiles were also examined by Rutherford backscattering spectrometry (RBS). Optical measurements include optical absorption, photoluminescence (PL) and infrared reflectance measurements. For the infrared reflectance spectrometry of samples containing Ge nanocrystals inside SiO_2 matrices, mid infrared measurements were made with a Bomen MB100 Fourier-transform spectrometer having a range of 450 to 5000 cm^{-1}. The samples were measured at an angle of incidence of 17°, unless otherwise indicated. All reflectance spectra were referenced to a silver mirror to provide a consistent normalization.

Si nanocrystals in SiO_2

Multiple implants of Si at different energies were performed to produce uniform concentration profiles of Si inside SiO_2. A TEM micrograph in Fig. 1 shows the Si nanocrystals are a few nanometers in size (in the 1-5 nm range, mostly ~ 2-3 nm) in the samples implanted with excess Si concentration of ~ 5 × 10^{21} cm^{-3} throughout a SiO_2 film on a Si substrate and annealed at 1100°C for 1 h. To achieve visible luminescence, the sizes of Si nanocrystals should be about a few nanometers in order to have sufficient quantum confinement effect [*4*].

Very strong PL has been observed in the samples containing Si nanocrystals about a few nanometers in size. Figure 2(a) shows PL spectra, excited by an Ar laser at 514.5 nm wavelength, from a sample implanted with Si at a dose of 1.5×10^{17} cm^{-2} before and after thermal annealing. There is some luminescence with a broad peak centered at ~ 650 nm from the as-implanted sample. These luminescence centers are presumably due to the ion-implantation damage of the SiO_2 matrix. The PL intensity is significantly reduced after the sample is annealed at 800°C. After annealing at a higher temperature of 1100°C, Si nanocrystals in sizes of a few nanometers have been formed and a very strong PL peak center at ~ 750 nm have been observed. This phenomenon is generally true for samples with a fixed implantation dose but annealed at different temperatures: the PL peak intensities are the highest for samples annealed at 1100°C.

Figure 2(b) shows PL spectra, measured using a SPEX Fluorolog 2 and excited at 300 nm, from samples implanted with flat Si concentration profiles of (A) 1×10^{21} cm^{-3}, (B) 5×10^{21} cm^{-3}, and (C) 2×10^{22} cm^{-3} and subsequently annealed at 1100°C (the detector's detection limit does not cover the whole range of the PL peaks). The only strong PL peak occurs around 750 nm. When measured using an Ar laser, excited at 514.5 nm, the PL peak from sample B has the highest intensity (Si nanoparticles in this sample are shown in Fig. 1). The PL peak is shifted toward a higher energy for sample A, which has been implanted with less Si, but shifted to a lower energy with much lower peak intensity for sample C which has much more Si implanted. TEM of sample C revealed that the Si nanocrystals are in the range from a few to several nanometers. Combining the results from PL and TEM from a number of samples, it is concluded that maximum PL intensity occurs in the samples annealed at 1100°C and containing Si nanocrystals with sizes about a few nanometers.

Strong optical absorption has been observed in the Si-nanocrystal-containing silica samples. The absorption bandgap energies measured are much larger than the PL peak energies observed [*10*]. When the implantation dose of Si is changed, the size distributions of Si nanocrystals change and the strong peak position can be shifted slightly, much less than the shift in the absorption energy. Large differences between luminescence and optical absorption energies have been reported for porous Si produced by electrochemical dissolution of Si [*11*]. Similarly, the optical properties of these Si-nanocrystals-in-SiO_2 systems can be explained by quantum confinement effects in Si nanocrystallites in association with surface/interface states. Theoretical calculations have also been reported to show the existence of self-trapped excitons on bonds of silicon crystallites [*12*]. The self-trapped state can be a stable situation for the excited states for very small crystallites.

Ge nanocrystals in SiO_2

The formation of nanocrystals of different materials can be very different under similar implantation and annealing conditions. The Ge nanocrystals can grow into much larger sizes than those of the Si nanocrystals. Supersaturated Ge starts to precipitate inside SiO_2 at temperatures as low as 300°C [*13*]. The size distribution of the nanocrystals is in general influenced by the implantation doses and annealing temperatures [*7*]. The sample annealed at 600°C contains Ge nanocrystals with sizes ranging between a few

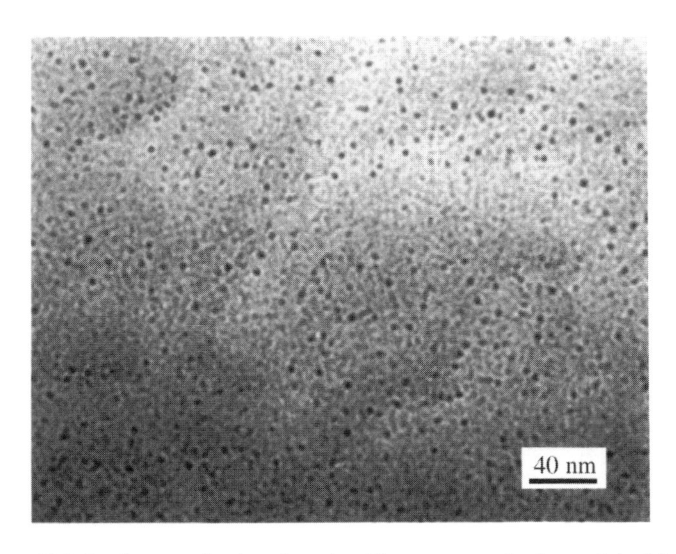

Figure 1. TEM micrograph showing the Si nanocrystals formed inside a SiO_2 matrix.

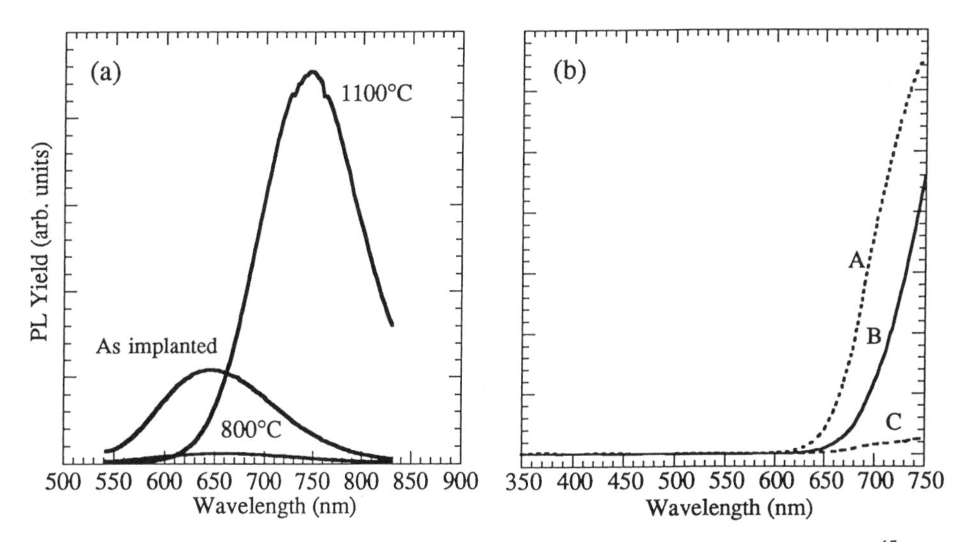

Figure 2. (a) PL spectra from a sample implanted with Si at a dose of 1.5×10^{17} cm^{-2} before and after thermal annealing at 800°C and 1100°C. PL was excited by an Ar laser at 514.5 nm wavelength. (b) PL spectra from samples implanted with flat Si concentration profiles of A: 1×10^{21} cm^{-3}, B: 5×10^{21} cm^{-3}, and C: 2×10^{22} cm^{-3}. These spectra were excited by a Xe lamp at 300 nm wavelength.

to several nanometers. Roughening of the nanocrystals is substantial in the high concentration areas when the annealing temperature is above 900°C.

A cross-sectional TEM image of a sample implanted with a dose of 3×10^{17} ions/cm^2 (implantation energy 500 keV) and annealed at 1000°C is shown in Fig. 3. At the middle of the implanted region, where the Ge concentration is the highest, the largest crystallites are ~ 25 nm in diameter. A wide range of particle sizes, from 1 to 25 nm, is observed in this sample. However, the most populated particle sizes over the whole implanted region are ~ 5 nm. When the sample is annealing at 1000C for a longer period of time, e.g., 17 hrs, the Ge particles do not seem to grow to be larger than 25 nm. The overall Ge concentration profile, measured by RBS, does not change much after annealing [7]. The Ge particles are randomly oriented, and spherical in shape, as is expected for an amorphous matrix. High-resolution electron microscopy (HREM) images of two Ge spheres from this sample are shown in Fig. 4. The Ge spheres are entirely crystallized. Some of them are single crystals, but crystal defects, such as twinning, are often observed in a large number of spheres, which could affect the optical properties. The PL intensities from the Ge-nanocrystal-containing samples are not nearly as strong as the PL intensity of the Si-nanocrystal-containing samples, which could be due to the fact that SiO2 can passivate the Si surface better than that of Ge or due to the defects in Ge nanocrystals. The role of defects in the visible luminescence of SiO2 films containing Ge nanocrystals is discussed in ref. [*14*]

It is known that Ge oxide is thermodynamically less stable than SiO_2 [*15*]. It should therefore be easier for Ge than Si to precipitate in SiO_2. While the initial nucleation of the nanocrystals depends on the thermodynamics of the material systems, roughening at higher temperatures appears to be related to the different melting temperatures of the implanted materials. The remarkable difference in sizes of the Ge and Si nanocrystals suggests that there is no rapid increase in crystallite size until the annealing temperature approaches the melting temperature of the implanted material. The annealing temperature (1000°C) for the Ge-nanocrystal sample shown in Fig. 3 is above the melting temperature of Ge (938.3°C), while the annealing temperature (1100°C) for the Si-nanocrystal samples shown in Fig. 1 is well below the melting temperature of Si (1414°C). Typical annealing temperatures in semiconductor processing do not reach as high as (or close to) the melting temperature of Si.

The implanted region, over a-few-hundred nanometers thick, is like a SiO_2-Ge composite film with Ge nanocrystals embedded inside the SiO_2 matrix. The existence of the matrix can influence the electronic and vibrational states in the nanocrystals, and the properties of the matrix can also be influenced by the inclusion of nanocrystals. The dielectric constant of a composite material can be calculated using the effective-medium theories, developed by Maxwell-Garnett [16] for spherical inclusions in the host medium and by Bruggman [17], the symmetric (in the two components) effective-medium theory. Infrared (IR) absorption spectra for a film with Ge spheres embedded in the SiO_2 matrix have been calculated using the effective medium theories [18]. A new peak, assigned to the so-called cavity mode, appears between the transverse optical (TO) and longitudinal optical (LO) mode (closer to the TO mode) in SiO_2 films when the Maxwell-Garnett theory is applied. There is no such new peak if using Bruggeman theory. The cavity mode predicted from Maxwell-Garnett theory was not

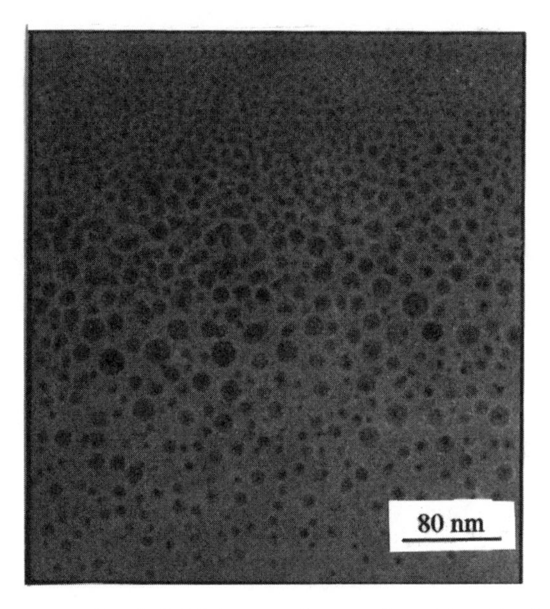

Figure 3. A cross-sectional TEM image of a sample with a Ge dose of 3×10^{17} cm^{-2} and annealed at 1000°C for 1 h.

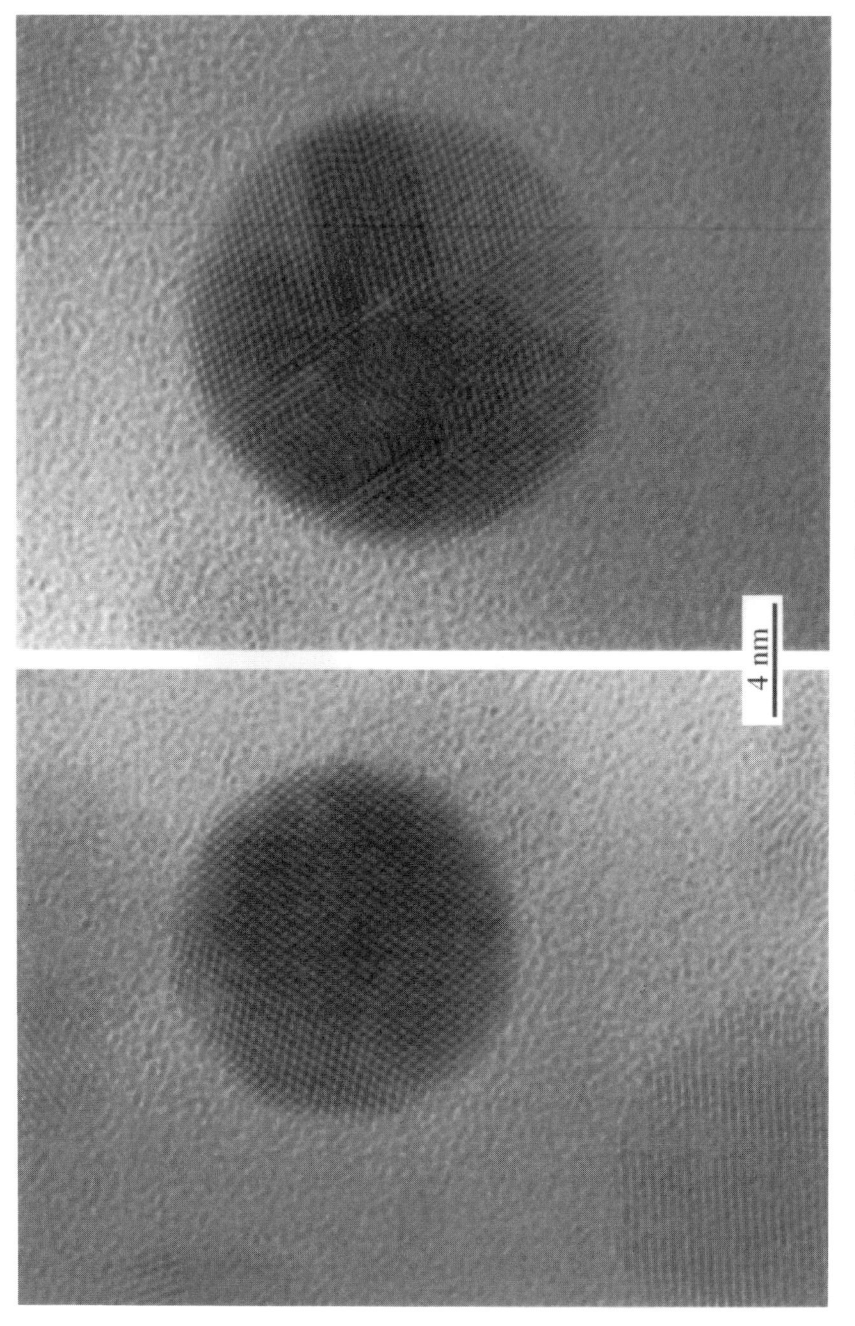

Figure 4. HREM images of two Ge spheres.

observed in the IR-absorption in composite films of SiO_2 and microcrystalline Ge prepared by the sputtering deposition and annealing method [*18*]. It was therefore concluded in ref. [*18*] that the influence of Ge microcrystals on the IR-absorption spectra can be well explained by the Bruggeman average dielectric function.

Infrared reflectance spectrometry is used to study the ion-implanted samples since the reflectance technique is more sensitive to the near-surface region where the nanocrystals are formed. Figure 5 shows the infrared reflectance spectra, measured with non-polarized light, recorded from samples implanted with Ge doses of (a) 6×10^{16} cm^{-2} and (b) 3×10^{17} cm^{-2} into a thermally grown SiO_2 film and annealed at various temperatures. The peak at ~ 1110 cm^{-1} is assigned to the TO mode of the Si-O stretch and the peak at a high wavenumber is assigned to the LO mode. The stretching vibration of Si-O dangling bonds, induced by ion implantation damage, could contribute to a shoulder peak (to the left of the TO peak) at 1025 cm^{-1}. The intensity of the dangling bond peak increases as the implantation dose increases, and decreases as the annealing temperatures increases since annealing will reduce the defect density. Therefore the intensity of the TO peak increases and shift toward 1110 cm^{-1} as the annealing temperature increases. This trend is easily observable from the samples implanted at a lower dose [Fig. 5(a)], and is more pronounced from the samples implanted with a higher dose [Fig. 5(b)]. For the sample implanted with a high dose, a new peak appears at ~ 1130 cm^{-1} after the sample has been annealed at 1000°C. Ge particles are relatively larger in this sample (a few to 25 nm in diameter, see Fig. 3). This new peak is assigned to the cavity mode predicted from the Maxwell-Garnett theory, which has not been observed in samples implanted with doses less than 3×10^{17} cm^{-2}. For this implantation dose and energy, the Ge volume fraction in the implanted region is about 10%. An inverse peak at 1260 cm^{-1} is assumed due to the SiO_2 thin film effect [*19*]. To avoid the complication due to thin film interference and the Si substrate absorption, Ge nanocrystals have also been formed in silica glass substrates by the ion implantation method. Figure 6 shows the IR-reflectance spectra from samples implanted with Ge doses of 1.5×10^{17} cm^{-2} and 3×10^{17} cm^{-2} and annealed at 1000°C, the condition the cavity mode is more substantial in the thin film samples. The spectra were measured with non-polarized, *p*-polarized and *s*-polarized light at 45° incidence. The spectra measured with obliquely incident *s*-polarized light contains only the contributions from the TO modes, while the specimen measured with the *p*-polarized light contains contributions from the LO modes as well as the TO modes (Berreman effect, [*20*]). The cavity mode is again observed in the samples implanted with the high dose of 3×10^{17} cm^{-2}, but not in the samples implanted with a Ge dose of 1.5×10^{17} cm^{-2}. This indicates that the cavity mode appears only in samples with enough Ge volume fractions. This is consistent with the Maxwell-Garnett theory.

GaAs nanocrystals in SiO_2

Nanocrystals of compounds or alloys can be formed by implanting the two constituent species sequentially at energies chosen to produce an overlap of the concentration profiles. If the constituents have strong chemical affinity for each other, precipitation and compound formation will take place during subsequent annealing. For the

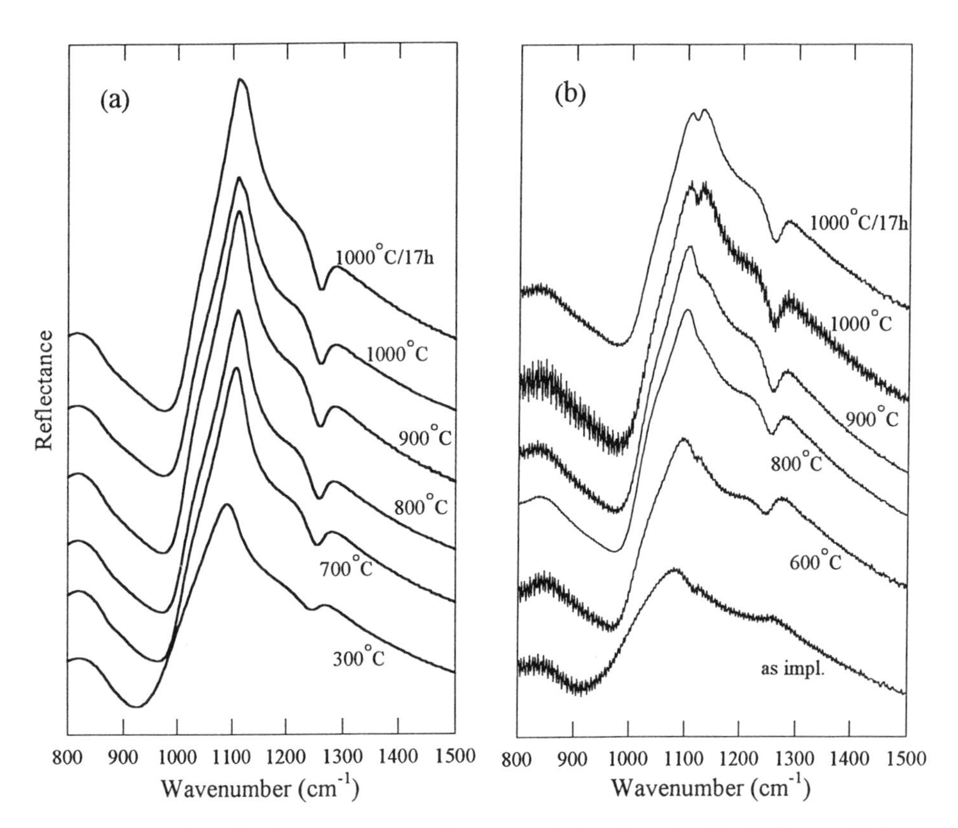

Figure 5. Infrared reflectance spectra from samples implanted with Ge doses of (a) 6×10^{16} cm^{-2} and (b) 3×10^{17} cm^{-2} into a thermally grown SiO_2 film and annealed at various temperatures.

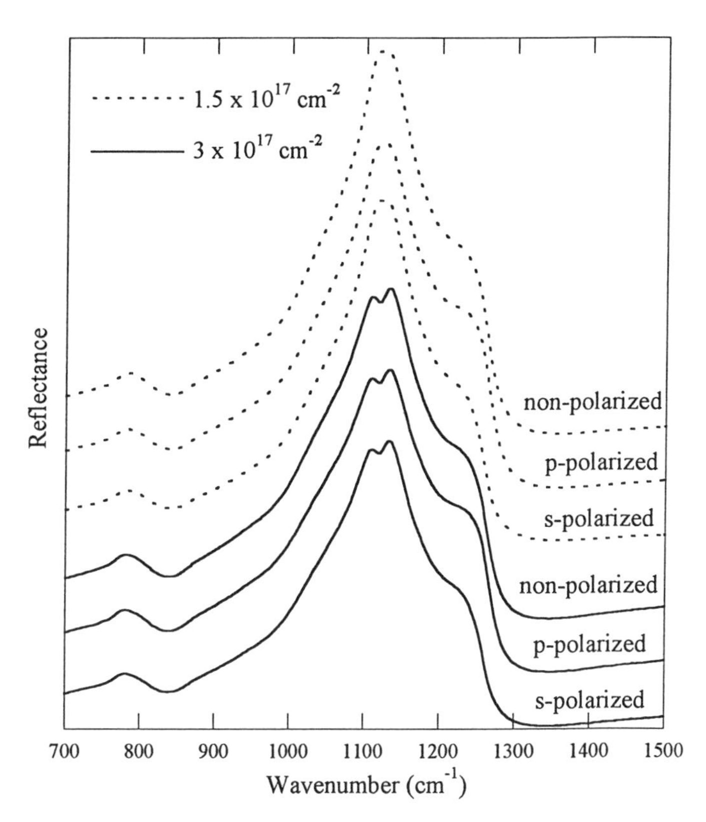

Figure 6. Infrared reflectance spectra from samples implanted with two different doses of Ge, 1.5×10^{17} cm^{-2} and 3×10^{17} cm^{-2}, into silica glass substrates and annealed at 1000°C for 1 h.

formation of compound semiconductors, such as GaAs, the sequence of Ga and As implantation was found to influence the nanocrystal formation dramatically [*21*]. Figure 7 shows cross-sectional TEM images from samples implanted with the equal amount of Ga and As, 1.5×10^{17} ions/cm2 each, but different implantation sequences and annealed at 1000°C for 1 h. The GaAs nanocrystals in Fig. 7(a) were formed in the sample implanted with Ga first and then As, and have sizes ranging from a few nanometers to ~ 30 nm. These nanocrystals are nearly spherical and randomly oriented with respect to each other, as expected for amorphous matrices. Some voids are observed in the region near the oxide surface. In the sample implanted with As and then Ga, the GaAs nanocrystals formed after annealing at the same temperature are much smaller with sizes in the range of 1 - 10 nm as shown in Fig. 7(b). The difference between the two samples shown in Fig. 7 (a) and (b) is very striking considering they have been through the same processing except the implantation sequence of Ga and As.

RBS measurements of the implanted ion profiles reveal that the As implanted in SiO_2 is thermally stable, while Ga implanted in SiO_2 is very mobile during implantation and annealing. Optical absorption spectra from samples implanted with Ga alone or As alone into silica glass wafers are shown in Fig. 8. A strong absorption peak at 220 nm has been observed in the as-implanted sample implanted with Ga only. This absorption peak is attributed to the surface plasmon resonance of Ga particles in SiO_2. Details on surface plasmon resonance of metal particles, such as Ag, in a dielectric matrix can be found elsewhere [*22*]. The optical density decreases after annealing at 1000°C, which is in agreement with the Ga loss observed in the RBS spectra. TEM from a sample implanted with Ga first and then As confirms the formation of Ga particles in the as-implanted stage. The optical absorption spectrum for the sample implanted with As only is virtually unchanged after annealing at 1000°C. The thermal stability of the As profile helps to interpret the thermal stability of the ion-concentration profile of As + Ga in the samples implanted with As first. When Ga ions are implanted after the implantation of As, they bond with the As atoms due to the chemical affinity between Ga and As. In the samples implanted with Ga first and then As, there is much more diffusion involved since Ga is very mobile in SiO_2. Consequently, the GaAs nanocrystals grow much bigger.

In the samples implanted with As first and then Ga, the GaAs nanocrystals formed after annealing are smaller than the bulk exciton radius. Therefore, quantum confinement effect is expected. An infrared reflectance spectrum is shown in Fig. 9 recorded from a sample implanted with equal amount, 1.0×10^{17} ions/cm^2, of As and then Ga and annealed at 1000°C for 1 h. A strong reflectance at 278 cm^{-1} has been observed. A Raman spectrum measured from the bulk GaAs is also plotted in Fig. 9. The single peak from the GaAs-nanocrystal-containing sample lies in between the transverse (267 cm^{-1}) and longitudinal (291 cm^{-1}) optical phonon modes of bulk GaAs. Observation of a similar single mode between the bulk transverse and longitudinal optical modes in the infrared spectrum has been reported for microcrystallites of MgO and GaP [*23*]. This peak is attributed to the excitation of surface phonon modes.

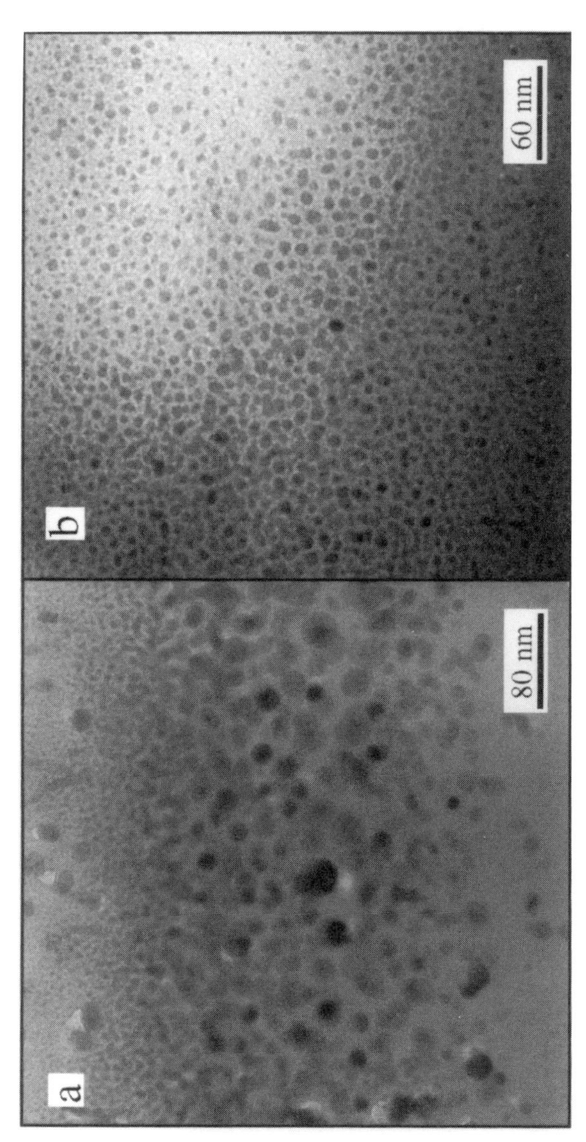

Figure 7. Cross-sectional TEM images from samples implanted with equal doses, 1.5×10^{17} cm^{-2}, of Ga and As, but different implantation sequences: (a) Ga first and then As and (b) As first and then Ga. Both samples have been annealed at 1000°C.

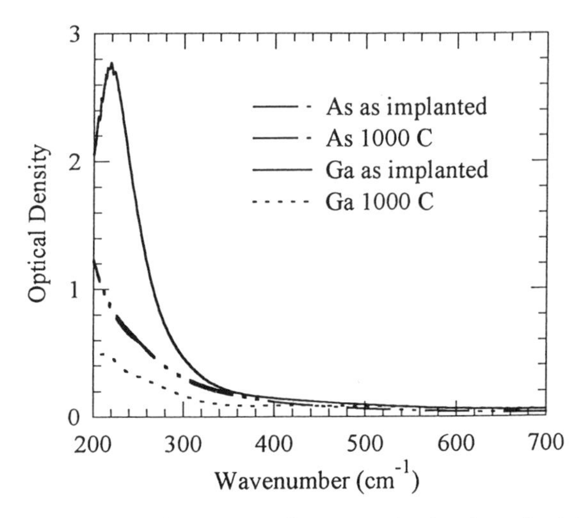

Figure 8. Optical absorption spectra from samples implanted with $1.0\times10^{17}/cm^2$ of Ga or As only, before and after annealing.

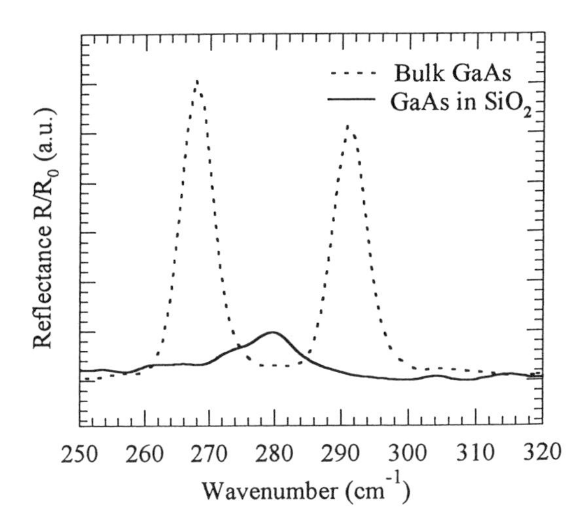

Figure 9. Infrared reflectance spectrum (solid curve) from a GaAs-nanocrystal-containing sample with the Raman spectrum (dashed curve) from bulk GaAs superimposed.

Conclusions

Semiconductor nanocrystals of Si, Ge and GaAs have been formed in SiO_2 by ion implantation and subsequent thermal annealing. The microstructure has been characterized extensively by TEM. A broad range of nanocrystal sizes can be produced through the control of ion implantation and annealing processes. Strong PL peaked at around 750 nm is observed in Si-nanocrystal-containing samples. The absorption bandgap energies measured are consistent with the quantum confinement effect, while the PL peak energies are considered to be associated with the surface/interface states. The dielectric properties of SiO_2 containing Ge nanocrystals have been studied by IR reflectance spectrometry. Cavity mode has been observed in the high dose sample, where Ge particles have grown into relatively large sizes, which is consistent with the Maxwell-Garnett theory. For the formation of compound semiconductor GaAs nanocrystals, it is demonstrated that the sequence of Ga and As ion implantation affects the size distributions of GaAs nanocrystals significantly. The nanocrystal sizes are much bigger in the samples with Ga implanted first than those with As implanted first. This phenomenon is explained by the different diffusion behaviors of Ga and As species. Optical absorption measurements show that Ga particles have already formed in the as-implanted stage with the absorption peak at 220 nm due to a surface plasmon resonance of metal particles in SiO_2. Single surface phonon mode has been observed from samples containing GaAs nanocrystals in the infrared reflectance measurement. We have also fabricated and characterized nanocrystals of other semiconductor materials in different matrices using the ion implantation technique, which is beyond the scope of this paper.

Acknowledgments

This research was sponsored by the Division of Materials Sciences, U.S. Department of Energy, under contract DE-AC05-96OR22464 with Lockheed Martin Energy Research Corp. and contract DE-FG05-94ER45521 with Fisk University.

Literature Cited

1. White, C. W.; McHargue, C. J.; Sklad, P. S.; Boatner, L. A.; Farlow, G. C. *Mater. Sci. Reports* **1989**, *4*, 41.
2. *Microcrystalline and Nanocrystalline Semiconductors*, Collins, R. W.; Tsai, C. C.; Hirose, M.; Koch, F.; Brus, L.; Mater. Res. Soc. Symp. Proc.; Materials Research Society: Pittsburgh, Pennsylvania, 1995, Vol. 358.
3. Alivisatos, A. P. *Science* **1996**, *271*, 933.
4. Takagahara, T.; Takeda K. *Phys. Rev. B* **1992**, *46*, 15578.
5. Brus, L. *Appl. Phys. A* **1991**, *53*, 465.
6. Zhu, J. G.; White, C. W.; Budai, J. D.; Withrow, S. P.; Chen, Y. *Mater. Res. Soc. Symp. Proc.* **1995**, *358*, 175.
7. Zhu, J. G.; White, C. W.; Budai, J. D.; Withrow, S. P.; Chen, Y. *J. Appl. Phys.* **1995**, *78*, 4386.

8. White, C. W.; Budai, J. D.; Zhu, J. G.; Withrow, S. P.; Hembree, D. H.; Henderson, D. O.; Ueda, A.; Tung, Y. S.; Mu, R.; Magruder, R. H. *J. Appl. Phys.* **1996**, *79*, 1876.
9. White, C. W.; Budai, J. D.; Zhu, J. G.; Withrow, S. P.; Aziz, M. J. *Appl. Phys. Lett.* **1996**, *68*, 2389.
10. White, C. W.; Budai, J. D.; Withrow, S. P.; Zhu, J. G.; Pennycook, S. J.; Zuhr, R. A.; Hembree, Jr., D. M.; Henderson, D. O.; Magruder, R. H.; Yacaman, M. J.; Mondragon, G.; Prawer, S.; *Nucl. Instr. Methods B.*, in press.
11. Lockwood, D. J. *Solid State Commun.* **1994**, *92*, 101.
12. Allan, G.; Delerue, C.; Lannoo, M.; to be published.
13. Maeda, Y. *Phys. Rev. B* **1995**, *51*, 1658.
14. Min, K. S.; Shcheglov, K. V.; Yang, C. M.; Atwater, H. A.; Brongersma, M. L.; Polman, A. *Appl. Phys. Lett.* **1996**, *68*, 2511.
15. LeGoues, F. K.; Rosenberg, R.; Nguyen, T.; Himpsel, F.; Meyerson, B. S. *J. Appl. Phys.* **1989**, *65*, 1724.
16. Garnett, J.C. M., Phil. *Trans. R. Soc. L.* **1904**, *203*, 385; ibid. **1906**, *205*, 237.
17. Bruggeman, D.A. G., *Ann. Physik. (Leipzig)* **1935**, *24*, 636.
18. Fujii, M.; Wada, M. *Phys. Rev. B* **1992**, *46*, 15930.
19. Yen, Y.-S.; and Wong, J. S., *J. Phys. Chem.* **1989**, *93*, 7208.
20. Berreman, D. E. *Phys. Rev.* **1963**, *130*, 2193.
21. Zhu, J. G.; White, C. W.; Wallis, D. J.; Budai, J. D.; Withrow, S. P.; Henderson, D. O. *Mat. Res. Soc. Symp. Proc.* **1996**, *396*, 447.
22. Arnold G. W.; Borders, J. A. *J. Appl. Phys.* **1977**, *48*, 1488.
23. Hayashi, S. *Jpn. J. Appl. Phys.* **1984**, *23*, 665.

Chapter 16

Synthesis and Handling of Single Sheets of a Covalent Monolayer Square Grid Polymer

Thomas F. Magnera, Jaroslav Pecka, Jaroslav Vacek, and Josef Michl[1]

Department of Chemistry and Biochemistry, University of Colorado, Boulder, CO 80309–0215

An ultrathin (0.7 nm × ~150 nm × ~150 nm) covalent grid consisting primarily of ~2.5-nm squares with ~1.5-nm square voids has been synthesized by two-dimensional cross-linking polymerization of lanthanum(III)bis[5,10,15,20-tetrakis(4-pyridyl)porphyrinate] anion with *p*-xylylene dibromide, constrained to proceed on a mercury surface under irreversible conditions. The grid was characterized by grazing incidence IR spectroscopy and, after transfer to the surface of highly ordered pyrolytic graphite, by STM imaging.

Covalent polymers with the topology of a grid would be of considerable intrinsic interest, and might also be useful for various applications. We report the synthesis, mercury to graphite surface transfer, IR spectrum, and STM images of 0.7 nm × ~150 nm × ~150 nm single sheets of a covalent ~2.5-nm square (~1.5-nm void) grid polymer consisting of porphyrin-based connectors and *p*-xylylene links. This first-generation polymer does not exhibit long-range order, but it is quite clear what needs to be done to secure it.

Molecular Grid Synthesis

For some time, we have been developing a set of molecular rigid rods and connectors, which we have referred to as a "molecular-size 'Tinkertoy' (1) construction kit". Our ultimate aim is to develop the ability to synthesize thin layers of solids with completely controlled robust covalent structure based on a scaffolding periodic in two (wide) dimensions and arbitrarily controlled in a third (thin) dimension. We refer to such solids as "designer solids"(2,3,4,5,6,7,8,9). The first target in the synthesis is the ground floor of the scaffolding, a single sheet of a regular grid. Single-layer two-dimensionally cross-linked covalent polymers have also been identified as desirable

[1]Corresponding author

synthetic targets by others (10,11). Single graphite sheets are of considerable interest as well (12).

To our knowledge, covalent grid polymers of subnanometer thickness have not been synthesized before, although much attention has been paid to two-dimensional polymerization in various constrained spaces such as oil-water interfaces (13,14,15), Langmuir monolayers (16,17,18,19,20) and multilayers (21,22,23,24), self-assembled monolayers (25), lipid bilayers, both vesicles (26,27,28,29,30,31,32) and cast assemblies (33,34,35), as well as liquid crystalline assemblies (36,37). The closest precedents are the polymeric "perforated membranes" made from functionalized calixarene multilayers (38,39,40,41) and the multilayer membrane prepared by interfacial polymerization of *meso*-tetrapyridylporphyrin quaternized by propargylation (42,43).

Our proposed (5,6,7,8,9) but previously unproven method of synthesis is two-dimensional rigid-rod cross-linking of point or star connectors (5,6,7) mounted on tentacled pedestals, oriented and confined to two dimensions by adsorption of the tentacles on a liquid surface, which provides firm adhesion combined with two-dimensional translational and rotational freedom. The star connector chosen is the sandwich anion lanthanum(III)bis[5,10,15,20-tetrakis(4-pyridyl)porphyrinate] (**1**), already known (8,9) to adhere firmly to mercury in the desired orientation, with porphyrin rings parallel to the surface. The upper deck is the connector proper and its pyridine rings are the arms for linear coupling. The lower one is the pedestal and its pyridines are the tentacles. We have improved our earlier synthesis of **1** (44) and find that treatment of its monolayer on clean mercury surface with *p*-xylylene dibromide (**2**) in absolute ethanol forms the desired grid, even though **2** is not entirely ideal as a linker. Its benzylic carbons are tetrahedral, preventing true linearity and permitting crankshaft-type motions, and under the reaction conditions it is slowly converted to a polymer. However, it was considered adequate for testing the feasibility of the concept.

Molecular Grid Spectroscopy

The structure of the grid has been proven by a combination of local (IR) and global (STM) structural probes, applied after removal of all residual **1** from the mercury surface bound grid by boiling in 3% ethanolic HCl. Both before and after transfer to the surface of highly ordered pyrolytic graphite (HOPG), grazing incidence IR (Figure 1) shows the absence of **1** and the presence of all the bands expected from the oriented grid according to conductor surface selection rules (45). Note the absence of in-plane polarized porphyrin bands. Our IR spectrum is very similar to that of the material obtained by coupling *p*-xylylene dichloride with a monolayer of perpendicularly stacked free-base 5,10,15,20-tetrakis(4'-pyridyl)porphyrin attached to a Si(100) surface via a silane (46).

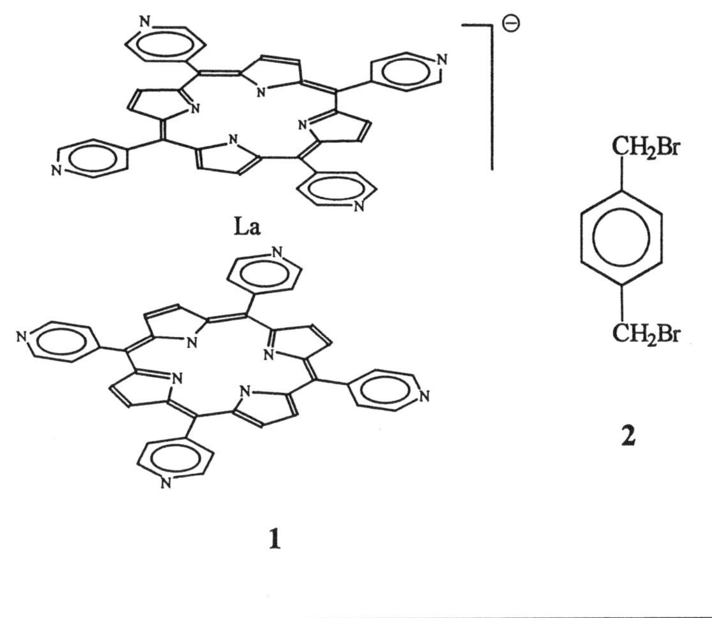

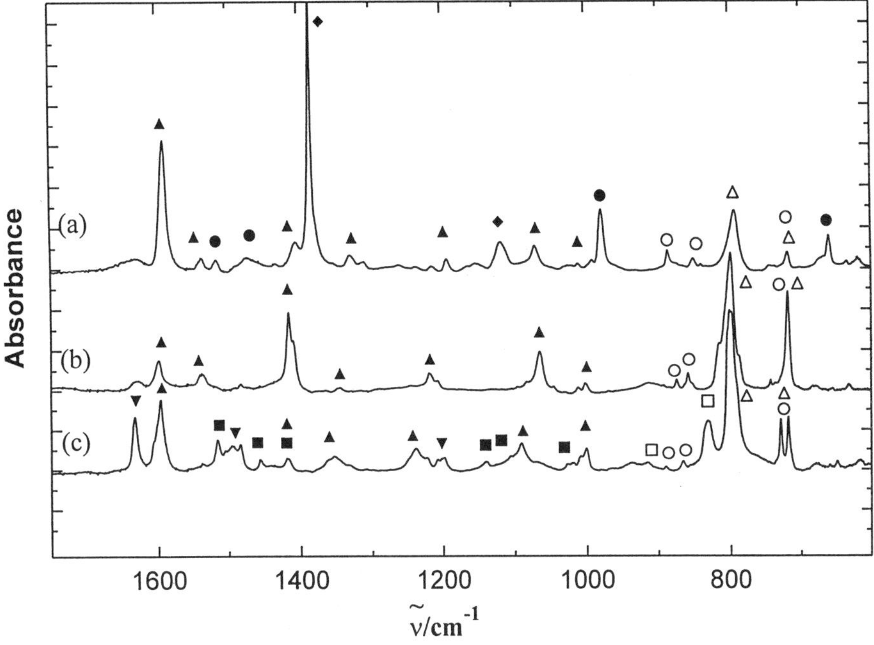

Figure 1. IR spectra: (a) n-Bu_4N^+ salt of **1** in KBr pellet; (b) Hg surface with adsorbed **1**; (c) Hg surface after coupling with **2**; ▲,Δ pyridine; ▼,∇ pyridinium; ●,O porphyrin; ■,□ *p*-xylylene (open symbols: locally out-of-plane polarized, closed symbols: locally in-plane polarized), ◆ tetra-*n*-butylammonium. Surface spectra at grazing incidence.

Our attempts at STM imaging of the grid directly on the mercury surface on which it was synthesized have failed in line with prior literature reports (47,48) that STM on liquid mercury is very difficult if not impossible. We have therefore decided to learn how to transfer the grid to another surface.

Molecular Grid Transfer and STM Imaging

Transfer of the grid to HOPG for STM imaging was accomplished by depositing a thin layer of polystyrene on the Hg surface by evaporation from a solution in THF, peeling off the resulting polymer sheet, placing it on a HOPG surface, swelling it with acetone, and dissolving the polystyrene in THF. IR intensities show that all of the grid is removed from the mercury surface, but only a fraction is actually successfully transferred to HOPG, since some of the domains are apparently flushed away by solvent in the polystyrene rinsing step.

A typical STM image (Figure 2) shows that the transferred grid is about 0.7 nm thick and consists of ~150 nm × ~150 nm sheets with the expected pattern of hollow squares of the anticipated size. Figure 3 presents a comparison with a segment of a

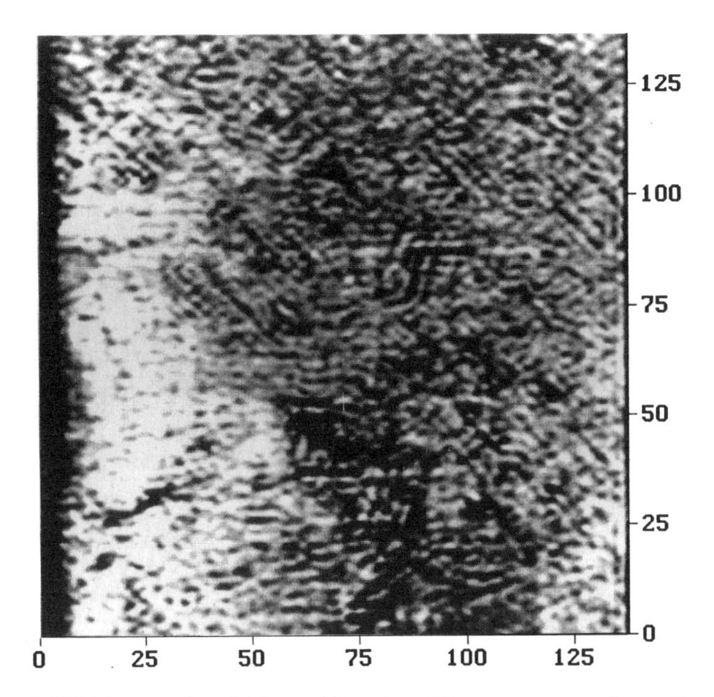

Figure 2. STM image of a grid formed from **1** and **2** on mercury surface, after transfer to HOPG. Pt/Ir tip, constant current mode (bias: -325 mV, current 131 pA).

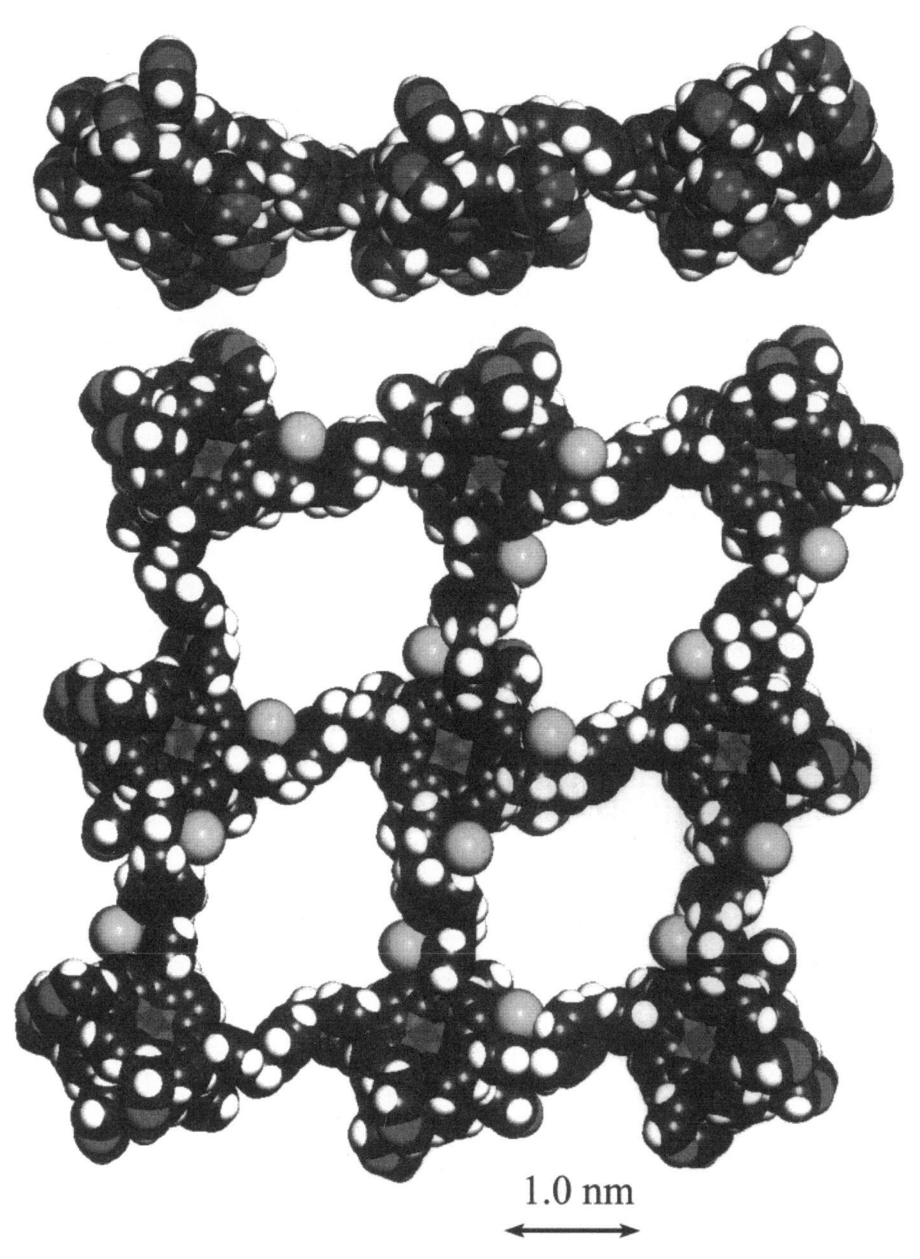

Figure 3. Calculated side (above) and top (below) view of a segment of a grid formed from **1** and **2**. Geometry optimized by molecular mechanics.

calculated grid structure as optimized by molecular mechanics (with chloride counterions, one of many potential energy minima) (49).

The imaged grid sheets are the first of a kind, but they have poor long-range order, contain many local irregularities, and are small. The shortcomings are attributed to the irreversibility of the coupling reaction under the conditions used, to imperfect linearity of the coupler, and to damage caused by the transfer, and they call for further effort. The irreversible nature of the coupling conditions used makes any repair of defects by annealing impossible. One way to secure a certain degree of reversibility would be to use *p*-xylylene diiodide instead of bromide. This cannot be done under open circuit conditions, since the diiodide is then reduced by mercury to the polymer of *p*-xylylene, but should be possible with appropriate potential control. The second problem with the *p*-xylylene coupler, its non-linearity, is harder to overcome, and we have begun experimenting with other coupling agents.

The present results suggest that the "molecular Tinkertoy (1)" approach to covalent giant-molecule grids is likely to succeed and they encourage us to continue work towards covalent multi-floor scaffoldings of designer solids. The single-floor grid sheets themselves may be of interest for fast size-based separations (38,39,40,41), growth of arrays of small particles (50), and other applications.

Acknowledgments. This work was supported by the Advanced Energy Project program of the U.S. Department of Energy (FG03-94ER12141). The STM instrument was funded by NSF grant CHE-9419335. We thank Professors R. Elber and A. Rappé for generous gifts of computer code.

Literature Cited

(1) Tinkertoy is a trademark of Playskool, Inc., Pawtucket, RI, and designates a children's toy construction set consisting of straight wooden sticks and other simple elements insertable into spool-like connectors.
(2) Kaszynski, P.; Michl, J. *J. Am. Chem. Soc.* **1988,** *110*, 5225.
(3) Michl, J.; Kaszynski, P.; Friedli, A. C.; Murthy, G. S.; Yang, H.-C.; Robinson, R. E.; McMurdie, N. D.; Kim, T. In *Strain and Its Implications in Organic Chemistry*; de Meijere, A.; Blechert, S., Eds.; NATO ASI Series, Vol. 273; Kluwer Academic Publishers: Dordrecht, The Netherlands, 1989; p 463.
(4) Kaszynski P.; Friedli, A.C.; Michl, J. *J. Am. Chem. Soc.* **1992**, *114*, 601.
(5) Michl, J. in *Applications of Organometallic Chemistry in the Preparation and Processing of Advanced Materials*, Harrod, J. F. and Laine, R. M., Eds., Kluwer, Dordrecht, The Netherlands, 1995, p. 243.
(6) Schöberl, U.; Magnera, T. F.; Harrison, R. M.; Fleischer, F.; Pflug, J. L.; Schwab, P. F. H.; Meng, X.; Lipiak, D.; Noll, B. C.; Allured, V. S.; Rudalevige, T.; Lee, S.; Michl, J. *J. Am. Chem. Soc.* **1997**, *119*, 3907.
(7) Harrison, R. M.; Brotin, T.; Noll, B. C.; Michl, J. *Organometallics*, in press.
(8) Harrison, R.M.; Magnera, T.F.; Vacek, J.; Michl, J. in *Modular Chemistry*, Michl, J., Ed., Kluwer, Dordrecht, The Netherlands, in press.
(9) Magnera, T. F.; Peslherbe, L. M.; Körblová, E.; Michl, J. *J. Organomet. Chem.*, in press.
(10) Diederich, F. *Nature* **1994**, *369*, 199.

(11) Bunz, U. H. F.; Wiegelmann-Kreiter, J. E. C. *Chem. Ber.* **1996**, *129*, 785, 1311.
(12) Ebbesen, T. W., in *Modular Chemistry*, Michl, J., Ed., Kluwer, Dordrecht, The Netherlands, in press.
(13) Dubault, A.; Casagrande, C.; Veyssié, M. *J. Phys. Chem.* **1975**, *79*, 2254.
(14) Rehage, H.; Schnabel, E.; Veyssié, M. *Makromol. Chem.* **1988**, *189*, 2395.
(15) Rehage, H.; Veyssié, M. *Angew. Chem.* **1990**, *102*, 497.
(16) Beredjick, N.; Burlant, W. J. *J. Polymer Sci.* **1970**, *A8*, 2807.
(17) Day, D.; Ringsdorf, H. *J. Polym. Sci. Polymer Lett.* **1979**, *16*, 205.
(18) Day, D.; Hub, H.-H.; Ringsdorf, H. *Isr. J. Chem.* **1979**, *18*, 325.
(19) Albrecht, O.; Laschewsky, A.; Ringsdorf, H. *J. Membrane Sci.* **1985**, *22*, 187.
(20) Bodalia, R. R.; Duran, R. S. *J. Am. Chem. Soc.* **1993**, *115*, 11467.
(21) Tieke, B.; Wegner, G.; Naegele; Ringsdorf, H. *Angew. Chem.* **1976**, *88*, 805.
(22) Tieke, B.; Lieser, G.; Wegner, G. *J. Polym. Sci. Polym. Chem. Ed.* **1979**, *17*, 1631.
(23) Regen, S. L.; Shin, J.-S.; Yamaguchi, K. *J. Am. Chem. Soc.* **1984**, *106*, 2446.
(24) Regen, S. L.; Shin, J.-S.; Hainfield, J. F.; Wall, J. S. *J. Am. Chem. Soc.* **1984**, *106*, 5756.
(25) Higashi, N.; Mori, T.; Niwa, M. *J. Chem. Soc. Chem. Commun.* **1990**, 225.
(26) Hub, H.; Hupfer, B.; Koch, H.; Ringsdorf, H. *Angew. Chem. Int. Ed. Engl.* **1980**, *19*, 938.
(27) Gros, L.; Ringsdorf, H.; Schupp, H. *Angew. Chem. Int. Ed. Engl.* **1981**, *20*, 305.
(28) Dorn, K.; Klingbiel, R. T.; Specht, D. P.; Tyminski, P. N.; Ringsdorf, H.; O'Brien, D. F. *J. Am. Chem. Soc.* **1984**, *106*, 1627.
(29) Kajiyama, T.; Kumano, A.; Takayanagi, M.; Kunitake, T. *Chem Lett.* **1984**, 915.
(30) Sakata, K.; Kunitake, T. *Chem. Lett.* **1989**, 2159.
(31) Higashi, N.; Mori, T.; Niwa, M. *Macromolecules* **1990**, *23*, 1475.
(32) Fendler, J. H.; Tundo, P. *Acc. Chem. Res.* **1984**, *17*, 3.
(33) Wu, J.; Harwell, J. H.; O'Rear, E. A. *Langmuir* **1987**, *3*, 531.
(34) Asakuma, S.; Okada, H.; Kunitake, T. *J. Am. Chem. Soc.* **1991**, *113*, 1749.
(35) Kuo, T.; O'Brien, D. F. *Langmuir* **1991**, *7*, 584.
(36) Stupp, S. I.; Son, S.; Lin, H. C.; Li, L. S. *Science* **1993**, *259*, 59.
(37) Stupp, S. I.; Son, S.; Li, L. S. Lin, H. C.; Keser, M. *J. Am. Chem. Soc.* **1995**, *117*, 5227.
(38) Markowitz, M. A.; Bielski, R.; Regen, S. L. *J. Am. Chem. Soc.* **1988**, *110*, 7545.
(39) Markowitz, M. A.; Janout, V.; Castner, D. G.; Regen, S. L. *J. Am. Chem. Soc.* **1989**, *111*, 8192.
(40) Conner, M.; Janout, V.; Regen, S. L. *J. Am. Chem. Soc.* **1993**, *115*, 1178.
(41) Lee, W.; Hendel, R. A.; Dedek, P.; Janout, V.; Regen S. L. *J. Am. Chem. Soc.* **1995**, *117*, 10599.
(42) Porteu, F.; Palacin, S.; Ruaudel-Teixier, A.; Barraud, A. *Makromol. Chem., Macromol. Symp.* **1991**, *46*, 37.
(43) Palacin, S.; Porteu, F.; Ruaudel-Teixier, A. *Thin Films* **1995**, *20*, 69.
(44) From 5,10,15,20-tetra(4'-pyridyl)porphine (619 mg, 1 mmol) in 1,2,4-trichlorobenzene (80 mL), ~10 mL of the solvent was distilled off under argon, lanthanum acetylacetonate hydrate (1.5 g, 3.3 mmol, dried 20 days over $Mg(ClO_4)_2$ at 0.02 Torr at rt) was added, and the mixture was refluxed and stirred for 30 min. This yielded an open-face lanthanum-porphyrin sandwich, convertible to unsymmetrical sandwiches. To obtain **1**, more 5,10,15,20-tetra(4'-pyridyl)porphine (619 mg, 1 mmol) was added and stirring and refluxing continued for another 16 h. The solvent was distilled off at 0.02 Torr, the residue was dissolved in methanol-chloroform (1:1) and chromatographed twice on thoroughly washed alumina, once with methanol-chloroform-water (20:20:1), and then with methanol-chloroform (from 1:3 to 1:1), eluting first some starting material (90 mg, 7%) and then **1** (1.08 g, 79%). The product contains a small amount of inorganic impurity extracted from the alumina but can be used in further synthesis. ^1H NMR (CD_3OD) d 6.5 - 7.5 (v.br., 8 H, pyridine), 8.21 (s, 16 H, pyrrole), 8.4-9.5 (m, v.br., 24 H, pyridine); ^{13}C NMR (CD_3OD) d 119.67, 131.12, 131.74, 147.77, 150.99, 154.52; ESI+ MS *m/z* 1373$[M]^+$, 687$[M+H]^{2+}$, 458$[M+H]^{3+}$.
(45) *Spectroscopy of Surfaces*, R. J. H. Clark and R. E. Hester, Eds., Wiley: New York, 1988.

(46) Li, D.; Buscher, C. T.; Swanson, B. I. *Chem. Mater.* **1994**, *6*, 803.
(47) Bruckner-Lea, C.; Janata, J.; Conroy, J.; Pungor, A.; Caldwell, K. *Langmuir* **1993**, *9*, 3612.
(48) Conroy, J. F. T.; Caldwell, K.; Bruckner-Lea, C.; Janata, J. *Electrochim. Acta* **1995**, *40*, 2927.
(49) The minimization was performed with a locally developed molecular mechanics and dynamics program based on the Universal Force Field and charge equilibration scheme proposed in Rappé, A. K., Casewit, C. J., Colwell, K. S., Goddard III, W. A., Skiff, W. M. *J. Am. Chem. Soc.* 1992, **114**, 10024 and Rappé, A. K., Goddard III, W. A. *J. Phys. Chem.* 1991, **95**, 3358, and incorporating large segments of code obtained from the program Moil 6.2 (Elber, R.; Roitberg, A.; Simmerling, C.; Goldstein, R.; Li, H.; Verkhivker, G.; Keasar, C.; Zhang, J.; Ulitsky, A., Hebrew University, 1994) and from a potential energy program written by Prof. A. Rappé (Colorado State University). The commercial Molgen 3.0 (Baričič, P.; Mackov, M.; Slone, J. E., Bratislava, Slovakia, 1993) program was used for building the initial geometry and the program Moil-View 8.5 (Simmerling, C., University of Illinois at Chicago, 1995) for molecular display.
(50) Fendler, J. H. *Membrane-Mimetic Approach to Advanced Materials*, Springer-Verlag: Berlin, 1994.

Chapter 17

Interfacial Electron Transfer Dynamics of Photosensitized Zinc Oxide Nanoclusters

Kei Murakoshi[1], Shozo Yanagida[1], Malcolm Capel[2], and Edward W. Castner, Jr.[3]

[1]Material and Life Science, Graduate School of Engineering, Osaka University, Suita, Osaka 565, Japan
[2]Biology and [3]Chemistry Departments, Brookhaven National Laboratory, Upton, NY 11973–5000

We have prepared and characterized photosensitized zinc oxide (ZnO) nanoclusters, dispersed in methanol, using carboxylated coumarin dyes for surface adsorption. Femtosecond time-resolved emission spectroscopy allows us to measure the photo-induced charge carrier injection rate constant from the adsorbed photosensitizer to the n-type semiconductor nanocluster. These results are compared with other photosensitized semiconductors.

Photosensitized nanoclusters of wide bandgap n-type semiconductors are promising systems for efficient solar energy conversion.[1] A photosensitizing dye adsorbed onto the surface of the semiconductor can inject an electron into the semiconductor, which could possibly be used directly in a solar photovoltaic device or stored in a battery. A system design goal is to select a dye/semiconductor pair for which the ground electronic state of the dye will lie in the bandgap between valence and conduction bands of the semiconductor, while the excited state of this dye will have an energy slightly higher than the sum of the conduction band edge and interfacial electron transfer reorganization energies. Several dyes have been used as photosensitizers, including organic laser dyes, porphyrin derivatives, and inorganic complexes of ruthenium, osmium, and iron. A number of semiconductor materials have been studied for this application, including TiO_2, SnO_2, SnS_2, and ZnO. For an overview of these studies of dye-sensitized electron-transfer at the electrode interface, we refer to the chapter by Willig.[2]

Zinc oxide electrodes photosensitized by a number of dyes were first reported by Gerischer and Tributsch.[3,4] Photosensitizers included riboflavin, eosin, rose bengal, rhodamine B, and cyanine dyes. Higher efficiency was achieved in similar systems by Tsubomura, *et al.*[5,6] using sintered porous disks of ZnO. This preparation was used in order to achieve a sufficiently high surface coverage of adsorbed dye on the ZnO electrode. Similar results using Rhodamine B and Ru^{II}-polypyridyl dye sensitized sintered-porous ZnO were also reported by Alonso, *et al.*[7] TiO_2 has also been extensively used as a photosensitizer, with the first report for single-crystal rutile photosensitized by $Ru(bpy)_3{}^{2+}$ and its derivatives given by Clark and Sutin.[8] A number of researchers have prepared and characterized nanocrystalline

ZnO, including Spanhel and Anderson,[9] Bahnemann,[10] and Cavaleri, *et al.*[11] The results mentioned here provide the foundation for the present study.

A number of different types of photosensitizers for metal-oxide semiconductors have been studied, including both organic and inorganic complex chromophores. For strong binding between the photosensitizer and the metal-oxide semiconductor surface, a carboxylate, phosphonate, or sulfate group on the photosensitizer molecule is needed[12-14]. In the experiments presented here, we use 7-aminocoumarin chromophores that have a carboxylate substituent at the 3-position. The carboxylate, perhaps in concert with the 2-position carbonyl oxygen, binds to one or more metal-cation surface sites. The three coumarins used in this study are shown in Scheme I.

Scheme I:

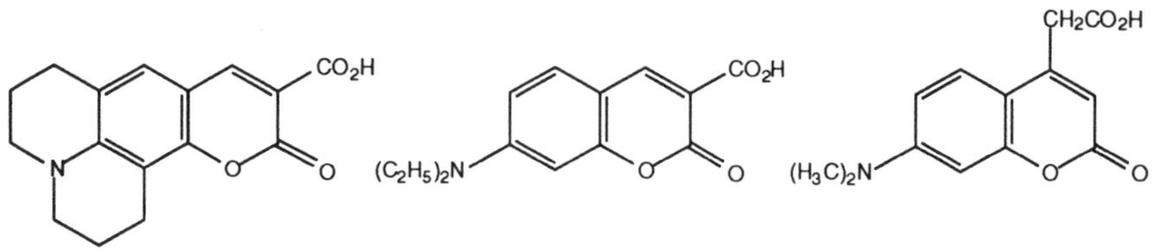

CO_2H CO_2H CH_2CO_2H

N O O $(C_2H_5)_2N$ O O $(H_3C)_2N$ O O

Coumarin 343 Coumarin D-1421 Coumarin D-126

Excited-states of coumarin molecules can be rapidly oxidized or reduced by quenchers of the appropriate electrochemical potential. In either case, typical emission lifetimes of 3 to 5 ns are shortened to sub-picosecond levels for homogeneous solution-phase bimolecular photo-induced electron-transfer.[15,16] Interfacial electron-transfer dynamics for coumarins adsorbed onto semiconductors can also be very rapid.[17,18] The ruthenium dye system used by Grätzel and co-workers[1,19] has recently been shown to undergo very rapid interfacial electron-transfer when adsorbed onto nanocrystalline TiO_2.[20]

Use of colloidal preparations of the semiconductor nanoclusters provides several advantages, including very high surface area for binding of photosensitizer, and straightforward solution-phase material preparation. The promising efficiency of such photoelectrochemical systems depends on very rapid forward electron-transfer from dye to semiconductor, rapid charge trapping, slow back electron-transfer, and a large absorption cross-section for the dye at appropriate visible wavelengths. A focus of our present work has been to characterize the very rapid forward electron-transfer rate. In aqueous photosensitized TiO_2 systems, we have measured forward photo-induced electron-transfer rate constants greater than 2.5×10^{13} s^{-1}.[18] By comparison, methanol dispersions of photosensitized TiO_2 and ZnO both show reaction rates constants that are substantially slower. For the methanol dispersions, the forward electron-transfer rate constants can exceed 10^{11} s^{-1}.

The general scheme for photosensitization of semiconductor nanoclusters is illustrated in Figure 1. On the left are closely spaced but discrete energy levels. These represent both the valence and conduction bands of the semiconductor nanocluster. The levels will be described neither by a continuum, as for a single crystal semiconductor, nor by the distinct and broadly spaced exciton levels (quantum confined) seen in other smaller diameter nanocrystalline materials such as CdS and CdSe. Other energy levels will lie in the bandgap region between valence and conduction band edges, including surface states and shallow and deep traps. To the right, the energy levels of the photosensitizing dye are indicated. It is desirable that the ground state labeled $S_{0,0}$ should lie between the valence and conduction band edges. The optical transition energy to an excited state, $E(S_{1,v} - S_{0,0})$ should be substantially less than the bandgap, to avoid optical excitation of the semiconductor.

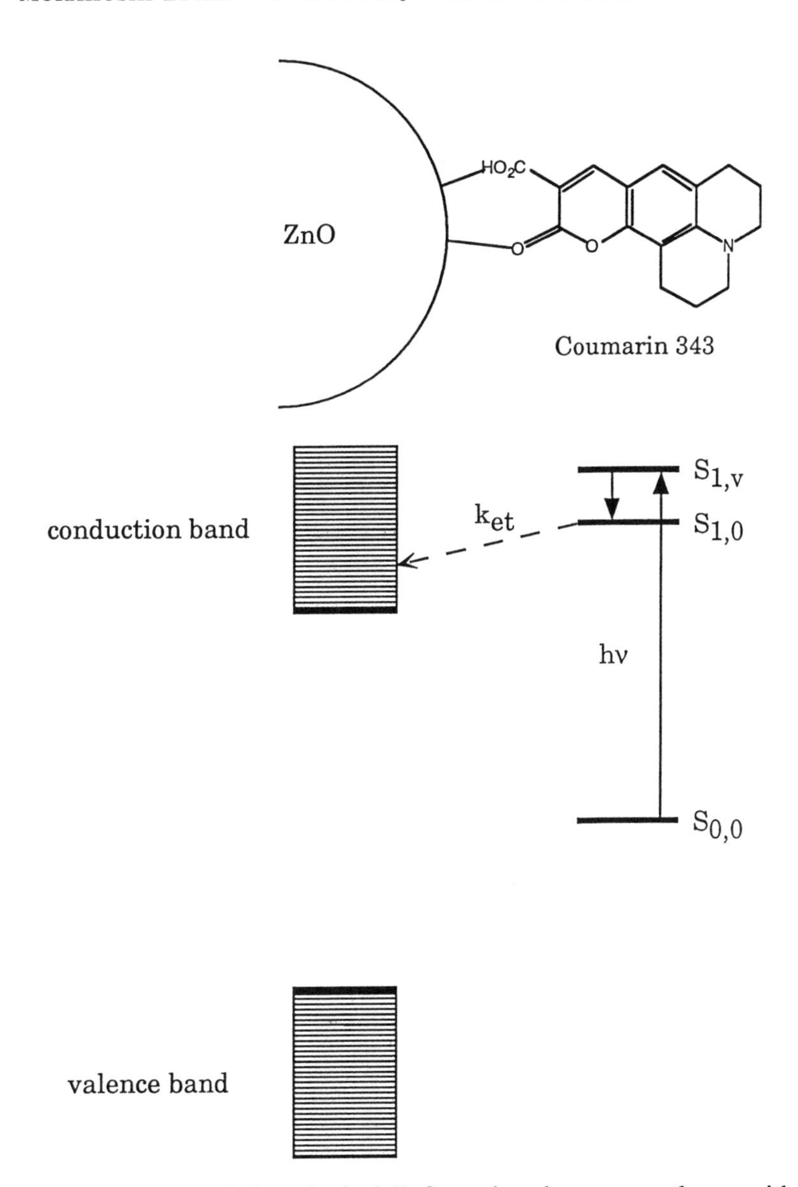

Figure 1. Diagram of the spherical ZnO semiconductor nanocluster with an adsorbed coumarin dye photosensitizer. The semiconductor energy levels are shown descriptively at the left. On the right, the relevant energy levels are shown for the neutral dye molecule. The optical transition between the ground electronic state and vibrationally populated excited electronic state is represented by the vertical arrow labeled hν ($S_{1,v} \leftarrow S_{0,0}$). The downward vertical arrow represents both intramolecular vibrational relaxation and dye-solvent dynamical interactions. The dashed arrow in the direction of the semiconductor conduction band represent the charge carrier (electron) injection into the n-type semiconductor.

There will be some competition between direct electron transfer from the dye excited state to the semiconductor and other radiative and non-radiative excited state relaxation pathways. Alternative relaxation channels include very rapid intramolecular vibrational relaxation and inertial solvent relaxation. A more accurate description would have the electron-transfer event taking place from any of the dye excited vibrational levels, at any point along the solvent relaxation coordinate, into several conduction band levels. If the interfacial electron transfer is to be adiabatic, then the vibrationally relaxed dye excited state $S_{1,0}$ must have an energy difference relative to the conduction band edge greater than the reorganization energy for the interfacial electron-transfer reaction, or 0.1 to 0.5 eV. More detailed discussion of these and other important details can be found in the review articles.[1,2]

Surface characterization of the binding of photosensitizers to metal-oxide surfaces is possible for single crystal sites in ultra-high vacuum (UHV).[21] However, it is more difficult to determine the binding geometry for the photosensitized nanocluster dispersed in methanol. Evidence from surface science studies of single crystals in UHV reveals that metal oxides, such as ZnO, adsorb an alkoxide coating via proton dissociation from alcohols. Previous studies on SnS_2 surfaces have found that the nature of the photosensitizer adsorption changes with increasing surface coverage.[22] In our own work, a parallel study of the same coumarins used to photosensitize TiO_2 showed substantially different steady-state and time-resolved spectra as a function of surface coverage.[18] Also apparent for ZnO is that the Zn^{2+} surface cationic site will bind carboxylic acids as carboxylates under UHV conditions.[21] Molecular modeling studies indicate that there exist several distinct binding geometries for the carboxylated dye on the semiconductor surface. Though direct detection of the nature of surface binding was unavailable from IR spectral data for the sols of photosensitized nanoclusters, dried films of Ru(II) polypyridyl-carboxylate dyes adsorbed on TiO_2 have previously been shown to have bind with both ester-like and chelate carboxylate geometries relative to the Ti^{4+} binding site.[23]

Despite the disclaimers above, we propose examples of four possible binding configurations that were observed to be feasible during molecular modeling. To properly model the surface binding, one would have to build a complex model incorporating at least two atomic layers at the ZnO nanocluster surface; calculations of the electronic structure of the ground, excited, and radical-cation energy levels of the photosensitizer and the band structure of the semiconductor (including quantum confinement); the effects of probable methoxide adsorption on nearby binding sites; and a realistic model for the solvent. Because such a model is presently an intractable computational problem, we instead examined crude models for the binding of coumarins D-1421 or 343 on ZnO, and deduced the proposed interactions shown in Figure 2. Though aromatic dye chromophores will bind directly (but weakly) to the semiconductor surfaces, the strength of the binding interaction is greatly increased if a specific chemical interaction between functional groups on the photosensitizer and the semiconductor nanocluster surface is used. Carboxylate and phosphonate groups have been used previously to tether dyes to TiO_2 and ZnO surfaces.[14,19] The coumarin dyes in Scheme I can interact with the Zn^{2+} cation sites on the surface in several ways. Besides a mono-ether type of linkage, more likely candidates for surface binding structures include a bridged or chelated structure for the carboxylate group, or a bridged or chelated geometry from the carbonyl at the coumarin ring 2-position and one of the 3-carboxylate oxygens. It is known that the distance between metal atoms on the Zn-atom surface (0001 face) of ZnO is 3.25 Å.[24] The distance between oxygen atoms on a carboxylate group of one of these coumarins is about 2.45 Å, whereas the distance between the 2-position carbonyl oxygen and a carboxylate oxygen on coumarin 343 or D-1421 is about 2.6 Å. Given these distances, the chelate rather than the bridge structures shown in Figure 2 are more likely. If one of the salicylate structures is in fact the most stable, this would explain why the third coumarin, D-126, does not seem to bind to the ZnO surface: D-126 cannot form the either of the salicylate geometries.

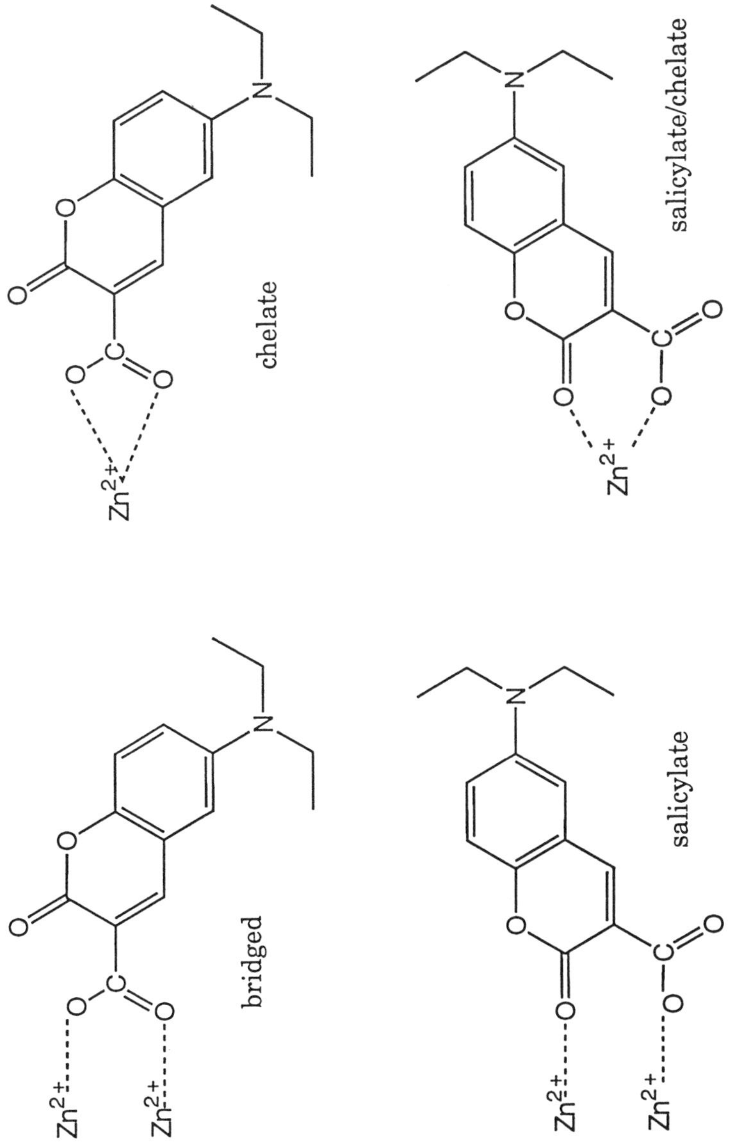

chelate

salicylate/chelate

bridged

salicylate

Figure 2. Models for binding of coumarin oxygens coordinated to Zn^{2+} cation sites on ZnO surface. Other models are feasible in addition to the four presented here.

Experimental

Materials. Zinc acetate dihydrate ($Zn(CH_3COO)_2 \cdot 2H_2O$, 99.9 %, Wako Pure Chemicals), lithium hydroxide monohydrate ($LiOH \cdot H_2O$, reagent grade, Wako Pure Chemicals), hydrochloric acid (HCl, 37 wt. %, reagent grade, Aldrich) were used as received. The solvents methanol, acetonitrile, and 2-propanol (spectral grade, Aldrich) were used without further purification. Ethanol for the preparation of zinc oxide (ZnO) nanocrystallites was purified by fractional distillation just before use.

The dyes coumarin 343 (laser grade, Acros Organics), 7-diethylamino-coumarin-3-carboxylic acid (D-1421, Molecular Probes Inc.), and 7-dimethylamino-coumarin-4-carboxylic acid (D-126, Molecular Probes Inc.) were used as received.

ZnO nanocrystallites were prepared under a dry nitrogen atmosphere following the method of Spanhel and Anderson.[9] An ethanol solution (500 ml) of $Zn(CH_3COO)_2 \cdot 2H_2O$ (0.1 M) was placed into a 1-L round-bottom flask fitted with a $CaCl_2$ moisture trap. The solution was boiled with stirring at 80 °C for 3 h. The condensate was collected continuously. After this procedure, 200 ml of reaction mixture and 300 ml of condensate were obtained. The suspended reaction solution (200 ml) was diluted to yield 500 ml of ethanol solution containing 0.1 M Zn. Then, 0.07 mol of $LiOH \cdot H_2O$ was added to the ethanol solution in an ultrasonic bath at 30 °C, resulting in transparent ethanol solution of ZnO nanocrystallites. The colloidal solution of ZnO nanocrystallites was evaporated at 50 °C and further dried *in vacuo* (*ca.* 1 mm Hg) at room temperature for 12 h, yielding a white powder of ZnO nanocrystallites. The resulting white powder of ZnO nanocrystallites was well-dispersible in dry ethanol and was stable more than 3 days. These results coincide with the literature report of Spanhel and Anderson.[9] Concentrations of ZnO ethanol dispersions were determined by absorption at 340 nm.

Apparatus. Absorption spectra were measured at room temperature on a Hitachi U-3300 spectrophotometer or a Hewlett Packard model 8452A photodiode array spectrophotometer. Emission spectra were measured at room temperature on a Hitachi Model 850 fluorescence spectrophotometer or a custom built multi-channel spectrophotometer. The excitation source for this latter emission spectrometer is from a 10 nm bandpass monochromator-filtered Xe lamp source, with front-surface fluorescence collection. In this instrument the emission was focused into a 0.5 meter focal length spectrometer, with detection by a 1152 x 256 element liquid-nitrogen/thermoelectric cooled CCD detector. Background-subtracted spectra were measured for 5 to 300 second accumulation times. High-resolution transmission electron microscopy (TEM) images and electron diffraction patterns were obtained on a Hitachi H-9000 instrument, operating at 300 kV. Small-angle x-ray scattering (SAXS) measurements were carried out at the National Synchrotron Light Source, beamline X12B.[25]

Electrochemical measurements were carried out using a Bioanalytical Systems potentiostat (BAS 100B). Reduction potentials of coumarin dyes were measured by cyclic voltammetry in acetonitrile solutions containing 0.2 M of tetraethylammonium perchlorate using a glassy carbon working electrode, measured versus SCE.

Ultrafast Spectroscopy. The fluorescence upconversion instrument will be described in a future publication.[26] The design of the instrument is nearly identical to that constructed by the Maroncelli group, described recently in detail.[27] This type of upconversion spectrometer with all-reflective fluorescence optics was described previously by Rosenthal, *et al.*[28] Important details of the Brookhaven instrument are described below.

For the upconversion experiment, 10 to 80 mW average power (82 MHz pulse repetition frequency) of second-harmonic laser light (from a Spectra-Physics Tsunami fs Ti:sapphire laser) was used to excite the sample at wavelengths in the 390-425 nm range. A Type I phase-matched nonlinear optical crystal (0.3 mm

LBO, lithium triborate, or 0.4 mm BBO, meta barium borate) was used for second-harmonic generation. Sample solutions were continuously flowed through a 1 mm path fused quartz flow cell. The emission was gated with near-infrared pulses at the fundamental wavelength (780-850) nm, by sum-frequency mixing fluorescence and laser fundamental in an angle-tuned, 0.4 mm path BBO nonlinear crystal. The upconverted (or sum-frequency) light was detected by a photon counter after passing through a 150 mm focal length monochromator (5 nm bandpass), tuned to the ultraviolet sum-frequency wavelength corresponding to emission at 425-600 nm. The sample was *excited* with 'magic-angle' polarization (linearly polarized with angle 54.7° from vertical) to eliminate polarization anisotropy contributions to the emission dynamics. The fluorescence was *polarization gated* by the BBO crystal to detect vertically polarized emission. Accumulation times of 1 to 20 seconds/point were used, with one to four emission transient scans summed to provide the final raw data set. After each scan, the scanning delay stage was reset to the peak signal position, and the signal intensity was checked to verify that the sample had not degraded, and that the laser pulse intensity had not changed.

The temporal instrument response function was measured by tuning the BBO sum-frequency crystal to phase-match the third-harmonic of second-harmonic excitation and the fundamental gating beams at 260-284 nm. The measured third-harmonic cross-correlation was in the range 130-160 fs (fwhm), and fit well to a Gaussian function, implying a pulse resolution of 90-115 fs. Nonlinear least-squares fitting was done with a program using the Levenberg-Marquardt algorithm, where a model function was iteratively convoluted and compared with the raw data. Best fits to the data were obtained using sums of exponential functions, where it is necessary to consider both rising and decaying components to find the best fit.

Emission lifetimes of the unbound photosensitizing dyes in solution were measured using a time-correlated single-photon-counting (TCSPC) instrument. This instrument is similar to a number of others,[29] except that femtosecond pulses from the Spectra-Physics Tsunami laser at 82 MHz were pulse selected at 4.1 MHz using an acousto-optic modulator instead of using a cavity-dumped picosecond dye laser. The 4.1 MHz pulse train was frequency doubled using a 1 mm Type I BBO crystal, providing $<<$ 1 mW of average power. The sample was contained in a 1 cm path fused silica cuvette, held in an Acton Research Corp. sample chamber attached to their SP-150 spectrometer equipped with a 1200 line/mm grating, for 5 nm bandpass. A Hamamatsu R3809-50U micro-channel plate photomultiplier (MCP-PMT) detected the emission with maximum count rates $<$40 kHz. A Philips electronics 6954-50X pre-amplifier was used, with a 3 dB/18 GHz 50 Ω SMA attenuator placed between the MCP-PMT and the pre-amp. A dichroic beamsplitter was used to remove the fundamental laser beam, which was then routed through a fiber-optic to a Newport D-100 ultrafast photodiode ($<$100 ps fwhm). EG&G/Ortec 9307 PicoTiming discriminators were used in both emission and laser pulse signal channels. Reverse counting methods were used: the signal from the MCP-PMT provided the start signal for the Oxford/Tennelec TC864 time-amplitude converter (TAC), while the D-100 photodiode signal provided the 4.1 MHz stop signal. The histogram of single-photon arrival times was generated by the Oxford PCA-Multiport multichannel analyzer. Using non-dairy coffee whitener, the measured instrument response is about 45 ps, fwhm. Analysis of the emission decays was carried out as for the fluorescence upconversion transients.

Results and Discussion

An atomic level characterization of the adsorbed dye on the ZnO nanocluster surface in methanol dispersion is not yet achievable. However, we have characterized the ZnO nanoclusters dispersed in methanol, before photosensitization. Steady-state absorption spectra for ZnO/MeOH dispersions show an onset at 350 nm (3.54 eV), with a peak at 330 nm (3.76 eV). This value of 3.54 eV for the ZnO bandgap should be compared with the bandgap of single crystal ZnO at 3.2 eV. These results are in good agreement with the spectral characterization of Spanhel and Anderson.[9] In a study of the size dependence of electron dynamics in ZnO nanoclusters, Cavaleri, *et al.*[11] used the model of Nosaka[30] to estimate the particle diameter as a function of the absorption onset energy. Extrapolating from the Cavaleri, *et al.*[11] curve for ZnO with a 2 eV potential well depth, we find that a diameter of 4 nm is predicted.

Our ZnO nanoclusters have been characterized by TEM and SAXS. TEM analysis revealed that ZnO consists of hexagonal ZnO nanocrystallites with a mean diameter of 4 nm. The SAXS data are shown in Figure 3. The raw data are shown in the top of Figure 3, with a Guinier plot of log intensity versus scattering wavevector squared shown in the bottom. A fit to the Guinier equation[31] resulted in a radius of gyration $R_g = 16.6$ Å. Because the particle size distribution is wide, we estimate that this radius has a distribution width greater than 15% of the mean. Converting to a particle diameter ($R_{sphere} = (5/3)^{0.5} R_g$) we find that the SAXS analysis provides a value of 43 Å, in good agreement with the estimates from the Nosaka model and the absorption onset. The Bohr radius for the lowest energy exciton in ZnO has been estimated to be 12.5 Å.[32] Thus, for 43 Å diameter particles, we expect partial quantum confinement, and this is manifested by the blue-shift of the bandgap relative to bulk ZnO, *vide supra*.

The x-ray analysis of the ZnO preparation done by Spanhel and Anderson[9] showed that the resulting nanoclusters have the wurtzite structure of bulk ZnO. Assuming the bulk density of 5.61 g cm^{-3} for the nanoclusters, we can deduce a number of other important parameters. Comparing the surface area of the nanocluster with that of a bulk spherical surface area, we find that the 43 Å nanoclusters have >160,000 times more surface area. This is an obvious reason why photosensitized nanocrystalline materials offer great promise over bulk semiconductors, because the available surface area for dye binding is so many orders of magnitude larger.

For the photosensitized nanoclusters, we have estimated the apparent association constant for coumarin 343 chemisorbed on ZnO to be $K_{app} = 1910\ M^{-1}$, obtained from the steady-state fluorescence quenching of coumarin 343 by ZnO. This value for K_{app} indicates strong binding of the dye to the ZnO surface. We can estimate the surface coverage from the relative dye concentrations and the surface area per nanocluster. Using the 43 Å diameter estimated from the SAXS data, we obtain surface areas and volumes per nanocluster of $A_s = 7.9 \times 10^{-13}\ cm^{-2}$ and $V = 6.5 \times 10^{-20}\ cm^{-3}$, respectively. Using the molecular weight of ZnO (81.37 g mol^{-1}) and the bulk density, we find that for a methanol dispersion of 10^{-2} M concentration, we have a ZnO volume of $1.45 \times 10^{-4}\ cm^{-3}$ in 1 cm^{-3} volume of dispersion. Dividing by the volume of a single nanocluster, we find that there are 3.5×10^{15} nanoclusters in 1 cm^{-3} of dispersion. Assuming 100% binding of the dye contained in the dispersion at 10^{-4} M concentration, we find 6.02×10^{16} dye molecules adsorbed onto 3.5×10^{15} nanoclusters, for an average surface coating of 17 dyes per nanocluster. Alternatively, we can estimate the fractional surface coverage. If we assume one of the salicylate type binding geometries shown in Figure 2, we can estimate the surface area covered by a coumarin dye molecule to be 3.4 Å thick for the aromatic ring, and about 8 Å across, for a subtended area per dye molecule of $2.7 \times 10^{-15}\ cm^{-3}$. Dividing by the surface area of one 43 Å diameter ZnO particle, and multiplying by 17 dyes/particle, we find that the surface coverage θ for 10^{-4} M dye/ 10^{-2} M ZnO would be $\theta = 0.06$. Our dye concentrations ranged from 90 to 170 μM for the optical experiments, so that we were working with surface coverages θ in the range of 5-10%.

Electrochemical characterization of the three coumarin dyes showed that D-126 has a substantially different electronic structure than coumarins 343 or D-1421.

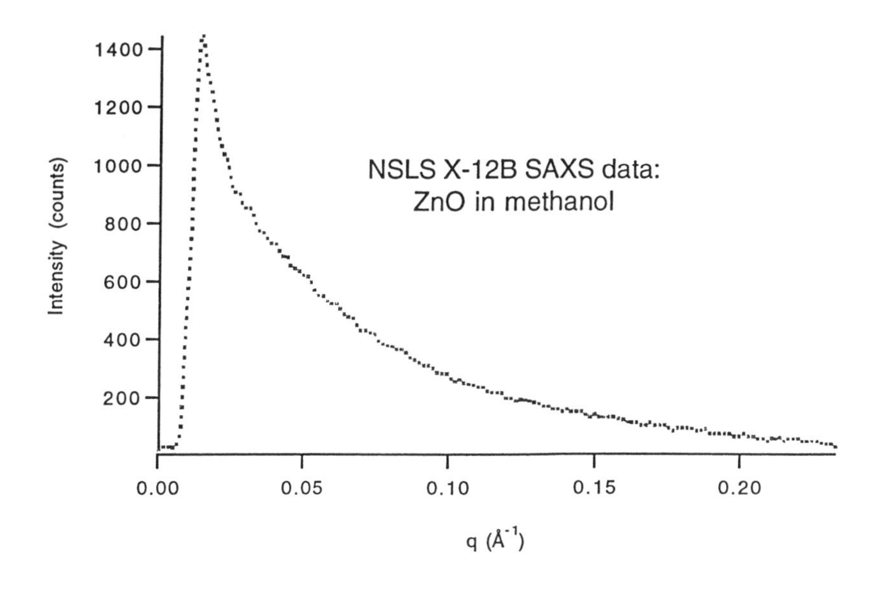

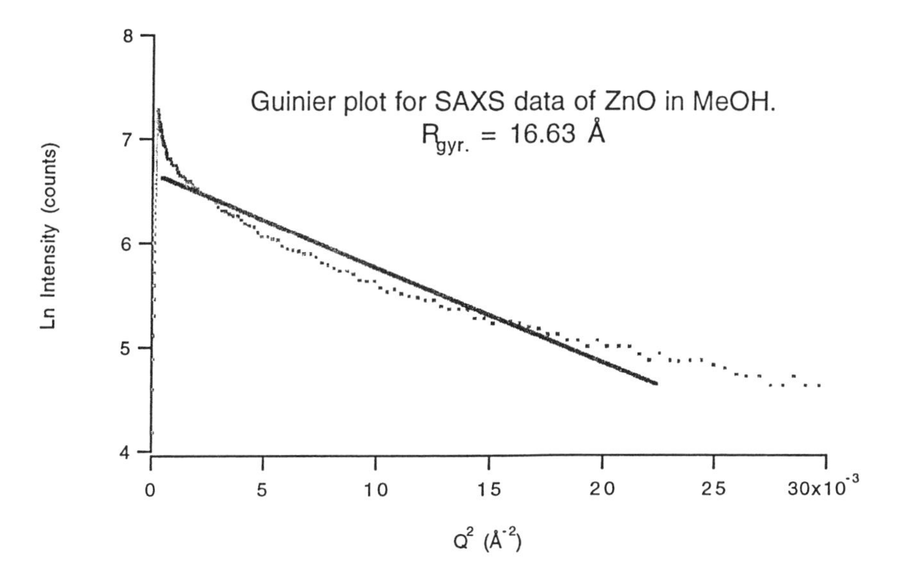

Figure 3. Small-angle x-ray scattering data for ZnO nanoclusters dispersed in methanol (not photosensitized). Data were obtained at beamline X12B at the National Synchrotron Light Source. Top: raw data of scattered intensity versus wavevector **q**. Bottom: Guinier plot of Ln intensity versus $\mathbf{q}^2$.

Quasi-reversible behavior was observed in the electrochemical oxidation of coumarins 343 and D-1421 as relatively clean forward and return waves at 0.99 and 1.20 V, respectively. Reduction of the dyes was irreversible. Although peak potentials were measured at -1.45 and -1.40 V vs. SCE in reduction forward waves for coumarins 343 and D-1421, reduction return waves were not observed. In the system of D-126, oxidation was irreversible (1.04 V), and reduction was quasi-reversible (-2.20 V). The direct determination for the redox potential of excited state to oxidized state of the dye molecules (U_{S^*/S^+}) was not carried out in this experiment. We assumed that the first reduction potentials ($U_{S^-/S}$) determined by the above electrochemical measurements correspond to U_{S^*/S^+} of the dyes under the present condition.

To assign the rapid excited state dye processes for the coumarin-sensitized ZnO dispersions observed in the femtosecond emission experiment, it is first necessary to know the values of the excited state lifetime in the absence of ZnO nanoclusters. The TCSPC emission lifetime data are given in Table I. While the approximately 4 ns lifetimes for the coumarins 343 and D-126 in methanol are unremarkable, the lifetime for the D-1421 dye is a factor of four shorter. This indicates that in the free dye some dynamical processes are contributing to the relevant excited state dynamics, leading to a more rapid non-radiative decay.

Table I. Fluorescence lifetimes of coumarins in methanol, measured by time-correlated single photon counting (TCSPC). Excitation 406 nm.

coumarin	Emission wavelength (nm)	Emission lifetime (ns)
343	475	3.74
D-1421	465	0.985
D-126	450	4.00

The time-integrated absorption and emission spectra for the dye/ZnO dispersions are shown in Figure 4. For the coumarin dyes 343 and D-1421, there is clear evidence of strong surface interaction as ZnO is added to the dye solution (not shown). At the concentrations used for the femtosecond emission experiments, there is a noticeable blue-shift in the absorption spectrum for coumarins 343 (30 nm) and D-1421 (16 nm) in ZnO dispersions. Only a very small shift of <2 nm is observed in the D-126/ZnO dispersion relative to the methanol-dye solution. For the C343/ZnO dispersion, emission quenching is clearly observed. Much less substantial quenching occurs for D-1421/ZnO dispersions. A factor of four less steady-state quenching would be expected because of the difference in excited state lifetimes between D-1421 and the other two coumarins, 343 and D-126. Though the absorption spectra of D-126 and D-126/ZnO show a much weaker interaction, there is an approximate 50% decrease in peak emission intensity for the D-126/ZnO system relative to the free dye solution. Extrapolating the charge carrier injection rate from time-integrated emission studies is likely to be incorrect, because the interfacial electron-transfer is but one of several possible decay pathways for the excited state of the photosensitizing dye. If the rates of competing processes are substantial, then only a direct measure of the excited state dynamics in the time domain will suffice to determine the interfacial electron-transfer dynamics directly.

The femtosecond time-resolved emission data shown in Figures 5 and 6 reveal the true excited state dynamics of the dye adsorbed on the ZnO nanoclusters. The fs

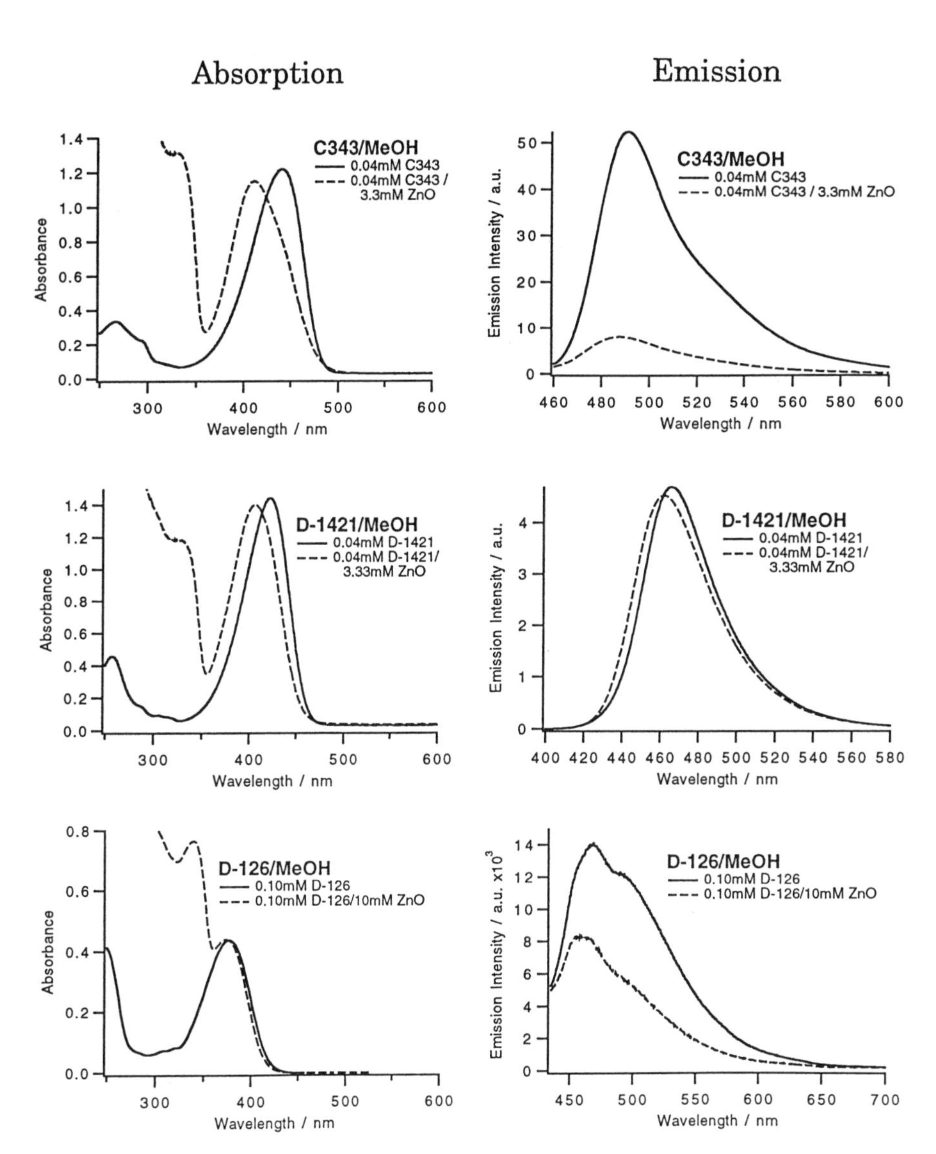

Figure 4. Absorption and emission spectra are shown for (top to bottom) coumarins 343, D-1421, and D-126 in methanol and in ZnO/methanol dispersions. The solid lines are for the spectra of coumarin/methanol solutions, while the dashed lines are for the coumarins/ZnO/methanol dispersions. Optical pathlength for absorption measurements of C343 and D-1421 were 10 mm; for D-126, 2mm. The uncorrected emission spectra were measured in 1.0 mm path cells, on the instrument built at Brookhaven.

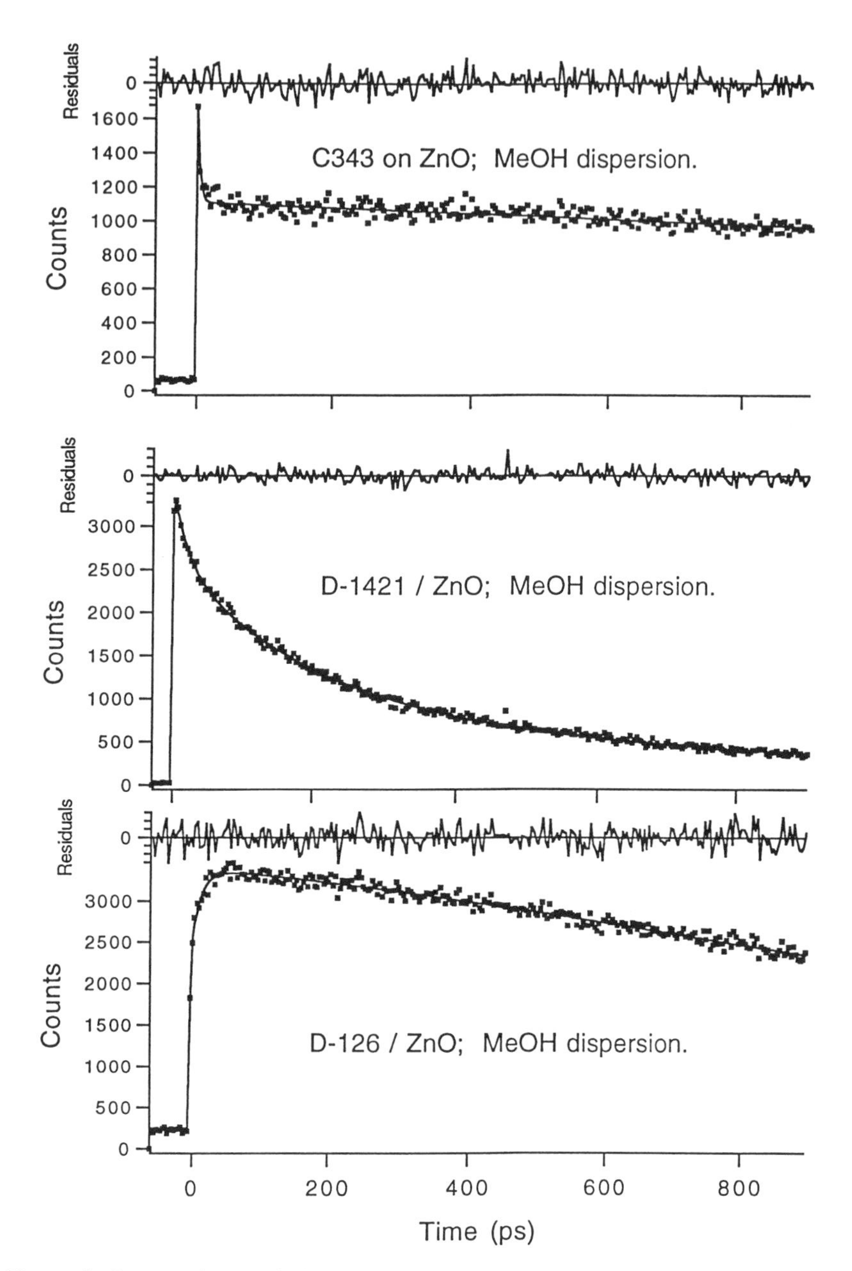

Figure 5. Longer-time emission dynamics measured by femtosecond fluorescence upconversion spectroscopy. Dots are fluorescence upconversion data; fit is a sum of two or three exponentials, and residuals for fit are shown above the plot of data and fit.

Top: [C343] = 1.7×10^{-4} M, [ZnO] = 10^{-2} M methanol dispersion. Excitation wavelength 393 nm; gated emission wavelength 480 nm.
Middle: [D-1421] = 9×10^{-5} M; [ZnO] = 10^{-2} M methanol dispersion. Excitation wavelength 393 nm; gated emission wavelength 480 nm.
Bottom: [D-126]=10^{-4} M; [ZnO] = 10^{-2} M methanol dispersion. Excitation wavelength 398 nm; gated emission wavelength 490 nm.

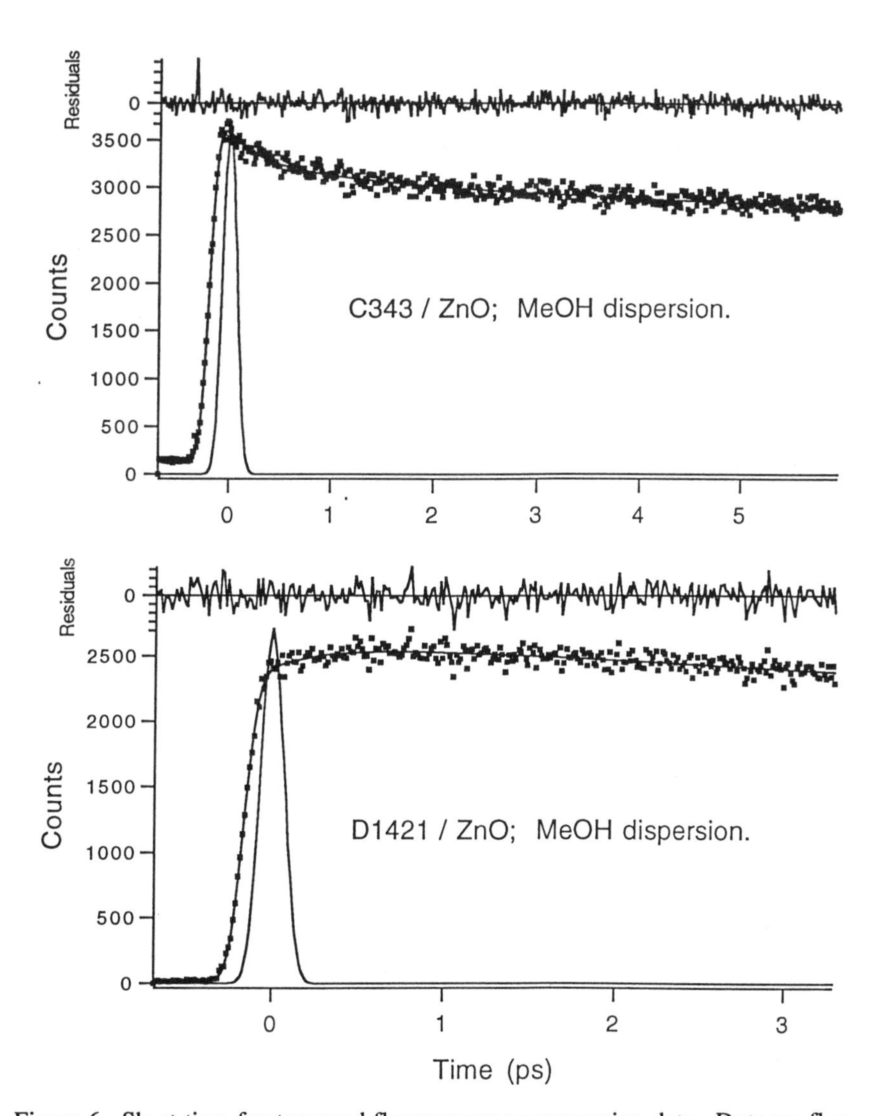

Figure 6. Short-time femtosecond fluorescence upconversion data. Dots are fluorescence upconversion data; third-harmonic cross-correlation (instrument response) is symmetric about zero time delay; fit is a sum of three exponentials, and residuals for fit are shown above the plot of data and fit. Excitation wavelength is 393 nm gated emission wavelength is 480 nm.
Top: [C343] = 1.7 x 10^{-4} M, [ZnO] = 10^{-2} M methanol dispersion.
Bottom: [D-1421] = 9 x 10^{-5} M; [ZnO] = 10^{-2} M methanol dispersion.

fluorescence upconversion spectrometer has a time window limited to < 1 ns because of the travel length of the scanning delay stage. Both long scans (Figure 5) and short scans (Figure 6) were measured for the photosensitized ZnO systems. The long scans were measured with a data point taken every 3.33 ps, while the short scans were measured with 13.34 fs per point.

In Figure 5 (top), a very rapid decay process in the excited-state of coumarin 343 adsorbed on ZnO is observed, with a bi-exponential fit giving a long lifetime of >6 ns, 43% amplitude, with a short component of 3.65 ps, 57% amplitude. The presence of a rapid component is unambiguous, but the fitted value of 3.65 ps must be viewed cautiously, because of the 3.33 ps/point data spacing. To characterize the rapid dynamics correctly, a scan with more densely spaced time points is needed, shown in Figure 6 (top). The results of the fit to this 6 ps window are shown in Table II. A multi-exponential decay for D-1421 on ZnO is observed on the long time-scale plot in Figure 5. A best fit with no constrained parameters was obtained, and then refined with equally good statistics by forcing the longest decay component equal to the value of the excited state lifetime of the free dye in methanol. Again, the fit parameters are found in Table II. The data for D-126 and ZnO nanoclusters in methanol is shown in Figure 5 (bottom). The dominant feature of this transient is surprising: a long rise-time, not a rapid decay. Careful observation of the data reveals a convex, rather than concave curvature to the decay at times from 100-900 ps, indicating that the rise is actually bi-exponential. The long decay time of 1.36 ns is a factor of three shorter than the emission lifetime of the free dye. The absorption spectra and electrochemistry data, combined with the rising emission dynamics transient, show that the coumarin D-126 is not acting at all like the ideal photosensitizer illustrated in Figure 1 and described in the Introduction.

Table II. Excited state dynamics of coumarins in ZnO / methanol dispersions, fit to a sum of up to three exponential decay components. A negative amplitude A_i indicates an exponential rise rather than a decay. The superscript ^{f} indicates that this parameter was fixed during the fitting procedure to match the measured emission lifetime in methanol.

Coumarin	A_1	τ_1 (ps)	A_2	τ_2 (ps)	A_3	τ_3 (ps)
343	0.1639	0.33	0.1494	3.47	0.6867	3740^f
D-1421	0.2576	19.6	0.5155	196.5	0.2269	985^f
D-126	-2.852	13.1	-3.076	489.2	1.000	1358.3

The ultrafast dynamics of coumarins 343 and D-1421 on ZnO are shown at high time resolution in Figure 6. In addition to the lifetime of about 3.5 ps, a substantially faster lifetime of 330 fs is required to fit the data for coumarin 343 on ZnO. No risetime could be fit to this data. This indicates that with our measured time resolution, if a dynamic process leading to a risetime were present, it must be faster than 25 fs. In contrast, the data for D-1421 on ZnO show the need for a 350 fs risetime in the fit, and a 34.9 ps decay (a value in between the 20 and 200 ps time constants fit for D-1421/ZnO in Figure 5, constrained by the 3 ps time window of this data set). The risetime is a lag-time prior to interfacial electron-transfer for D-1421 sensitized ZnO. It is unlikely that this risetime results from vibrational relaxation for coumarin D-1421/ZnO; coumarin chromophores of similar size have intramolecular vibrational relaxation occurring in <50 fs.[27] A possible explanation is that since we are observing the emission dynamics on the red side of the spectrum, the risetime could result from dynamical solvation leading to a time-dependent fluorescence Stokes shift. The ultrafast solvation dynamics of a related molecule in methanol solution, coumarin 153, have been studied in detail.[27,28] The experimental time correlation function for the energy relaxation occurring as a result of fluctuations of the methanol solvent has four exponential time constants: 30 fs, 280 fs, 3.2 ps, and 15.3 ps. The shortest time constant results from librational fluctuations of the hydrogen-bonded hydroxy group. The 280 fs value is assigned to collision-induced fluctuations in the solvent. The longer two time constants of 3.2 and 15.3 ps result from diffusive reorientation of the methanol molecules. All of these solvent fluctuation processes are expected to affect rapid electron-transfer reactions, though with a different weighting factor than would be observed for the free dye molecule.

We are confident that the very rapid excited-state dynamical processes observed in the coumarins adsorbed onto ZnO result from electron-transfer to the ZnO nanocluster. In a number of systems, other organic dye photosensitizers adsorbed on ZnO electrodes have produced substantial photocurrent.[3-7] Coumarin 343 has been shown to be an excited-state electron donor when adsorbed on TiO_2 by two methods. First, current proportional to light flux is obtained from a coumarin 343 sensitized TiO_2 photoelectrochemical cell. Second, transient absorption measurements observe the radical-cation species of coumarin 343.[33,34] However, we note that our assignment of the excited state decay processes for these coumarin photosensitized ZnO systems is based on inference from the prior work; we have not yet measured photocurrent or transient absorption from coumarin/ZnO systems.

The analysis of the time-resolved emission dynamics provides us with the interfacial electron-transfer rate constants. The effective rate constant for electron transfer is the inverse of the weighted average lifetime:

$$k_{ET} = 1/\langle\tau\rangle; \qquad \langle\tau\rangle = \frac{a_1}{a_1+a_2}\tau_1 + \frac{a_2}{a_1+a_2}\tau_2 .$$

For ZnO dispersions photosensitized by coumarin 343, we obtain $k_{ET} = 5.5 \times 10^{11}\ s^{-1}$
from the 0.33 and 3.47 ps components of the fast decay. The interfacial electron-transfer for D-1421 sensitized ZnO is much slower. If we assume that both the 20 and 197 ps decay components arise from interfacial electron-transfer, then $k_{ET} = 7.3 \times 10^9\ s^{-1}$. A possible explanation for the substantially slower electron-transfer to ZnO from coumarin D-1421 relative to 343 is that weaker interaction between ZnO surface and the dye molecule leads to the interfacial electron-transfer described as a non-adiabatic 'normal' Marcus regime photoreaction.

One factor in understanding the relatively slow interfacial electron-transfer in both these cases may be the effect of methanol on the ZnO nanocluster surface. Single crystal studies show a range of binding behavior for methanol, especially for methoxide groups attached to the Zn^{2+} sites on the surfaces.[21] Because of the LiOH used in the synthetic procedure, we may expect methoxide binding to be far more likely than methanol H-bonding. For the case of coumarin 343, we have evidence that the dye is tightly bound from the value of $K_{app} = 1910\ M^{-1}$. However, the likelihood of surface binding methoxide on ZnO in methanol dispersion means that the electronic structure of particles may be substantially modified. In nanocrystalline systems, such a change of surface structure significantly modifies surface electronic structure as well as the energies of whole particle, i.e., band structure. The adsorption of negatively charged methoxide should cause a shift of the conduction band edge to negative potential, leading to slower injection. As mentioned in the introduction, we find that the charge injection rate to photosensitized TiO_2 is larger than the rate for injection into ZnO, using the same photosensitizer. Redmond, *et al.* have estimated that the density of states for the conduction band of TiO_2 has been estimated to be 113 times the density of states for ZnO.[32] The theories of Marcus and Levich describing interfacial electron-transfer[2] would then predict faster electron-transfer for photosensitized TiO_2 versus ZnO, as observed.

Rapid interfacial electron-transfer in photosensitized semiconductors has long been rationalized by a scheme such as described in Figure 1.[19,35] Time-resolved spectroscopy of the dye sensitizer excited states have been carried out for tin disulfide (SnS_2). Willig, *et al.* measured charge carrier injection from cresyl violet sensitized SnS_2 with a dye excited state decay less than their 10 ps instrument-limited response, and assigned this to interfacial electron transfer into the SnS_2 conduction band.[22] A similar explanation was given by Lanzafame, *et al.*, who used oxazine dye as a photosensitizer for SnS_2. Though the excited state decay they measured was limited by the 40 ps instrument response, by careful calibration of the time versus intensity quenching in this experiment, they estimated that the interfacial electron transfer occurred in 40 fs.[36]

Very rapid adiabatic interfacial electron-transfer has recently been observed for nanocrystalline photosensitized TiO_2 films or dispersions in at least three different laboratories. Tachibana, *et al.* studied the Ru^{II} dye used in the successful Grätzel-group photocell, adsorbed onto nanocrystalline TiO_2 films.[20] They observe rapid interfacial electron transfer with equally weighted time constants of 0.15 and 1.2 ps, leading to an effective rate constant of at least $k_{ET} = 1.5 \times 10^{12}\ s^{-1}$. These authors are careful to point out that the 150 fs time constant is instrument limited, and that the true rate may be even faster. Rehm, *et al.* studied coumarin 343 photosensitized TiO_2, dispersed in 95% water/5% acetone solvent, and found an excited state decay of <190 fs (instrument limited), corresponding to an adiabatic rate constant of k_{ET} exceeding $5.3 \times 10^{12}\ s^{-1}$.[17] We have studied the same three coumarins reported on here, adsorbed on 53 Å diameter TiO_2 nanoclusters in aqueous dispersions at pH 4. Our femtosecond emission results show dominant rapid decay time constants of 42 fs (instrument limited) and 85 fs, for coumarins 343 and D-1421, respectively.[18] These time constants correspond to adiabatic interfacial electron transfer rate constants of at least 2.4×10^{13} and $1.2 \times 10^{13}\ s^{-1}$.

Summary and Future Directions

We find that coumarins 343 and D-1421 successfully photosensitize ZnO nanoclusters in methanolic dispersions, but that D-126 does not. The electron-transfer rates are rapid, but indicate an electronic coupling that is between non-adiabatic and adiabatic. The similarities between these dyes on ZnO and on TiO_2 in methanol indicate that a non-alkanol solvent, especially aqueous dispersion, may well lead to the type of extremely rapid charge carrier injection dynamics found for photosensitized TiO_2 in water. The ZnO nanoclusters used in this study were not stable in water. However, an alternative preparation by Bahnemann provides 50 Å diameter wurtzite ZnO nanoclusters that are stable in aqueous dispersion.[10] It will be of great interest to measure the photo-induced interfacial electron-transfer for photosensitizers such as coumarin 343 or Ru^{II}-polypyridyl dyes to learn whether extremely rapid, adiabatic charge carrier injection becomes possible with aqueous nanocrystalline ZnO.

Acknowledgments

The research carried out at Brookhaven National Laboratory (E.W.C.) was funded under contract DE-AC02-76CH00016 with the U.S. Department of Energy and supported by its Division of Chemical Sciences, Office of Basic Energy Sciences. K.M and S.Y. wish to thank the Monbusho International Scientific Research Program: Joint Resarch for support. E.W.C. wishes to thank Prof. S. H. Courtney for a loan of the nonlinear least-squares fitting program used in analyzing the fs fluorescence and TCSPC data, and Dr. M.-L. Horng and Prof. M. Maroncelli at Penn State for invaluable advice during the setup of the femtosecond fluorescence upconversion spectrometer.

References

(1) Hagfeldt, A.; Grätzel, M., *Chem. Rev.* **1995**, *95*, 49-68.
(2) Willig, F. In *Surface Electron Transfer Processes*; R. J. D. Miller, G. L. McLendon, A. J. Nozik, W. Schmickler and F. Willig, Ed.; VCH: New York, NY, 1995; pp 167-309.
(3) Gerischer, H.; Tributsch, H., *Ber. Bunsenges. Phys. Chem.* **1968**, *72*, 437-445.
(4) Tributsch, H.; Gerischer, H., *Ber. Bunsenges. Phys. Chem.* **1968**, *73*, 251-260.
(5) Tsubomura, H.; Matsumura, M.; Nomura, Y.; Amamiya, T., *Nature* **1976**, *261*, 402-403.
(6) Matsumura, M.; Nomura, Y.; Tsubomura, H., *Bull. Chem. Soc. Jpn.* **1977**, *50*, 2533-2537.
(7) Alonso, N.; Beley, V. M.; Chartier, P.; Ern, V., *Revue Phys. Appl.* **1981**, *16*, 5-10.
(8) Clark, W. D. K.; Sutin, N., *J.A.C.S.* **1977**, *99*, 4676-4682.
(9) Spanhel, L.; Anderson, M. A., *J.A.C.S.* **1991**, *113*, 2826-2833.
(10) Bahnemann, D. W., *Israel Journal of Chemistry* **1993**, *33*, 115-136.
(11) Cavaleri, J. J.; Skinner, D. E.; Colombo, J., D. Phillip; Bowman, R. M., *J. Chem. Phys.* **1995**, *103*, 5378-5386.
(12) Dare-Edwards, M. P.; Goodenough, J. B.; Hamnett, A.; Seddon, K. R.; Wright, R. D., *Journal of the Chemical Society: Faraday Discussions* **1980**, *70*, 285-298.
(13) Péchy, P.; Rotzinger, F. P.; Nazeeruddin, M. K.; Kohle, O.; Zakeeruddin, S. M.; Humphry-Baker, R.; Grätzel, M., *J. Chem. Soc., Chem. Commun.* **1995**, 65-66.
(14) Yan, S.; Hupp, J. T., *J. Phys. Chem.* **1996**, *100*, 6867-6870.
(15) Jones, G. I.; Griffin, S. F.; Choi, C.-y.; Bergmark, W. R., *J. Org. Chem.* **1984**, *49*, 2705-2708.
(16) Yoshihara, K.; Tominaga, K.; Nagasawa, Y., *Bull. Chem. Soc. Jpn.* **1995**, *68*, 696-712.
(17) Rehm, J. M.; McLendon, G. L.; Nagasawa, Y.; Yoshihara, K.; Moser, J.; Grätzel, M., *J. Phys. Chem.* **1996**, *100*, 9577-9578.
(18) Murakoshi, K.; Yanagida, S.; Castner, E. W., Jr., *to be submitted to JACS* **1997**,
(19) Nazeeruddin, M. K.; Kay, A.; Rodicio, I.; Humphry-Baker, R.; Müller, E.; Liska, P.; Vlachopoulos, N.; Grätzel, M., *J.A.C.S.* **1993**, *115*, 6382-6390.
(20) Tachibana, Y.; Moser, J. E.; Grätzel, M.; Klug, D. R.; Durrant, J. R., *J. Phys. Chem.* **1996**, *100*, 20056-20062.
(21) Heinrich, V. E.; Cox, P. A., *The Surface Science of Metal Oxides;* 1 ed.; Cambridge University Press: Cambridge, UK, 1994, p 278.
(22) Willig, F.; Eichberger, R.; Sundaresan, N. S.; Parkinson, B. A., *J.A.C.S.* **1990**, *112*, 2702-2707.

(23) Murakoshi, K.; Kano, G.; Wada, Y.; Yanagida, S.; Miyazaki, H.; Matsumoto, M.; Murasawa, S., *Journal of Electroanalytical Chemistry* **1995**, *396*, 27-34.
(24) Petrie, W. T.; Vohs, J. M., *Surface Science* **1991**, *245*, 315-323.
(25) Capel, M., http://crim12b.nsls.bnl.gov/x12b.htm:
WWW reference for NSLS Beamline X12B, Upton, NY, for 1997.
(26) Castner, E. W., Jr., **1997**, *in preparation*,
(27) Horng, M. L.; Gardecki, J. A.; Papazyan, A.; Maroncelli, M., *J. Phys. Chem.* **1995**, *99*, 17311-17337.
(28) Rosenthal, S. J.; Jiminez, R.; Fleming, G. R.; Kumar, P. V.; Maroncelli, M., *Journal of Molecular Liquids* **1994**, *60*, 25-56.
(29) Chapman, C. F.; Fee, R. S.; Maroncelli, M., *J. Phys. Chem.* **1990**, *94*, 4929-4935.
(30) Nosaka, Y., *J. Phys. Chem.* **1991**, *95*, 5054-5058.
(31) Glatter, O.; Kratky, O., *Small Angle X-ray Scattering*; Academic Press: New York, NY, 1982, pp 515.
(32) Redmond, G.; O'Keefe, A.; Burgess, C.; MacHale, C.; Fitzmaurice, D., *J. Phys. Chem.* **1993**, *97*, 11081-11086.
(33) Moser, J. E.; Grätzel, M., *Chem. Phys.* **1993**, *176*, 493-500.
(34) Castner, E. W., Jr.; Murakoshi, K., unpublished results.
(35) O'Regan, B.; Grätzel, M., *Nature* **1991**, *353*, 737-739.
(36) Lanzafame, J. M.; Miller, R. J. D.; Muenter, A. A.; Parkinson, B. A., *J. Phys. Chem.* **1992**, *96*, 2820-2826.

Chapter 18

Nickel–Aluminum Alloy Clusters: Structural and Dynamical Properties

J. Jellinek and E. B. Krissinel[1]

Chemistry Division, Argonne National Laboratory, Argonne, IL 60439

Structural and dynamical properties of mixed Ni_nAl_m alloy clusters mimicked by a many-body potential are studied computationally for all the possible compositions n and m such that n+m=13. It is shown that the manifold of the usually very large number of the different possible structural forms can be systematized by introducing classes of structures corresponding to the same concentration of the components, geometry and type of the central atom. General definitions of mixing energy and mixing coefficient are introduced, and it is shown that the energy ordering of the structural forms within each class is governed by the mixing coefficient. The peculiarities of the solid-to-liquid-like transition are described as a function of the concentration of the two types of atoms. These peculiarities are correlated with and explained in terms of the energy spectra of the structural forms. Class-dependent features of the dynamics are described and analyzed.

Understanding of properties of different materials and, even more so, design and fabrication of novel materials require a microscopic (i.e., atomic level) description of the fundamental characteristics of these systems. The field of atomic and molecular clusters grew out of an effort to provide such a description. Clusters are a tool for arriving at understanding of how properties of individual atoms and molecules (gas phase) evolve into those of solids and liquids (condensed phase) as the number of atoms or molecules in a cluster increases. It has been realized along the way that clusters themselves form a class of extremely interesting and important, from the applied point of view, objects. Cluster research evolved into a sovereign discipline with its own methodologies and techniques. At its present stage it already begins to have a noticeable impact on bordering disciplines, such as materials science, chemistry and physics, and on a variety of modern technologies, such as microelectronics, catalysis, nucleation and growth of particles and layers, etc. One of the well-known ways of obtaining materials with superior properties is to mix different elements. Alloys (mixes of metals) are known to the mankind for millennia

[1]Current address: Institute for Water and Environmental Problems, 105 Papanintsev Str., Barnaul 656099, Russia

(at least from the bronze age). A promising novel approach to design and fabrication of materials with predefined mechanical, thermal, electrical, optical, magnetic and other properties emerges in the cluster field. The idea is to use clusters with special properties as building blocks for manufacturing so-called cluster-assembled (*1-4*) and nanophase (*5-7*) materials. It is clear that different (in size, structure, type and relative concentration of the component elements, etc.) clusters will have to be used to manufacture cluster-assembled and nanophase materials tailor-made to meet special-purpose needs and specifications. Clusters of mixed composition stand the best chance of becoming the optimal building blocks of such novel materials. In this respect, alloy clusters are of especial interest because they are natural "laboratories" both for furthering our understanding of the traditional (bulk) alloys and for searching for building blocks of novel metal-cluster based materials.

Here we give a description and analysis of some of the structural and dynamical properties of two-component metal clusters (*8-18*). We use as a paradigm the Ni_nAl_m system with all the possible values of n and m such that n+m=13. In the next section, we specify the potential energy function representing the microscopic (interatomic) interactions and describe the necessary theoretical and computational tools. Then we present an analysis of the structural properties of the mixed clusters and formulate a classification scheme, which allows for systematization and rationalization of the energy ordering of the different structural forms of the alloy 13-mer. The discussion of the structural and energy characteristics is followed by a description of the dynamical properties of the mixed clusters. It is shown that these properties correlate with and can be understood in terms of the classes of structural forms. A brief summary is given in conclusion.

Theory and Methodology

All the structural and dynamical properties of alloy clusters, as of any physical system, are ultimately defined by the interactions between the constituent atoms. We mimic these interactions in a mixed 13-atom Ni-Al cluster by a Gupta-like potential of the form (*19*)

$$V=\sum_{i=1}^{N}\left\{\sum_{j=1(j\neq i)}^{N} A_{ij}\exp\left(-p_{ij}\left(\frac{r_{ij}}{r_{ij}^{o}}-1\right)\right)-\left[\sum_{j=1(j\neq i)}^{N}\xi_{ij}^{2}\exp\left(-2q_{ij}\left(\frac{r_{ij}}{r_{ij}^{o}}-1\right)\right)\right]^{1/2}\right\}, \qquad (1)$$

where N is the total number of atoms in the cluster; A_{ij}, ξ_{ij}, p_{ij}, q_{ij}, and r_{ij}^{o} are adjustable parameters, the values of which depend on the type of the i-th and the j-th atoms; and r_{ij} is the distance between the i-th and the j-th atoms. The values of the parameters for the Ni-Ni, Al-Al and Ni-Al pairs of atoms are adopted from ref. (*19*) and are listed in Table I. These parameters are fitted in such a way that in the cases n=0 or m=0 the potential describes the pure Al_{13} or Ni_{13} clusters, respectively [for details see ref. (*19*)]. As has been established in earlier studies, the lowest energy structure of both Ni_{13} (*20-24*) and Al_{13} (*16*) is an icosahedron. The distance from the center of the icosahedron to its surface, as obtained from the potential, equation (1), is 2.43 Å and 2.66 Å for Ni_{13} and Al_{13}, respectively. These values are in good agreement with those obtained in other studies (*25-32*). The average binding energy per atom is 2.67 eV in Ni_{13} and 2.60 eV in Al_{13}. The corresponding values obtained in electronic structure calculations vary from 2.91 to 4.26 eV in Ni_{13} (*25-27*) and from 2.77 to 3.21 eV in Al_{13} (*28-32*).

A possible way to obtain different structural forms of the mixed Ni_nAl_m, n+m=13, clusters is to perform simulated thermal quenchings from a variety of initial configurations. These can be generated, for example, by a sufficiently high energy and sufficiently long molecular dynamics trajectory that accomplishes a sufficiently complete sampling of the configuration space of the cluster of interest. A distinguishing feature of two-component clusters is that the total number of their different structural forms is much larger than that corresponding to their one-component counterparts. The reason for this is that the structural forms of mixed clusters can differ not only by the geometry (packing), as in the case of one-component clusters, but also by different distributions of the two types of atoms between the sites of a chosen geometry. We retain the term *isomers* to label the different geometric forms of a mixed cluster of a given size and introduce the term *homotops* to label a subset of the structures of a cluster with a given composition that belong to the same isomer, but differ by the distribution of the two types of atoms between the sites of that isomer. The total number of the different possible structural forms (isomers and homotops) of mixed clusters is usually very large. For example, the number of ways, in which one can distribute six Al atoms and seven Ni atoms (or seven Al atoms and six Ni atoms) in a single isomer of a 13-atom cluster is $13!/(6!\cdot7!)=1716$, although, because of the possible symmetry of an isomer, not all the 1716 distributions necessarily form different homotops. Instead of performing a comprehensive search for the possible structures of Ni_nAl_m through the thermal quenching procedure mentioned above, we carried out a search for all the homotopic forms corresponding to six isomers. These isomers were chosen as the first six lowest energy structures of the pure Al_{13} (Figure 1). (The search for the structures of Al_{13} was performed using the simulated thermal quenching technique; it produced a total of 129 different isomers.) For each composition, i.e. fixed n and m, n+m=13, all the possible replacements of n Al atoms by n Ni atoms were considered in each of the six chosen isomers of Al_{13}. The so formed mixed Ni_nAl_m clusters were then allowed to relax to their corresponding equilibrium configurations. The technical details of the applied relaxation procedure are given in ref. (*14*). We consider all the stable equilibrium structures that differ only by small local relaxations in the positions of the individual atoms, but have the same geometric packing, as the same isomer. A normal mode analysis was applied to each resulting configuration to filter out those structures that represent saddles, rather then minima, of the corresponding potential energy surfaces. In general, not necessarily all the stable isomeric forms of the pure clusters survive as equilibrium geometries of the corresponding mixed clusters (at least, not necessarily for all the compositions of the latter), and, vice versa, some stable isomers of the mixed clusters may not survive as equilibrium geometries of their one-component counterparts [cf. refs. (*14,16*)].

An important structural aspect of two-component clusters is mixing vs segregation of the component materials and the quantification of the degree of mixing. It has been suggested (*11*) to use as a measure of the degree of mixing the total number of bonds between first neighbor atoms of unlike type. The shortcomings of this measure in establishing a correlation between the preference for mixing (or segregation) and the configurational energy of a mixed cluster are obvious. Such a correlation exists only when and if 1) the total interaction energy can be represented as a sum of the pairwise interactions (pairwise-additivity) and 2) the interatomic interactions do not extend beyond the first nearest neighbors of each atom. Neither of the two assumptions holds for many systems. The inherently many-body character of the interatomic interactions in metals is one of the most distinguishing features of these systems.

A physically meaningful and general definition of the mixing energy V_{mix} of a two-component cluster A_nB_m in a given configuration, as specified by the coordinates of its atoms, where A and B are the types of the atoms, is given by (*14*)

Table I. Parameters of the Gupta-Like Potential

Parameters	Ni-Ni	Al-Al	Ni-Al
A (eV)	0.0376	0.1221	0.0563
ξ (eV)	1.070	1.316	1.2349
p	16.999	8.612	14.997
q	1.189	2.516	1.2823
r^o (Å)	2.4911	2.8638	2.5222

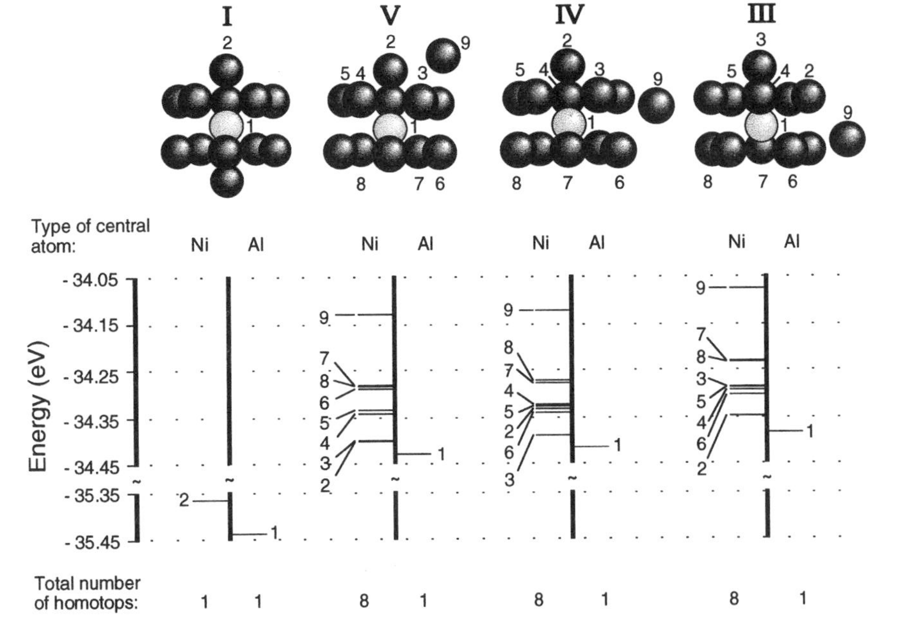

Figure 1. Four isomeric forms and energy diagrams of the corresponding to them homotops for the $Ni_{12}Al$ cluster (the darker spheres represent Ni, the grays Al). The diagrams are separated into classes defined by the type of the central atom. The total number of homotops in each class is also shown. Each isomer is depicted by its lowest energy homotop. The isomers are shown in the order of increasing energy and are labeled by roman numerals, which indicate the isomer of Al_{13} with the same geometry (see text). The number labeling each energy level represents the position of the Al atom in the homotop corresponding to that level.

$$V_{mix} = V_{A_nB_m} - \left[V_{A_n}^{(A_nA_m)} + V_{B_m}^{(B_nB_m)} \right], \tag{2}$$

where $V_{A_nB_m}$ is the energy of the A_nB_m cluster, $V_{A_n}^{(A_nA_m)}$ is the energy of the A_n subcluster in the A_nA_m cluster, and $V_{B_m}^{(B_nB_m)}$ is the energy of the B_m subcluster in the B_nB_m cluster. In the above definition the one-component A_nA_m and B_nB_m clusters have the same configuration as A_nB_m, and the configurations of the n-atom and of the m-atom subclusters are the same in A_nB_m, A_nA_m, and B_nB_m. Since

$$V_{A_nB_m} = V_{A_n}^{(A_nB_m)} + V_{B_m}^{(A_nB_m)}, \tag{3}$$

where $V_{A_n}^{(A_nB_m)}$ and $V_{B_m}^{(A_nB_m)}$ are the energies of the A_n and B_m subclusters in the A_nB_m cluster, the mixing energy can be written as

$$V_{mix} = \left(V_{A_n}^{(A_nB_m)} - V_{A_n}^{(A_nA_m)} \right) + \left(V_{B_m}^{(A_nB_m)} - V_{B_m}^{(B_nB_m)} \right). \tag{4}$$

V_{mix} is, therefore, a measure of the total change in energy experienced by the A_n and B_m subclusters (or, alternatively, their individual atoms) when they are removed from the one-component A_nA_m and B_nB_m clusters, respectively, and are brought together to form the A_nB_m cluster. It is clear that for pairwise-additive potentials which include only first-neighbor interactions our definition of V_{mix} becomes equivalent, more precisely, proportional, to the number of bonds between first-neighbor atoms of unlike type. (The proportionality is obtained rigorously only if one neglects the possible small differences, caused by the relaxations, in the distances between the nearest neighbor atoms of unlike type.) The (global) mixing coefficient M can be defined as (*14*)

$$M = \frac{V_{mix}}{V} \cdot 100\%. \tag{5}$$

The dynamical features of the clusters are derived from microcanonical (i.e., constant energy) molecular dynamics simulations. Long trajectories (up to $8 \cdot 10^6$ steps with a step size of 2 fs) were run on a fine grid covering a broad energy range for all the possible compositions of the mixed Ni-Al 13-mer. The trajectories were generated using the velocity version of the Verlet propagator (*33*). The initial coordinates and velocities of the atoms were chosen so as to supply no total linear and angular momenta to the cluster. The total energy is conserved, even along the longest trajectories, within 0.01 %. The cluster temperature in a given run is defined as

$$T = \frac{2\langle E_k \rangle_t}{(3N-6)k}, \tag{6}$$

where E_k is the total (internal) kinetic energy, k is the Boltzmann constant, and $\langle \; \rangle_t$ stands for time averaging over the entire trajectory. The dynamical features of the clusters are analyzed in terms of the following quantities: 1) caloric curve, i.e., time-averaged kinetic energy (per atom) as a function of total energy (per atom); 2) relative root-mean-square (rms) bond length fluctuation δ,

$$\delta = \frac{2}{N(N-1)}\sum_{i<j}\frac{\left(\left\langle r_{ij}^2\right\rangle_t - \left\langle r_{ij}\right\rangle_t^2\right)^{\frac{1}{2}}}{\left\langle r_{ij}\right\rangle_t}, \tag{7}$$

as a function of total energy (per atom); and 3) specific heat (per atom) (*34*),

$$C = \left[N - N\left(1 - \frac{2}{3N-6}\right)\left\langle E_k\right\rangle_t \left\langle E_k^{-1}\right\rangle_t\right]^{-1}, \tag{8}$$

as a function of total energy (per atom).

Structural Properties

The number of homotops corresponding to a given isomer depends on the geometry (degree of its symmetry) and the composition of the isomer. As a rule, a larger number of homotops corresponds to an isomer 1) of lower symmetry and 2) with a more equal concentration of the two types of atoms. For example, 30, 805, 887, 900, 137 and 43 different stable homotops correspond, respectively, to the six isomers of Ni_7Al_6 sequenced in order of increasing energy as defined by the most stable homotop of each isomer. The ordering of the isomers by the energy of the corresponding most stable homotop produces a sequence which is, in general, different from that for the isomers of the pure Al_{13} and which changes with the concentration of the Ni and Al atoms in the 13-mer. However, for all n and m, n+m=13, including the cases of n=0 or m=0, the lowest energy geometry of Ni_nAl_m is the icosahedron. Figures 1-4 display the first four (of the six considered) isomers, each represented by its lowest energy homotop, for the cases of $Ni_{12}Al$, Ni_7Al_6, Ni_6Al_7 and $NiAl_{12}$, respectively. The figures also show the homotop energy diagrams (either in the form of individual energy levels or distributions of these levels), as well as the total number of homotops corresponding to each isomer. The energy diagrams clearly display the features mentioned above. They also show that $Ni_{12}Al$ prefers Al as the central atom, whereas Ni_7Al_6, Ni_6Al_7 and $NiAl_{12}$ prefer Ni as the central atom. In fact, n=12, m=1 is the only composition of the Ni_nAl_m 13-mer for which it is energetically favorable to place an Al atom in the center of the cluster. With the exception of the case of $Ni_{12}Al$, the homotop energy spectra of the individual isomers are bimodal. As is clear from the figures, the lower energy branches of the distributions correspond to homotops with Ni in the center, whereas the higher energy branches represent homotops with Al in the center. The branches become better separated as the number of the Al atoms in the mixed 13-mer increases. In the case of $NiAl_{12}$ the lower energy branch degenerates into a single energy level - that of the homotop with Ni in the center. This level is separated by a large gap from the manifold of levels corresponding to homotops with Al in the center. The degree of bimodality of the homotop energy spectra of a given isomer can be quantified by the difference between the energies of the two most stable homotops of this isomer, representing the branches with Ni and Al, respectively, as the central atom. Figure 5 shows the graph of this difference as a function of the number m of the Al atoms for the icosahedral isomer of the mixed 13-mer.

The features discussed above suggest that one can introduce a classification of the manifold of the structural forms of mixed clusters of a given size and chosen materials (which in the case of the six isomers of the mixed Ni-Al 13-mer considered here includes 13603 such stable structures) based on the isomeric form, composition, and the type of the central atom. What defines then the energy ordering of the homotops within each class specified by these three attributes? The answer to this question is - the mixing coefficient. Figure 6 displays the graphs of the mixing coefficient as a function of the homotop energy for four classes representing two

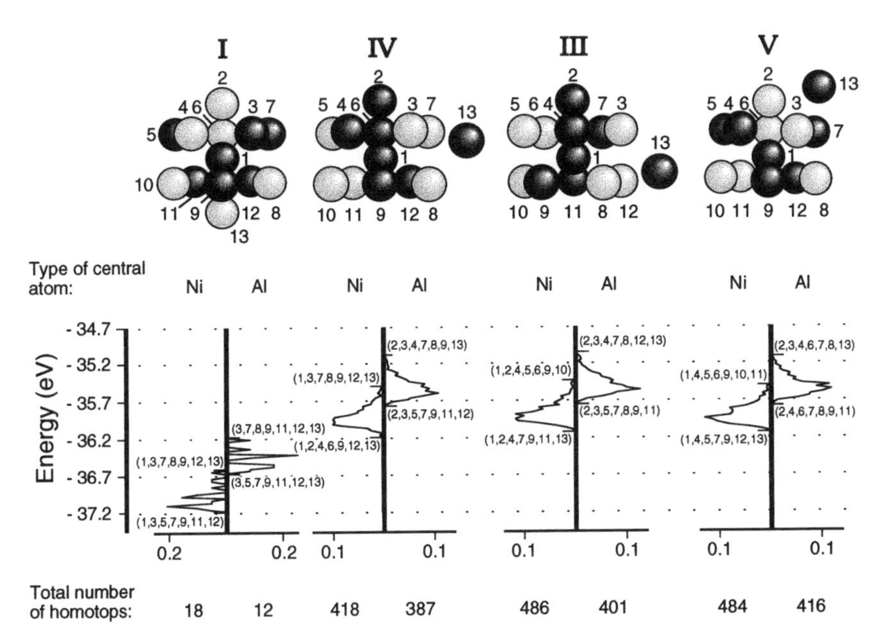

Figure 2. The same as Figure 1, but for Ni_7Al_6. Because of the large number of homotops that correspond to each class, distributions of the homotop energy levels are shown instead of the levels themselves. The distributions are obtained with an energy box size of 0.03 eV. The homotops of the lowest and the highest energy in each class are labeled. The numbers forming a label of a homotop represent the positions of the Ni atoms in it.

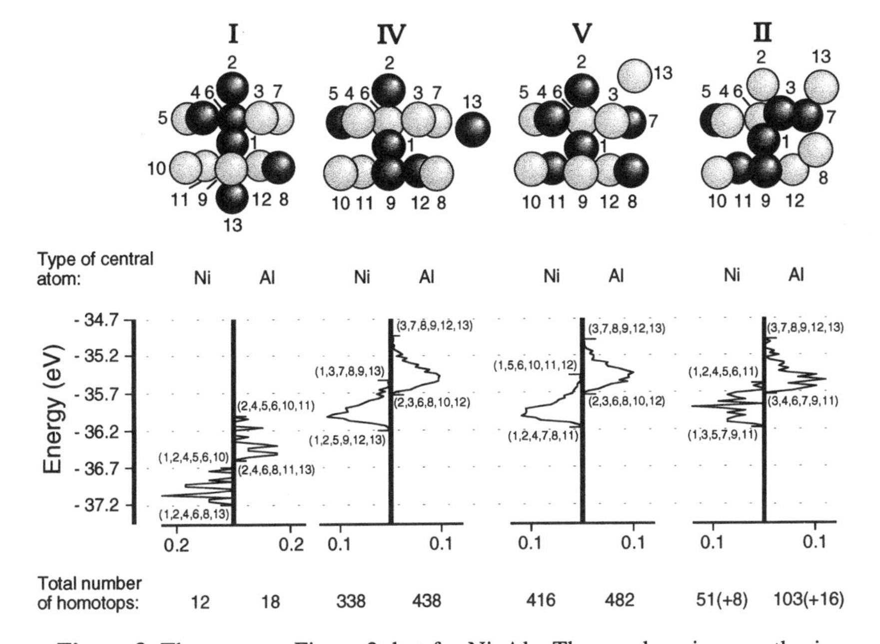

Figure 3. The same as Figure 2, but for Ni_6Al_7. The numbers in parenthesis indicate the number of additional stationary homotops that correspond to saddles, rather than to minima, of the potential energy surface.

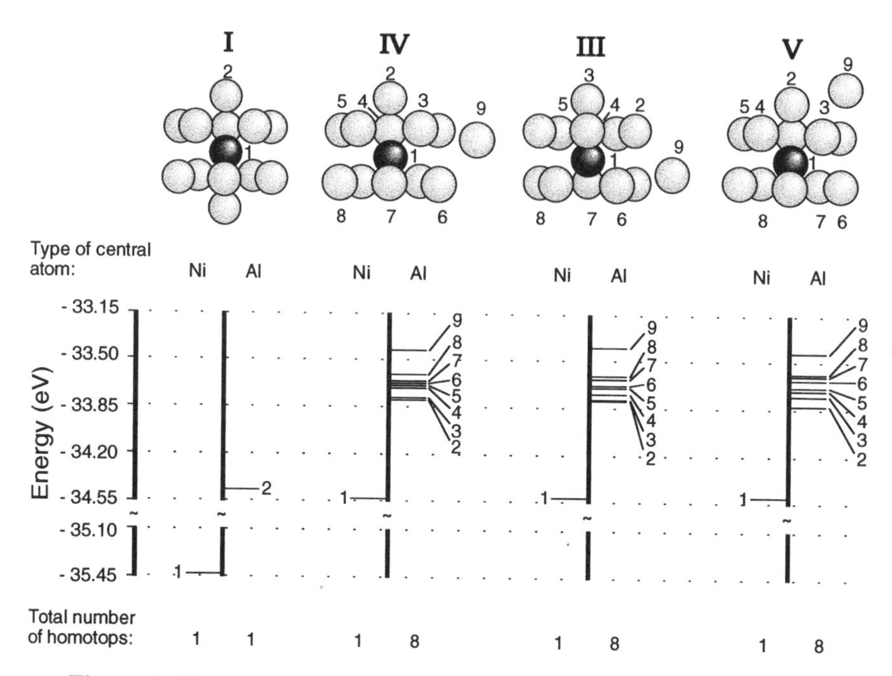

Figure 4. The same as Figure 1, but for $NiAl_{12}$. The number labeling each energy level represents the position of the Ni atom in the homotop corresponding to that level.

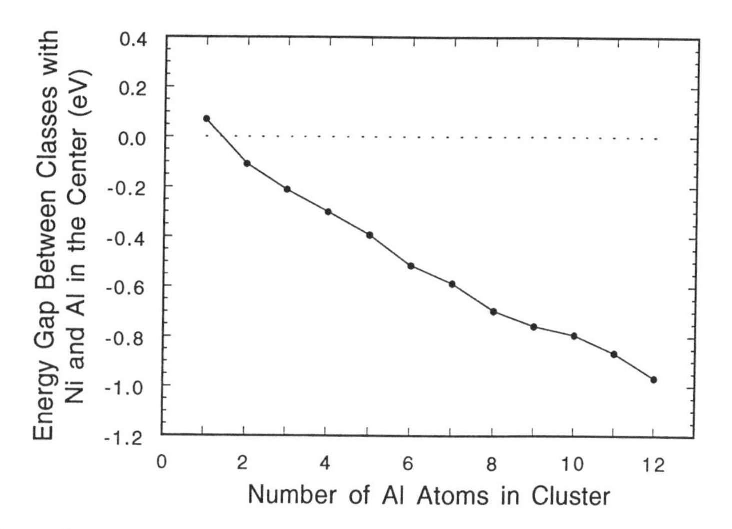

Figure 5. The energy difference between two icosahedral homotops, one of which is the most stable structure in the class with Ni in the center, and the other with Al in the center, as a function of the number of Al atoms in the cluster (see text).

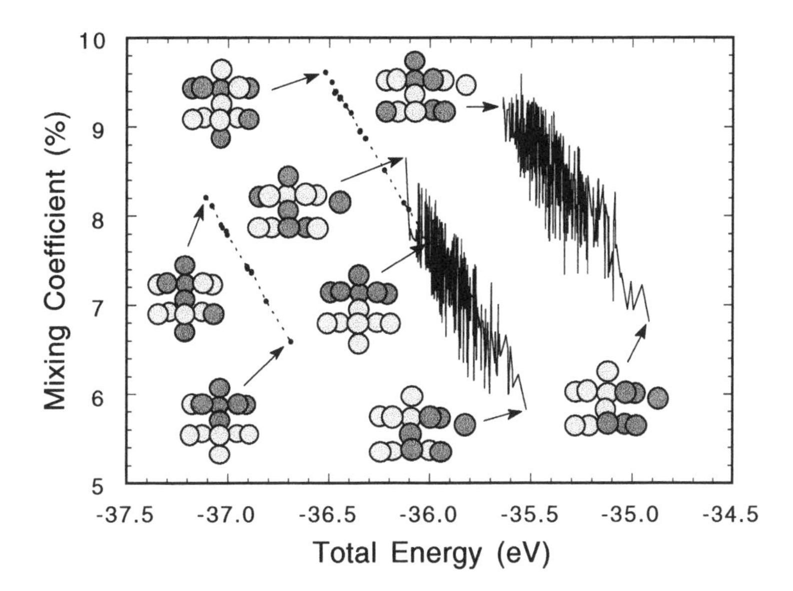

Figure 6. Mixing coefficient as a function of the homotop energy for four classes of Ni_6Al_7. The pictures show the lowest and the highest energy homotops of each class. The darker spheres represent Ni, the grays Al.

isomers of the Ni_6Al_7 cluster. The feature to notice is the monotonicity of the graphs, which is either absolute (exact), as in the case of the icosahedron, or global (overall), as in the case of the second isomer. The higher the degree of mixing of a homotop, the lower its energy within its class. The monotonicity is exact when the energy gap between neighboring homotops within a class is larger then the energy changes caused by the local relaxations in the structures of the individual homotops. The reason for the local violation of the monotonicity in the case of the second isomer is that the energy gap between neighboring homotops in the two classes corresponding to it is very small: 338 homotops with Ni in the center and 438 homotops with Al in the center fall into energy ranges with width of 0.557 eV and 0.722 eV, respectively. The defining role of the mixing coefficient as an ordering index in a given class of homotops holds for all isomers and all compositions of the mixed Ni-Al 13-mer.

All the structural and energy properties of the Ni_nAl_m clusters discussed above are, of course, imbedded in the parameters, cf. Table I, of the potential, equation (1). A model which transparently displays and explains the role of the *homo* (i.e., Ni-Ni and Al-Al) vs *hetero* (i.e., Ni-Al) interactions in defining these properties is discussed elsewhere (*17,18*).

Dynamical Properties

A sample of the dynamical properties of the clusters is given in Figure. 7, which displays graphs of the caloric curve, rms bond length fluctuation δ, and specific heat C, all considered as a function of the cluster energy, for three compositions of the mixed Ni-Al 13-mer. The icosahedral isomer is chosen to represent the zero-temperature geometries of the clusters, and two cases - one corresponding to Ni and the other to Al as the initial central atom (each is represented by the homotop of the lowest energy in the corresponding class) - are considered for each composition. The global features of the graphs are typical of those corresponding to a solid-to-liquid-like transition in clusters induced by increase of their internal energy [cf., e.g., refs. (*22-24,35-37*)]. The details - stages and mechanism(s) - of this transition, however, depend on the class as defined by the zero-temperature structure of the cluster.

Consider first the cases when the zero-temperature icosahedra have the energetically preferred type of atom in the center (cases A in Figure. 7). Whereas two abrupt changes characterize the δ graphs of the $Ni_{12}Al$ and Ni_7Al_6 clusters, only one such change is present in the δ graph of the $NiAl_{12}$ cluster. In contrast, the specific heat graphs display only one peak for the $Ni_{12}Al$ and Ni_7Al_6 clusters, but two peaks for the $NiAl_{12}$ cluster. These peculiarities of the dynamical features can be correlated with and explained in terms of the peculiarities of homotop energy spectra associated with the individual isomers of different compositions (cf. Figures. 1-4). As the energy of a mixed cluster is increased, it begins to sample its higher energy isomeric and homotopic forms. The order in which additional structural forms become accessible depends on their energy and/or the height of the barriers separating them from the other structures. Any homotopic rearrangement within the icosahedral geometry would require overcoming very high barriers. It is energetically easier to convert the complete icosahedron into geometric forms of higher energy. The cluster, therefore, first undergoes isomerization transitions that destroy the completeness of the icosahedral shell. The isomerizations that require the least extra energy are those that involve only the surface atoms. These "surface isomerizations" leave the central atom in the central position, and they are the ones that are responsible for the first abrupt change in the δ graphs. As the energy of the $Ni_{12}Al$ and Ni_7Al_6 clusters with the preferred type of the atom in the center is increased, they sample, through surface isomerizations and homotopic transitions, an increasing number of structural forms. The reason for this is that in these clusters the most stable homotop of each isomer with incomplete shell is separated from the higher energy homotops, either in the same or the other class of either the same or similar isomers, only by small or at most moderate energy gaps (cf. Figures 1 and 2). Further increase of the energy (by about 0.1 eV/atom over that needed to trigger surface isomerizations and homotopic transitions) results in that also global structural changes, i.e., ones that involve permutations between the central atom and a surface atom, become observable on the time scale of our simulations. In $Ni_{12}Al$ such permutations imply replacing the Al atom by a Ni atom in the central position. In Ni_7Al_6 these permutations may or may not change the central Ni by an Al. It is these global structural rearrangements that cause the second abrupt change in the δ graphs. When the preferred central atom in $Ni_{12}Al$ and Ni_7Al_6 gets replaced by an atom of the other type, the clusters sample also their icosahedral homotop(s) of higher energy (i.e., those with Ni in the center of $Ni_{12}Al$ and with Al in the center of Ni_7Al_6); both homotops of the icosahedral $Ni_{12}Al$ and most of the icosahedral Ni_7Al_6 have energies that are lower than those of the homotopic forms of the corresponding higher energy isomers of these clusters. Further increase of the energy leads to an increase in the rate of the surface and global structural transitions and eventually produces a liquidlike state [*22-24,35-37*] of the $Ni_{12}Al$ and Ni_7Al_6. The single peak in the graphs of the specific heat of these clusters is the signatures of transition to this state.

The overall sequence in which the different structural transitions (surface and global) become observable in the $NiAl_{12}$ cluster with Ni as the initial central atom is the same as described above for the $Ni_{12}Al$ and Ni_7Al_6 clusters. The factor that makes the dynamical behavior of $NiAl_{12}$, as exhibited by the graphs of Fig. 7, different from that of either $Ni_{12}Al$ or Ni_7Al_6 is the large energy gap separating the most stable homotop of its higher energy isomers from the higher energy homotop(s) of these isomers. The abrupt change in the δ graph of the $NiAl_{12}$ cluster is caused by surface isomerizations, which convert the lowest energy homotop of the icosahedral isomer into the lowest energy homotops of the higher energy isomers. The global isomerizations and homotopic transitions (i.e., the ones that replace the Ni atom by an Al atom in the center) remain unobservable on the time scale of our simulations until an extra energy of about 0.18 eV is deposited into the cluster. Before the global isomerizations become observable, however, the extra energy makes the rate of the surface isomerizations so high that the cluster attains the state of surface melting. The twelve Al surface atoms form a highly mobile shell, which is not necessarily completely closed at all times, but which overall encages the single Ni atom. The signature of surface melting is the first peak in the corresponding graph of the specific heat. The minimum between the two peaks in this graph corresponds to the energy at which the replacement of Ni by Al in the central position becomes observable on the time scale of our simulation runs. The rapid increase of the rms bond length fluctuation due to surface melting masks the distinct signature of the onset of global rearrangements in the corresponding δ graph. The second peak in the graph of the specific heat corresponds to complete melting of the $NiAl_{12}$ cluster. In the state of complete melting the Ni atom executes frequent excursions to the surface of the cluster.

Comparison of pairs of graphs A and B of a given quantity in Figure 7 corresponding to the same composition but different types of the central atom in the zero-temperature structure of the mixed 13-mer allows one to analyze the class-dependence of the dynamics as defined by the initial type of the central atom. Inspection of the graphs shows that for each composition there is an energy above which the two dynamics, one associated with Ni and the other with Al as the initial central atom, become identical. Above this energy the clusters loose memory of the fact that their zero-temperature icosahedral structure may have been prepared with the energetically less favorable type of atom in the center. Clearly, this composition-dependent energy is one at which structural changes that alter the type of the central atom are feasible. An important and tantalizing observation is that this energy, as defined by the graphs B - they correspond to dynamics associated with the classes with the less preferred type of the central atom, - is lower than the energy at which global structural changes become observable, as defined by the corresponding graphs A (cf. the discussion above). This "paradox" is apparent, rather than real, in that it is based on long but finite observation (simulation) times, and it can be explained either by different extents ("volumes") of the parts of the configuration space ("basins of

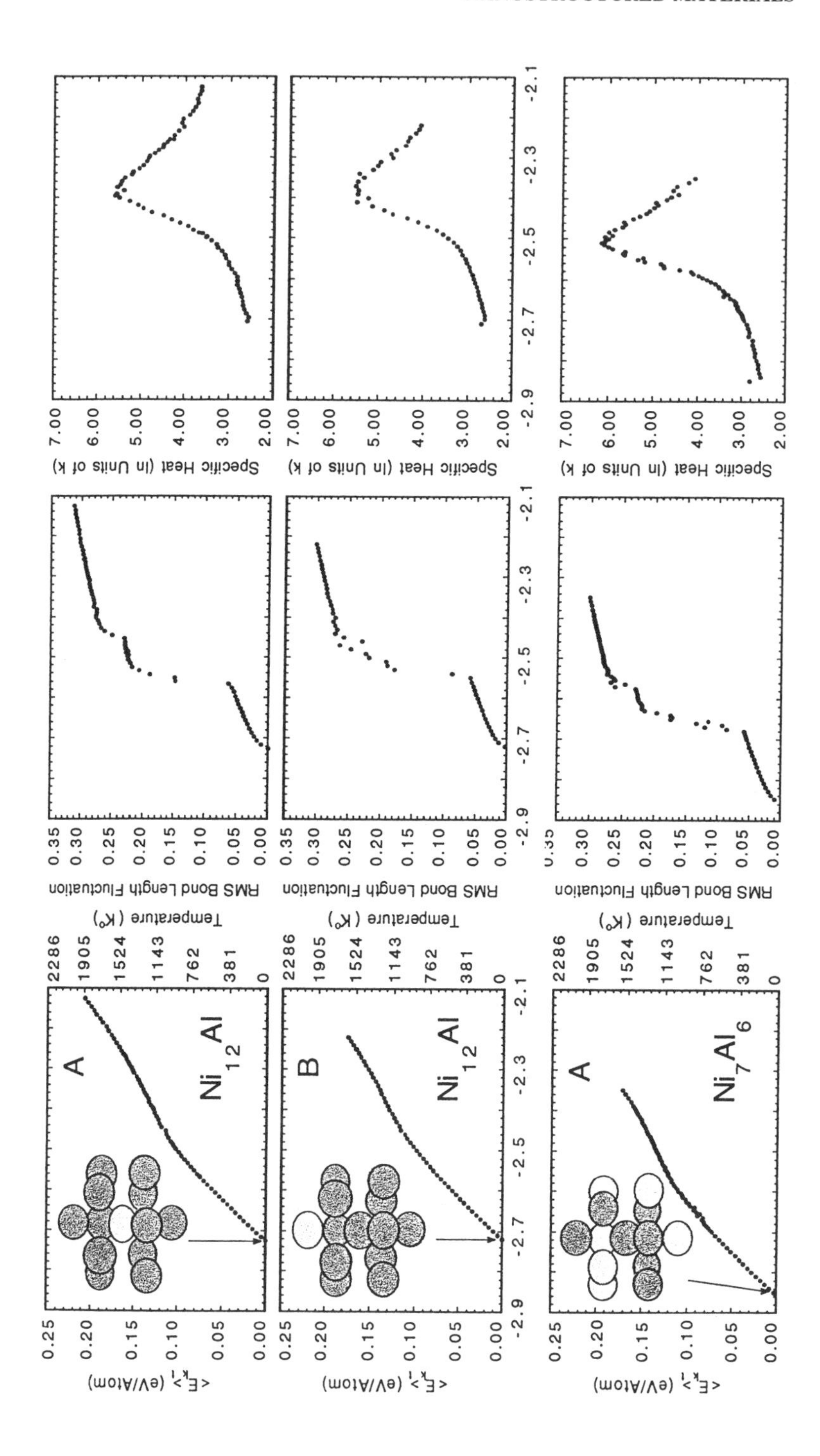

A
$Ni_{12}Al$
B
$Ni_{12}Al$
A
Ni_7Al_6
$<E_{k}>_{t}$ (eV/Atom)
Temperature (K°)
RMS Bond Length Fluctuation
Specific Heat (in Units of k)

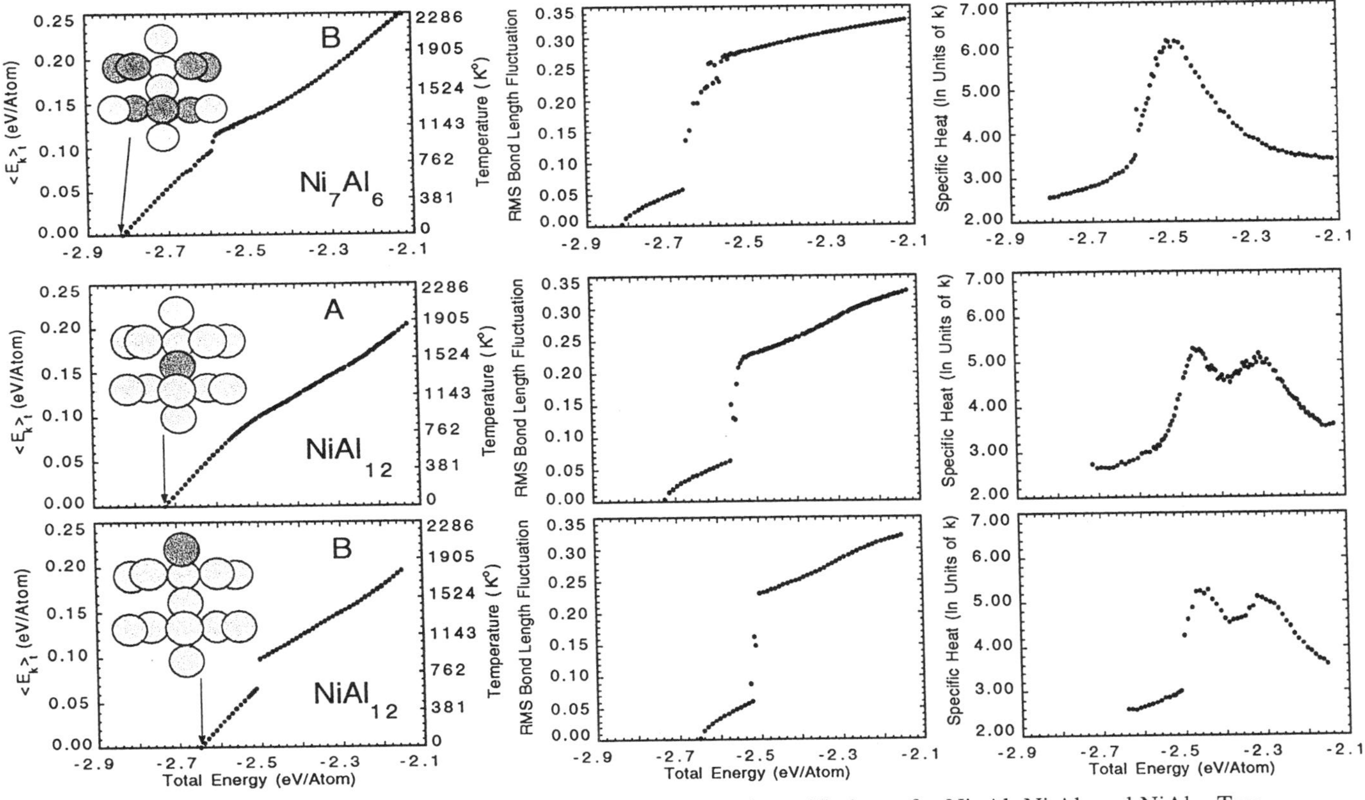

Figure 7. Caloric curves, rms bond length fluctuations and specific heats for $Ni_{12}Al$, Ni_7Al_6 and $NiAl_{12}$. Two cases are considered for each composition—one corresponds to the class with Ni in the center, the other with Al in the center. The pictures represent the lowest energy homotop in each class. These homotops are used as the initial zero-temperature structures of the corresponding dynamics.

attraction") associated with the different structural forms (especially those with the more preferred type of the central atom, on the one hand, and those with the less preferred, on the other), or by peculiarities of the topography of the corresponding potential energy surfaces that make the transition over a barrier in a given direction more probable than in the reverse direction, or both.

The way the clusters loose memory of the initial less preferred central atom depends on their composition (cf. graphs B of Figure 7). This composition dependence also can be explained through the features of the corresponding homotop energy spectra. When one energizes an icosahedral $NiAl_{12}$ with Al initially in the central position, the surface isomerizations become observable at the energy of about -2.52 eV/atom, which is approximately 0.04 eV/atom higher than the corresponding energy in the case when Ni is initially in the central position (cf. graphs B and A for $NiAl_{12}$ in Figure 7). Very little additional energy (less than 0.02 eV/atom) leads then to a replacement of the initially central Al atom by the energetically more preferred Ni atom. After this replacement the cluster behaves as if it were originally prepared with Ni in the center - the A and B graphs become identical. The first replacement of Al by Ni in the central position takes place early on along the trajectory of the corresponding energy, and after it occurred the cluster never revisits in the course of the same trajectory structures with Al in the center. The homotops with Ni in the center completely dominate the dynamics, and, as a consequence, the time-averaged kinetic energy of the cluster drastically increases because the configurational energies of these homotops are much lower than those of homotops with Al in the center. The change is so abrupt that it appears as a discontinuity in the B-labeled graphs of the caloric curve and specific heat of the $NiAl_{12}$ cluster. Similar arguments explain also the quite rapid but smaller increase in the B-labeled caloric curve of the Ni_7Al_6 cluster. In this case, however, the change in the time-averaged kinetic energy is not as drastic, and it translates into an additional sharp local peak (represented by a single point), rather than a discontinuity, at the energy of about -2.59 eV/atom in the B-labeled graph of the specific heat of this cluster. The dynamical replacement of an initial Al atom by a more preferred Ni atom in the center of Ni_7Al_6 triggers a less abrupt change in the time-averaged kinetic energy because 1) the homotops of this cluster with Ni in the center are separated from the corresponding (i.e., belonging to the same isomer) homotops with Al in the center by smaller energy gaps, and 2) the number of homotopic forms participating in the switchover from Al to Ni as the central atom increases gradually as the cluster energy is increased. In the B-labeled δ graph of the Ni_7Al_6 cluster this is reflected by a less distinct (as compared to the corresponding A graph) representation of the two changes associated with the onsets of the surface and of the global structural transformations, respectively. In the case of $Ni_{12}Al$, the small difference between the configurational energies of the homotops with Ni in the center and the corresponding homotops with Al in the center makes the replacement of an initial Ni central atom by the more favorable Al atom essentially unnoticeable, as gauged by the B-labeled graphs of the caloric curve and specific heat of this cluster. The only noticeable effect of this replacement is some blurring of the distinction between the two rapid increases in the B-labeled δ graph of $Ni_{12}Al$ (as compared to the corresponding A graph), which is caused by inclusion in the sampling of additional homotops participating in the switchover from Ni to Al in the central position as the energy of the cluster is increased.

Summary

Mixed metal (alloy) clusters are systems with an enormous, yet unexplored, potential for fabrication of novel materials and other applications. The richness of their properties, - not only structural and dynamical considered in this study, - but also other (e.g., electronic, optical, magnetic, chemical, etc.), and the ability to control and shape these properties through rational selection of their sizes, types and concentrations of their component elements, as well as conditions (e.g., temperature, etc.) of their preparation, make them ideal for arriving at materials tailor-made to unique and demanding specifications. For this, however, an understanding of the fundamental properties of these systems and of how these properties change with the "control variables", e.g., those listed above, is necessary. Theoretical and numerical simulation studies are the means of arriving at a microscopic level of description and understanding of these properties. The two major ingredients of such studies are 1) adequate description of the many-body interatomic interactions and 2) conceptually adequate and efficient analysis tools. An example of the latter is the concept of mixing energy and mixing coefficient, as implemented for systems characterized by interactions that are of inherently many-body character. The finding that the usually very large manifolds of the possible different structural forms of mixed clusters can be subdivided into classes, within which the energy of the different structural forms changes monotonically with the degree of mixing of the component elements, suggests in the case of the Ni-Al clusters considered here that annealing will produce clusters with a higher degree of mixing of the Ni and Al atoms, whereas rapid thermal quenching will preferentially yield clusters with higher degree of segregation of the two metals. Clusters of the same size, as well as type and concentration of the component elements, but different degree of mixing of these elements, most certainly exhibit differences in many, if not all, of their physical and chemical properties. It is the very rich spectrum of these properties that makes two-component clusters, especially alloy clusters, so challenging and attractive both from the fundamental cognitive and applied points of view.

Acknowledgments

This work was performed under the auspices of the Office of Basic Energy Sciences, Division of Chemical Science, US-DOE, under Contract Number W-31-109-ENG-38. E. B. K. was also supported by the NIS-IPP Program.

References

1. *Clusters and Cluster-Assembled Materials;* Averback, R. S.; Bernholc, J.; Nelson, D. L., Eds.; Materials Research Society Symposium Proceedings; Materials Research Society: Pittsburgh, PA, 1991, Vol. 206; and references therein.
2. Guo, B. C.; Kerns, K. P.; Castleman, A. W., Jr. *Science* **1992,** *255,* 1411.
3. Khanna, S. N.; Jena, P. *Phys. Rev. Lett.* **1992,** *69,* 1664.
4. Jena, P.; Khanna, S. N. *Material Science & Engeneering A* **1996,** *217,* 218.
5. *Nanophase and Nanocomposite Materials;* Komarneni, S.; Parker, J. C.; Thomas, G. J., Eds.; Materials Research Society Symposium Proceedings; Materials Research Society: Pittsburgh, PA, 1993, Vol. 286.
6. *Molecularly Designed Ultrafine/Nanostructured Materials;* Gonsalves, K. E., Ed.; Materials Research Society Symposium Proceedings; Materials Research Society: Pittsburgh, PA, 1994, Vol. 351.
7. *Nanophase Materials, Synthesis, Properties, Applications;* Hadjipanayis, G. C.; Siegel, R. W., Eds.; NATO ASI, Series E: Applied Sciences; Kluwer Academic Publishers: Dordrecht, 1994, Vol. 260.
8. Rothlisberger, U.; Adreoni, W. *Chem. Phys. Lett.* **1992,** *198,* 478.
9. Yannouleas, C.; Jena, P.; Khanna, S. N. *Phys. Rev. B* **1992,** *46,* 9751.
10. Cheng, H.-P.; Barnett, R. N.; Landmann, U. *Phys. Rev. B* **1993,** *48,* 1820.
11. López, G. E.; Freeman, D. L. *J. Chem. Phys.* **1993,** *98,* 1428.
12. Bol, A.; Alonso, J. A.; López, J. M. *Int. J. Quantum Chem.* **1995,** *56,* 839.
13. López, M. J.; Marcos, P. A.; Alonso, J. A. *J. Chem. Phys.* **1996,** *104,* 1056.
14. Jellinek, J.; Krissinel', E. B. *Chem. Phys. Lett.* **1996,** *258,* 283.
15. Rey, C.; Garcia-Rodeja, J.; Gallego, L. J. *Phys. Rev. B* **1996,** *54,* 2942.
16. Krissinel', E. B.; Jellinek, J. *Intern. J. Quantum Chem.* **1997,** *62,* 185.
17. Krissinel', E. B.; Jellinek, J. *Chem. Phys. Lett.* **1997,** *in press.*
18. Jellinek, J.; Krissinel', E. B. *to be published.*

19. Cleri, F.; Rosato, V. *Phys. Rev. B* **1993,** *48,* 22.
20. Raghavan, K.; Stave, M. S.; DePristo, A. *J. Chem. Phys.* **1989,** *91,* 1904.
21. Stave, M. S.; DePristo, A. *J. Chem. Phys.* **1992,** *97,* 3386.
22. Jellinek, J.; Garzón, I. L. *Z. Phys. D* **1991,** *20,* 239.
23. Garzón, I. L.; Jellinek, J. In *Physics and Chemistry of Finite Systems: From Clusters to Crystals;* Jena, J.; Khanna, S. N.; Rao, B. K., Eds.; Kluwer Academic Publishers: Dordrecht, 1992, Vol. 1.; pp 405-410.
24. Jellinek, J. In *Metal-Ligand Interactions;* Russo, N.; Salahub, D. R., Eds.; Kluwer Academic Publishers: Dordrecht, 1996, pp 325-360.
25. Pacchioni, G.; Chung, S.-C.; Krüger, S.; Rösch, N. *Chem. Phys.* **1994,** *184,* 125.
26. Reuse, F. A.; Khanna, S. N. *Chem. Phys. Lett.* **1995,** *234,* 77.
27. Lathiotakis, N. N.; Andriotis, A. N.; Menon, M.; Connolly, J. *J. Chem. Phys.* **1996,** *104,* 992.
28. Yi, J.-Y.; Oh, D. J.; Bernholc, J.; Car, R. *Chem. Phys. Lett.* **1990,** *174,* 461.
29. Cheng, H.-P.; Berry, R. S.; Whetten, R. L. *Phys. Rev. B* **1991,** *43,* 10647.
30. Pederson, M. R. In *Physics and Chemistry of Finite Systems: From Clusters to Crystals;* Jena, P.; Khanna, S. N.; Rao, B. K., Eds.; Kluwer Academic Publishers: Dordrecht, 1992, Vol. 2.; pp 861-866.
31. Gong, X. G.; Kumar, V. *Phys. Rev. Lett.* **1993,** *70,* 2078.
32. Khanna, S. N.; Jena, P. *Chem. Phys. Lett.* **1994,** *219,* 479.
33. Swope, W. C.; Andersen, H. C.; Berens, P. H.; Wilson, K. R. *J. Chem. Phys.* **1982,** *76,* 637.
34. Pearson, E. M.; Halicioglu, T.; Tiller, W. A. *Phys. Rev. A* **1985,** *32,* 3030.
35. Jellinek, J.; Beck, T. L.; Berry, R. S. *J. Chem. Phys.* **1986**, *84*, 2783.
36. Berry, R. S.; Beck, T. L.; Davis, H. L.; Jellinek, J. In *Advances in Chemical Physics;* Prigogine, I.; Rice, S. A., Eds.; Wiley-Interscience: New York, 1988, Vol. 70, Part 2.; pp 75-138.
37. Heidenreich, A.; Jortner, J.; Oref, I. *J. Chem. Phys.* **1992**, *97,* 197.

INDEXES

Author Index

Affiliation Index

Subject Index